Tributes
Volume 42

Abstract Consequence and Logics
Essays in Honor of Edelcio G. de Souza

Volume 33
Logic and Computation. Essays in Honour of Amílcar Sernadas
Carlos Caleiro, Fransciso Dionísio, Paula Gouveia, Paulo Mateus and João Rasga, eds.

Volume 34
Models: Concepts, Theory, Logic, Reasoning, and Semantics. Essays Dedicated to Klaus-Dieter Schewe on the Occasion of his 60th Birthday
Atif Mashkoor, Qing Wang and Bernhrd Thalheim, eds.

Volume 35
Language, Evolution and Mind. Essays in Honour of Anne Reboul
Pierre Saint-Germier, ed.

Volume 36
Logic, Philosophy of Mathematics and their History.
Essays in Honor of W. W. Tait
Erich H. Reck, ed.

Volume 37
Argumentation-based Proofs of Endearment. Essays in Honor of Guillermo R. Simari on the Occasion of his 70th Birthday
Carlos I. Chesñevar, Marcelo A. Falappa, Eduardo Fermé, Alejandro J. García, Ana G. Maguitman, Diego C. Martínez, Maria Vanina Martinez, Ricardo O. Rodríguez, a Gerardo I. Simari, eds.

Volume 38
Logic, Intelligence and Artifices. Tributes to Tarcísio H. C. Pequeno
Jean-Yves Béziau, Francicleber Ferreira, Ana Teresa Martins and
Marcelino Pequeno, eds.

Volume 39
Word Recognition, Morphology and Lexical Reading. Essays in Honour of Cristina Burani
Simone Sulpizio, Laura Barca, Silvia Primativo and Lisa S. Arduino, eds

Volume 40
Natural Arguments. A Tribute to John Woods
Dov Gabbay, Lorenzo Magnani, Woosuk Park and Ahti-Veikko Pietarinen, eds.

Volume 41
On Kreisel's Interests. On the Foundations of Logic and Mathematics
Paul Weingartner and Hans-Peter Leeb, eds.

Volume 42
Abstract Consequence and Logics. Essays in Honor of Edelcio G. de Souza
Alexandre Costa-Leite, ed.

Volume 43
Judgements and Truth. Essays in Honour of Jan Woleński
Andrew Schumann, ed.

Tributes Series Editor
Dov Gabbay   dov.gabbay@kcl.ac.uk

Abstract Consequence and Logics
Essays in Honor of Edelcio G. de Souza

edited by

Alexandre Costa-Leite

© Individual authors and College Publications 2020. All rights reserved.

ISBN 978-1-84890-342-5

College Publications
Scientific Director: Dov Gabbay
Managing Director: Jane Spurr

http://www.collegepublications.co.uk

Cover design by Laraine Welch

All rights reserved. No part of this publication may be reproduced, stored in a retrieval system or transmitted in any form, or by any means, electronic, mechanical, photocopying, recording or otherwise without prior permission, in writing, from the publisher.

Contents

Introduction

Alexandre Costa-Leite
On Edelcio G. de Souza

Part 1
Abstraction, unity and logic

3 Jean-Yves Beziau
   Logical structures from a model-theoretical viewpoint

17 Gerhard Schurz
   Universal translatability: optimality-based
   justification of (not necessarily) classical logic

37 Roderick Batchelor
   Abstract logic with vocables

67 Juliano Maranhão
   An abstract definition of normative system

79 Newton C. A. da Costa and Decio Krause
   Suppes predicate for classes of structures and
   the notion of transportability

99 Patrícia Del Nero Velasco
   On a reconstruction of the valuation concept

Part 2
Categories, logics and arithmetic

115 Vladimir L. Vasyukov
    Internal logic of the $H - B$ topos

135 Marcelo E. Coniglio
    On categorial combination of logics

173 Walter Carnielli and David Fuenmayor
   GÖDEL'S INCOMPLETENESS THEOREMS FROM A PARACONSISTENT PERSPECTIVE

199 Edgar L. B. Almeida and Rodrigo A. Freire
   ON EXISTENCE IN ARITHMETIC

PART 3
NON-CLASSICAL INFERENCES

221 Arnon Avron
   A NOTE ON SEMI-IMPLICATION WITH NEGATION

227 Diana Costa and Manuel A. Martins
   A ROADMAP OF PARACONSISTENT HYBRID LOGICS

243 Hércules de Araujo Feitosa, Angela Pereira Rodrigues Moreira and Marcelo Reicher Soares
   A RELATIONAL MODEL FOR THE LOGIC OF DEDUCTION

251 Andrew Schumann
   FROM PRAGMATIC TRUTHS TO EMOTIONAL TRUTHS

263 Hilan Bensusan and Gregory Carneiro
   PARACONSISTENTIZATION THROUGH ANTIMONOTONICITY: TOWARDS A LOGIC OF SUPPLEMENT

PART 4
PHILOSOPHY AND HISTORY OF LOGIC

277 Diogo H. B. Dias
   HANS HAHN AND THE FOUNDATIONS OF MATHEMATICS

289 Cassiano Terra Rodrigues
   A FIRST SURVEY OF CHARLES S. PEIRCE'S CONTRIBUTIONS TO LOGIC: FROM RELATIVES TO QUANTIFICATION

301 Jonas R. B. Arenhart and Sanderson Molick
   ON THE VERY IDEA OF CHOOSING A LOGIC: THE ROLE OF THE BACKGROUND LOGIC

321 **Lorenzzo Frade** and **Abilio Rodrigues**
SOME REMARKS ON LOGICAL REALISM AND LOGICAL PLURALISM

341 **Duško Prelević**
MODAL RATIONALISM, LOGICAL PLURALISM, AND
THE METAPHYSICAL FOUNDATION OF LOGIC

359 **Fabien Schang**
QUASI-CONCEPTS OF LOGIC

On Edelcio G. de Souza

Edelcio Gonçalves de Souza was born in the village of Bauru (in the State of São Paulo, Brazil) in the beginning of the sixties, precisely at the year 1960, May 22. In 1963, his family moved to the city of São Paulo, the place where Edelcio grow up, studied, works and lives up to recent days.

In 1979, he started undergraduate studies in Physics at the *Universidade de São Paulo* (USP), but he decided to change his path and moved towards Philosophy in the eighties. Edelcio got an undergraduate degree and a PhD in the area of Philosophy (USP) supervised by the Brazilian philosopher and logician Newton da Costa. Since then, Edelcio's researches are in the intersection of philosophy of science, mathematics and logic. Particularly, his investigations, up to now, are in topics such as the structure of scientific theories, category theory, model theory and abstract logics.

During his career, he had research stays at *Stanford University* and *Miami University*. He worked in many Brazilian universities such as *Pontifícia Universidade Católica de São Paulo* (where he worked for 21 years), *Fundacão Escola de Sociologia e Política do Estado de São Paulo* and also at the first center of philosophical research in Brazil: *Faculdade de São Bento*. In 2013, Edelcio started his activities as professor of Logic at the *Universidade de São Paulo*.

Edelcio is a enthusiastic supporter of the soccer team *Sociedade Esportiva Palmeiras* and an Aikido Master who is an strong admirer of Brazilian cuisine, especially *feijoada* and *churrasco*. He is married with Vanessa Boarati and they have two children: Helena and Maria Luiza.

Edelcio G. de Souza has a plurality of interests from general abstract logic to the notion of *quasi-truth*. Departing from

Bueno, O; de Souza, E.G. (1996). The concept of quasi-truth. *Logique et Analyse*, 39(153/154), pp. 183-199;

de Souza, E. G. (2000). Multideductive logic and the theoretic-formal unification of physical theories. *Synthese*, 125, pp.253-262.

Edelcio still tries to model and represent formally the essential philosophical idea of *quasi-truth* (developed by his professor Newton da Costa) in the domain of the philosophy of science. This research attempts to establish the formal limits of any empirical scientific knowledge about reality. Besides that, considering *multideductive logic* and its role in abstraction and unification of theories in general, he naturally goes to the domain of model theory. His steps in this area are substantial and they were basically conducted with the mathematician Alexandre Augusto Martins Rodrigues (*in memoriam*). Some representative articles of their collaboration are

Rodrigues, A.A.M; Miranda Filho, R.C; de Souza, E.G. (2006). Invariance and set-theoretical operations in first-order structures. *Reports in Mathematical Logic*, 40, pp.207-213;

de Souza, E.G; Rodrigues, A.A.M. (2017). On extensions of isomorphisms of substructures. *South American Journal of Logic*, 3(1), pp.123-130.

A next very natural level also towards generic abstraction is to consider Tarskian consequence operators as a source of logical and philosophical investigation. Connecting abstract logics, category theory and paraconsistency, we (Edelcio, myself and Diogo H.B. Dias) were able to study in detail a particular technique of paraconsistentization (i.e. different ways which can be used to transform any explosive logic into a paraconsistent formalism). The results appeared in

de Souza, E.G; Costa-Leite, A; Dias, D.H.B. (2016). On a paraconsistentization functor in the category of consequence structures. *Journal of Applied Non-Classical Logics*, 26(3), pp.240-250;

de Souza, E.G; Costa-Leite, A; Dias, D.H.B. (2019). Paradeduction in axiomatic formal systems. *Logique et Analyse*, 62(6), pp.161-176.

The title of the book *Abstract Consequence and Logics* intends to capture and unify Edelcio's main interests in logic, mathematics and philosophy. Thanks to Prof. Dov Gabbay and to Jane Spurr for all outstanding work they have done and, of course, they are still doing in *College Publications* by means of *avant-garde* logical, philosophical and scientific publications. Thanks also to all contributors of this volume for providing original and excellent articles to celebrate Edelcio's 60th birthday. The year 2020 has been very weird and complicated for all of us and it shows precisely the contradictory and paradoxical condition of human life in Earth: at the same time that soon we will be probably launching *James Webb Space Telescope* (NASA) and other special technologies, which looks like amazing from our current primary technological perspective, we, as species, are desperately (and anxiously) searching for a (successful) vaccine against coronavirus. It was in this inconsistent situation, in world quarantine, that Edelcio's 60 years old arrived and in this context it was celebrated: *obrigado por ser esse querido amigo, Tio Ed, saúde e paz, um forte abraço*!

Brasília, September 15, 2020
Alexandre Costa-Leite
Departamento de Filosofia
Universidade de Brasília
costaleite@unb.br

Figure 1: Edelcio G. de Souza at the *III Colóquio UnB-USP de Lógica e Filosofia da Lógica*, Universidade de Brasília, Brasília, 2018.

Part 1

Abstraction, unity and logic

Logical structures from a model-theoretical viewpoint

Jean-Yves Beziau

Federal University of Rio de Janeiro (UFRJ),
Brazilian Research Council (CNPq), Brazil

Dedicated to Edelcio Gonçalves de Souza for his 60th birthday

Abstract

We first explain what it means to consider logics as structures. In a second part we discuss the relation between structures and axioms, explaining in particular what axiomatization from a model-theoretical perspective is. We then go on by discussing the place of logical structures among other mathematical structures and by giving an outlook on the varied universe of logical structures. After that we deal with axioms for logical structures, in a first part in an abstract setting, in a second part dealing with negation. We end by saying a few words about Edelcio.

1 Logics as structures

It is usual nowadays to consider a logic as a structure of type $\mathcal{L} = \langle \mathbb{F}; \vdash \rangle$ where
• $\mathbb{F}$ is a set of objects, called *formulas*. Sets of formulas are called *theories*.
• $\vdash$ is a binary relation between theories and formulas, i.e. $\vdash \subseteq \mathcal{P}(\mathbb{F}) \times \mathbb{F}$, called *consequence relation*.

The idea is to consider this kind of structure in the same way as other mathematical structures. A structure than can be seen as a *model* of some *axioms*, similarly for example to a structure of order $\mathcal{O} = \langle \mathbb{O}; < \rangle$ where
• $\mathbb{O}$ is a set of objects.
• $<$ is a binary relation between objects i.e., $< \subseteq \mathbb{O} \times \mathbb{O}$, called *order relation*.

The situation for logical structures is a bit ambiguous, tricky, mysterious because there is an interplay between the method used and the

objects under study. These objects are logics and the method used is part of logic, namely *model theory*. There are four mammals of mathematical logic, in order of appearance:
- Set Theory
- Proof Theory
- Recursion Theory
- Model Theory.

In modern times there has been a proliferation of *logical systems*, that can be simply called *logics* and that we consider here as *logical structures*. The study of logical systems can be called *metalogic*. It is performed using the four mammals of modern logic. Since 1993 the present author has promoted *universal logic* [3], not as one system among the jungle of logical systems, not even as a super system. Universal logic is a general theory of all these logical systems, in a way similar to *universal algebra*, which is a general theory of *algebraic systems*, or simply *algebras*. And, like in universal algebra, the idea is to consider these systems as mathematical structures. Universal logic is part of metalogic or/and a way to approach metalogic, using in particular model theory, but it can also be developed using for example category theory.

One ambiguity we are facing here is that the word *theory* is used in three different ways:
- When we are talking about model theory, the word is used in the sense of a *general scientific field*, like relativity theory, the theory of evolution or number theory.
- In model theory, a *set of axioms* that characterizes a given class of structures, is called a theory, for example a set of of axioms for lattices. This is different from *Lattice Theory*, which is the study of all the different kinds of lattices and the way they can be axiomatized.
- In universal logic we are considering structures where a *set of objects* is called a theory. This is not the case when dealing with a structure whose elements are, for example, numbers.

2 Structures and axiomatization

Model theory does not reduce to the study of logical structures, it deals with any kind of structures. There is no canonical definition of model theory. In a general perspective, we can say that model theory studies the relations between structures and axioms.

Given a class of structures, we may want to *axiomatize* it, by finding some axioms whose models are exactly the structures of this class. On the other hand, given some axioms, we can investigate the class of structures that are models of these axioms.

What is a structure? We can reply to this question in the same way as we can reply to the question *What is a cat?* by pointing at our favorite cat Miaou. Let us therefore first start with an ostensive reply, by pointing at a famous structure, the structure $\mathcal{N} = \langle \mathbb{N}, < \rangle$ where
- $\mathbb{N}$ is the set of natural numbers.
- $<$ is the relation of strict order between natural numbers.

In some sense it is quite easy to understand what it is, a 7-year old child can understand it. Natural numbers such as 0, 1, 2, 3, 4, 5, are well-known and also one can understand what a big number like 7.794.798.739 (the number of human beings on Earth, right now) is. All numbers have a name, it is not like dogs. And if we ask if $7689 < 987$ we know how to answer. We don't even need a calculator (curiously calculators generally don't make this kind of operation, maybe they think there is no operation to perform here).

A more complicated story is to find some axioms which characterize this structure. What does this mean? An order relation is transitive and anti-symmetric:
- If $a < b$ and $b < c$ then $a < c$.
- If $a < b$ then $b \not< a$.

But the relation of strict order on natural numbers does not reduce to these axioms, or, to put it the other way round, such axioms are not enough to characterize it. An additional axiom is for example the following:
- Given any number a, there is a number b such that $a < b$.

This can be expressed in a more colloquial way as:
- There is not greatest natural number.

And in a more formal way as:
- $\forall x \exists y \; xRy$.

Note that in both cases the symbol "$<$" was sent to the sky. Its presence is only in the middle way, which is generally the way of the mathematician by contrast to the butcher and the logician.

One may want to find a set of axioms that exhausts all properties of the relation $<$ of the structure $\mathcal{N} = \langle \mathbb{N}; < \rangle$. From a structuralist point of view, this also means that it characterizes the natural numbers themselves. The numbers are nothing else than their relations, a number like 9 has no inner nature, what it has, is, its position. The structuralist approach was strongly promoted by Bourbaki [11].

It is not possible to axiomatize in first-order model theory the structure $\mathcal{N} = \langle \mathbb{N}; < \rangle$. Any set of axioms expressed in first-order logic has models which are different from the structure $\mathcal{N}$. This result is due to Skolem [18]. This is an application of the compactness theorem, according to which if every finite subtheory of a theory has a model, this theory has a model.

We will not enter here in the details of such kind of result: its relation with Gödel's first incompleteness theorem and so on. But we take this example to emphasize three important characteristics of axiomatization from a model-theoretical perspective:

• Model-theoretical axiomatization is not the same as proof-theoretical axiomatization, i.e. to derive some theorems from some basic principles, called axioms.

• In the perspective of model-theory, axioms are specific cases of theories, they are finite or recursive sets of formulas.

• Axioms, as well as theories, are generally expressed in a specific formal language, the most famous one being the language of first-order logic.

Having made these clarifications, we will in the next sections present logics as structures in a model-theoretical way, studying the relation between these structures and some axioms. We will stay in the middle way of ordinary mathematics, not specifying, not formalizing too much, the language we are using for expressing the axioms. We just want to point out that if this would be formalized, it would not be naturally formalized in the language of first-order logic, because the central concept of logical structures, the notion of consequence relation, is a relation between sets and objects, typically a second-order relation, by contrast to first-order relations which are only between objects.

We conclude this section emphasizing two points. The first-point is that using logic to talk about logic can be done in a fruitful and intelligent way. Linguists use language to talk about languages, this is not a problem, there are no vicious circles if the perspective is clearly understood. For example it should be clear that general linguistics, the theory of all languages, is not itself a super language. It is expressed and developed using languages, there is no priority of a given language for doing that. The second point is that what we are doing here is not fundamentally new, it is in the line of the Polish school of logic: connected to some works of Tarski (consequence operator), Roman Suszko (abstract logic) or Helena Rasiowa and Roman Sikorski *The mathematics of metamathematics* (see [4], [6] and [9]). The difference between the approach presented here, *Universal Logic*, is that we are *thematizing* the notion of logical structure and clearly differentiating these structures from other mathematical structures, as we will explain in the next section.

3 Logical structures within the family of mathematical structures

Let us come back to our starting point:

Logical structures
A logic is a structure $\mathcal{L} = \langle \mathbb{F}; \vdash \rangle$ where
• $\mathbb{F}$ is a set of objects, called *formulas*. Sets of formulas are called *theories*.
• $\vdash$ is a binary relation between theories and formulas, i.e. $\vdash \subseteq \mathcal{P}(\mathbb{F}) \times \mathbb{F}$, called *consequence relation*.

Logical structures are part of the family of mathematical structures. Using the biological hierarchical distinction between *family*, *genus*, and *species*, we consider that logical structures are a specific genus of structures. Let us examine three other genera of the family.

Order structures
A *structure of order* is a structure $\mathcal{O} = \langle \mathbb{O}; < \rangle$ where
• $\mathbb{O}$ is a set of objects.
• $<$ is a binary relation between objects i.e. $< \subseteq \mathbb{O} \times \mathbb{O}$, called *order relation*.

Algebraic structures
An *algebraic structure* is a structure $\mathcal{A} = \langle \mathbb{A}; f_{(i \in I)} \rangle$ where
• $\mathbb{A}$ is a set of objects.
• $f_{(i \in I)}$ is a collection of functions defined on $\mathbb{A}$.

Topological structures
A *topological structure* is a structure $\mathcal{T} = \langle \mathbb{P}; \mathfrak{T} \rangle$ where
• $\mathbb{P}$ is a set of objects, called *points*.
• $\mathfrak{T}$ is a set of subsets of $\mathbb{P}$ called a *topology*.

All these four genera of structures are similar in the sense that they are made of a pair. The left part of the pairs are all the same, this is a naked set. But there is a variation on the right side, this is the essence of the genus. The right part is often called the *signature*. Two structures having different signatures are of different genera.

An order relation is not considered as of the same genus as an algebra because its signature is a binary relation, whether in the case of an algebra the signature is made of functions. The signature of an algebra may vary: it can be only one binary function or one unary function together with two binary functions, etc. So we have different species of algebras. A group is not of the same species as a ring.

It is not necessarily easy to make the difference between species and genera of structures. For example if in the definition of logical structures, we replace the signature by a binary relation on the Cartesian product of the the power set of formulas, i.e. $\Vdash \subseteq \mathcal{P}(\mathbb{F}) \times \mathcal{P}(\mathbb{F})$, can we say that we still are in the same genus?

Also a structure of a particular type can be equivalent to a structure of a different type. A striking example is a result of Stone showing that a Boolean ring is the same as a distributive complemented lattice (see [19]). A Boolean structure can be presented as a structure of order or as an algebra, or as a mix.

The equivalence between structures of different types has been conceptualized in model-theory with the notion of *expansion*. Two structures are equivalent if they have a common expansion by definition up to isomorphism.

There is also a well-known correspondence between the notion of Boolean algebra and logical structures: by factoring classical propositional logic, we get a Boolean algebra. Logical structures can be "viewed" as algebras, but this is not always the case, and it is only one point of view.

4 The diversity of logical structures

There are different ways to define a logical structure. First of all let us consider three variations of the signature:

Tautological logical structures
A logic is a structure $\mathcal{L} = \langle \mathbb{F}; \mathbb{T} \rangle$ where
- $\mathbb{F}$ is a set of objects, called *formulas*. Sets of formulas are called

theories.
- $\mathbb{T}$ is a set of formulas called *tautologies*.

Consequence logical structures
A logic is a structure $\mathcal{L} = \langle \mathbb{F}; \vdash \rangle$ where
- $\mathbb{F}$ is a set of objects, called *formulas*. Sets of formulas are called *theories*.
- $\vdash$ is a binary relation between theories and formulas, i.e. $\vdash \subseteq \mathcal{P}(\mathbb{F}) \times \mathbb{F}$, called *consequence relation*.

Multiple-conclusion logical structures
A logic is a structure $\mathcal{L} = \langle \mathbb{F}; \Vdash \rangle$ where
- $\mathbb{F}$ is a set of objects, called *formulas*. Sets of formulas are called *theories*.
- $\Vdash$ is a binary relation between theories and theories: $\Vdash \subseteq \mathcal{P}(\mathbb{F}) \times \mathcal{P}(\mathbb{F})$, called *multiple-consequence relation*.

The first formulation corresponds to how logical systems were originally conceived at the beginning of the 20th century. The second approach was mainly promoted in Poland but using a different set-up (see [20]), which is the following one:

Consequence Operator
A logic is a structure $\mathcal{L} = \langle \mathbb{F}; Cn \rangle$ where
- $\mathbb{F}$ is a set of objects, called *formulas*. Sets of formulas are called *theories*.
- Cn is a binary function from theories to theories, i.e. from $\mathcal{P}(\mathbb{F})$ to $\mathcal{P}(\mathbb{F})$, called *consequence operator*.

These two set-ups are equivalent independently of any axioms. Multiple-conclusion logical structures were developed only in the 1970s (see [17]), although one may say that they already showed up in the case of Gentzen's sequent-calculus [14], but this is rather ambiguous as we will explain in the next section.

Let us point out that we have presented all these variations on the one hand without specifying the structure of the set $\mathbb{F}$, on the other hand without stating some axioms for the "thing" which appears in the signature. This is typical of the universal logic approach we have been developing, but focusing on the second type of structures, i.e. consequence logical structures. The spirit of *axiomatic emptiness* (cf. [7]) can however also be applied to other types of logical structures. Regarding the dressing of the naked set $\mathbb{F}$, there are various ways to proceed also independent of the signature. A typical dressing, for propositional logic,

is to consider the domain of the structure as follows (cf. [15]):
- $\mathcal{F}$ is an absolutely free algebra $\langle \mathbb{F}; \wedge, \vee, \neg \rangle$ whose domain $\mathbb{F}$ is generated by the functions $\wedge, \vee, \neg$ from a set of atomic formulas $\mathbb{A} \subseteq \mathbb{F}$.

We have then a mix of two kinds of structures, by putting within a logical structure an algebra. This is what Bourbaki called a *carrefour de structures*.

The model-theoretical axiomatic methodology for logical structures does not mean that we need to fix a set of axioms. It is similar to universal algebra. There are good reasons not to fix a set of axioms, both philosophically and theoretically. Let us just consider the theoretical aspect here. Among all logical structures, it is interesting to consider the two extreme cases:
- Nothing is a consequence of nothing, i.e. $\vdash = \emptyset$
- Everything is a consequence of everything, i.e. $\vdash = \mathcal{P}(\mathbb{F}) \times \mathbb{F}$

And also it is interesting to consider that the opposite of classical logic, i.e. the set-theoretical complement of the consequence relation of this logic, is a logical structure. This is what we have called *anti-classical logic* [10]. It is not possible to host all these structures, in the universe of logical structures, if we are working with a specific set of axioms.

Similarly to the universal algebra approach the universal logic approach does not mean that we will not consider axioms. But axioms are always relative and are a way to classify and study the relations between different logical structures, to navigate within the ocean of logical structures

5 Axioms for abstract logical structures

The most famous axioms for logical structures are the three following Tarski's axioms (cf. [5]):
- $a \vdash a$ (*Reflexivity*)
- If $T \vdash a$ and $T \subseteq U$ then $U \vdash a$ (*Monotonicity*)
- If $T \vdash a$ and $U, a \vdash b$ then $T, U \vdash b$ (*Transitivity*)

We call *Tarskian logic*, a logical structure obeying these axioms. These axioms were originally presented by Alfred Tarski but in a different way because, he was working with a consequence operator, not with a consequence relation. There are lots of different equivalent ways to present these axioms, even within the same type of structures. For

example sometimes reflexivity is presented as
- $T, a \vdash a$ (*Extended reflexivity*)

If we are a minimalist, it is not necessary to present it in this way because in fact this can be deduced from the three above axioms: by reflexivity we have $a \vdash a$, and since $a \subseteq \{T\} \cup a$, applying monotonicity we have $T, a \vdash a$. Since $a \vdash a$ is a particular case of $T, a \vdash a$, i.e. the case when $T = \emptyset$, the axiom $a \vdash a$ is equivalent to the axiom $T, a \vdash a$ modulo monotonicity. This means that if we replace the axiom of reflexivity by extended reflexivity we define the same class of logical structures.

We have shown that extended reflexivity is deducible from reflexivity using monotonicity. Is this a proof? Yes! But an informal proof as standard mathematicians are doing, when dealing with order structures, algebraic structures, etc. Such kind of proof cannot be easily translated into first-order logic, like in fact most of mathematical proofs, in particular here because we are using second-order structures, but this can be translated into first-order logic for example via set theory. Although we are aware that this is not straightforward and that some complications may show up, we are, to start with, not interested to work on the formalization in first-order logic of such proofs.

What is important for us here is to make a clear distinction between this structural approach and a proof-theoretical approach such as sequent calculus. There can be some confusions, which are in particular generated by terminology and symbolism. For example the above Tarski's axioms look like the so-called *structural* rules of sequent calculus. But are they the same? Can we identify the cut-rule with transitivity? This would be highly misleading.

Gentzen's sequent system LK generates a logic which is the same as the logic generated by LK^-, i.e. LK wihtout the cut-rule. The logical structure generated by LK^- is the same classical logic as the one generated by LK. The consequence relation generated by the cut-free system LK^- is transitive! This is what shows the cut-elimination theorem.

What is interesting however is that we can develop informal proofs about logical structures which are inspired or directly imported from sequent calculus and vice-versa, we can import informal proofs about logical structures within sequent calculus. This is important because this can secure an "algorithmic" aspect. Note however that even if everybody agrees about the computable aspect of first-order classical sequent-calculus, this does not mean that the metatheory of this system is itself formalized.

But the idea we are discussing here it the applications of model-theoretical methods to the study of logical structures. Let us give a fairly simple example. Consider for example the following axiom:
- If $T \vdash a$ then there is To finite, $To \subseteq T$ such that $To \vdash a$ (*Compactness*)

We can show that this axiom is not a consequence of Tarski's axioms by giving an example of a logical structure which verifies Tarski's axioms but not the axiom of compactness. An example of such a structure is second-order classical logic considered from a standard model-theoretical way.

6 Axioms for logics of pure negation

The *law* or *principle* of non-contradiction was traditionally considered as a basic principle of logic. It was considered either as a law of thought or a law of reality, or both. We will not discuss these pataphysical questions here. There are even more extravagant people considering that this law has to be rejected. We will also not comments this kind of extravagance. What is important for us here is to show how we can have a new and hopefully better understanding of this law using the model-theoretical approach, independently of wanting to approve or reject it.

Boole formulated the law of non-contradiction as $x(1-x) = 0$ and showed how to deduce it from $x^2 = x$, which for this reason he considered as the fundamental law of thought (see [8]). This is a purely algebraic approach in the sense that he is using functions and equalities. But we don't consider algebra as a panacea for mathematics, there are other mathematical structures, and moreover we consider that a different approach provides a better understanding.

We consider negation as a unary function $\neg$. The second step is to consider this function on a naked set, with only this function, so we have the following structure: $\mathcal{F} = \langle \mathbb{F}; \neg \rangle$. The third step, which is properly original. is to consider the structure: $\mathcal{LPN} = \langle \mathcal{F}; \vdash \rangle$. $\mathcal{LPN}$ is an acronym for *Logic of Pure Negation*. And then we consider axioms for this negation.

There are here two important features directly connected with the spirit of universal logic:
- We can work with an algebra which is not necessarily an absolutely

free algebra. We may have an algebra where $\neg\neg a$ is a, or even $\neg a$ is a.
• The axioms for the consequence relation are not absolute. Axioms for negation can be considered independently of axioms for the consequence relation.

We can consider the following axiom:

$$\text{Given } T \text{ and } a, \text{ for any } x: T, a, \neg a \vdash x$$

independently of Tarski's axioms for consequence relation. And also we can consider the relations between the above axiom and the following second one

$$\text{Given } T \text{ and } a: T, \neg a \vdash x, \text{ for any } x \text{ iff } T \vdash a$$

according to or not according to such or such axiom for the consequence relation.

We have shown that it is possible to deduce all axioms for negation from this second axiom, modulo Tarski's axioms (see [2]). Can we call this axiom, the axiom of non-contradiction? To answer this question it is important to study the relation between this axiom and the following principle:

$$a \text{ is true iff } \neg a \text{ is false,}$$

To do so we need to connect a theory of truth and falsity with general abstract logic. This has been done by Newton da Costa with his theory of valuation on which we have been working together (see [12]).

7 Dedication and acknowledgments

I am glad to dedicate this paper to Edelcio whom I have known since 1991. I met Newton da Costa in Paris in January 1991 and he invited me to come to work with him for one year at the University of São Paulo. Arriving at São Paulo's airport in August 1991 da Costa was there together with Edelcio, who was one of his students. I stayed at Edelcio's flat for a few days in the district of *Campos Elísios* (*Champs-Élysées*) and then he took me to a residence in the campus of the university.

Since then I have continuously been in touch with Edelcio, for example taking part to the jury of his PhD Student Patricia del Nero Velasco [16]. And we share some common interest: chess, Italian food and logical structures of course. That's why I decided to choose this topic for the present paper. In particular Edelcio took part to the *1st World Congress on Universal Logic* (UNILOG'2005) that I organized in Montreux in 2005 with the help of Alexandre Costa-Leite, the editor of this volume, who was doing a PhD with me at this time at the University of Neuchâtel in Switzerland [13].

Edelcio, together with Alexandre and Hilan Bensusan, wrote a paper for the *Festschrift* volume of my 50th birthday: "Logics and their galaxies" [1]. I am glad to reward him by the present paper.

Edelcio on the way to Marmot's paradise during UNILOG'05

References

[1] H.Bensusan, A.Costa-Leite and, E.G de Souza, "Logics and their galaxies", in A.Koslow and A.Buchbaum (eds), *The Road to Universal Logic - Vol 2*, Birkhäuser, Basel, 2015, pp.243-252.

[2] J.-Y.Beziau, "Théorie législative de la négation pure", *Logique et Analyse*, 147-148 (1994), pp.209-225.

[3] J.-Y. Beziau, "Universal Logic", in *Logica '94 - Proceedings of the 8th International Symposium*, T.Childers and O.Majers (eds), Czech Academy of Science, Prague, 1994, pp.73–93.

[4] J.-Y.Beziau, "From consequence operator to universal logic: a survey of general abstract logic", in *Logica Universalis: Towards a general theory of logic*, Birkhäuser, Basel, 2005, pp.3-17.

[5] J.-Y. Beziau, "Les axiomes de Tarski", in R.Pouivet and M.Rebuschi (eds), *La philosophie en Pologne 1918-1939*, Vrin, Paris, 2006, pp.135-149.

[6] J.-Y. Beziau, "13 Questions about universal logic", *Bulletin of the Section of Logic*, 35 (2006), pp.133-150.

[7] J.-Y.Beziau, "What is a logic ? - Towards axiomatic emptiness", *Logical Investigations*, vol.16 (2010), pp.272-279.

[8] J.-Y.Beziau, "Is the Principle of Contradiction a Consequence of $x^2 = x$?", *Logica Universalis*, vol.12 (2018), pp.55-81.

[9] J.-Y.Beziau, "Metalogic, Schopenhauer and Universal Logic", in J.Lemanski (ed), *Language, Logic, and Mathematics in Schopenhauer*, Birkhäuser, Basel, 2020, pp.207-257.

[10] J.-Y.Beziau and A.Buchsbaum, "Let us be Antilogical: Anti-Classical Logic as a Logic", in A.Moktefi, A.Moretti and F.Schang (eds), *Soyons logiques / Let us be Logical*, College Publication, London, 2016, pp.1-10.

[11] N.Bourbaki, "The architecture of mathematics", *American Mathematical Monthly*, 57 (1950), pp.221-232.

[12] N.C.A. da Costa and J.-Y.Beziau, "Théorie de la valuation", *Logique et Analyse*, 146 (1994), pp.95-117.

[13] A.Costa-Leite, *Interactions of metaphysical and epistemic concepts*, PhD Thesis, University of Neuchâtel, 2007.

[14] G.Gentzen, "Untersuchungen über das logische Schließen. I", *Mathematische Zeitschrift*, **39** (1935), 176–210.

[15] J.Łoś and R.Suszko, 1958, "Remarks on sentential logics", *Indigationes Mathematicae*, **20**, 177–183.

[16] P. del Nero Velasco, *Sobre uma reconstrução do conceito de valoração*, PhD, Pontifical Universiy of São Paulo, Brazil, 2004.

[17] D.J.Shoesmith and T.J.Smiley, *Multiple-Conclusion Logic*, Cambridge University Press, Cambridge, 1978.

[18] T.Skolem, "Über die Nicht-charakterisierbarkeit der Zahlenreihe mittels endlich oder abzählbar unendlich vieler Aussagen mit ausschliesslich Zahlenvariablen", *Fundamenta Mathematicae*, 23 (1934), pp.150-161.

[19] M.Stone, "Subsomption of the theory of Boolean algebras under the theory of rings", *Proceedings of National Academy of Sciences*, 21 (1935), pp.103-105.

[20] A.Tarski, "Remarques sur les notions fondamentales de la méthodologie des mathématiques", *Annales de la Société Polonaise de Mathématiques*, 7 (1929), pp.270-272.

UNIVERSAL TRANSLATABILITY: OPTIMALITY-BASED JUSTIFICATION OF (NOT NECESSARILY) CLASSICAL LOGIC

Gerhard Schurz

Duesseldorf Center for Logic and Philosophy of Science (DCLPS),
Department of Philosophy, Heinrich Heine University Duesseldorf, Germany

Abstract

In order to prove the validity of logical rules, one has to assume these rules in one's semantic, or metalogic. But how is a non-circular justification of a logical system possible? The question becomes especially pressing insofar in present time a variety of *non-classical* alternatives to classical logics have been developed. Is the threatening situation of an epistemic circle or infinite regress unavoidable? The situation seems hopeless. Yet, in this paper I suggest a positive solution to the problem based on the fact that logical systems are *translatable* into each other. I propose a translation method based on introducing additional concepts into the language of classical logic. Based on this method I demonstrate that all finite multi-valued logics - and I conjecture all non-classical logics - can be translated into classical logic. If this argument is correct, it shows that classical logic is *optimal* in the following sense: by using it we cannot lose, because if another logic turns out to have advantages for certain purposes, we can translate and thus embed it into classical logic. This optimality argument does not exclude that there can be other, non-classical logics that are likewise optimal in the explained sense.

1 Introduction: The significance of optimality-based justifications for foundation-theoretic epistemology

In other writings (Schurz (2008a), (2018a), (2019)) I have defended a 'modernized' version of an internalist and foundation-theoretic epistemology. Within this epistemological framework the class of 'basic' beliefs that are considered as 'immediately evident' or not in need of further justification is minimalistic, consisting only of analytical and introspective beliefs. Moreover, circular justifications are rejected because they are demonstrably epistemically worthless. This is demonstrated by the fact that with circular justification both a proposition and its negation may be proved. For example, both the rule of induction and the rule of counter-induction can be circularly 'pseudo-justified' (Salmon (1957), 46). More drastic examples of circular 'justifications' of obviously irrational rules are given in Achinstein (1974) and Schurz (2019), sec. 2.4 and 3.3.

In a foundation-theoretical framework of the described sort, the 'epistemic load' that has to be carried by deductive, inductive or abductive reasoning is high. Therefore the justification of the truth-conduciveness of these inferences - in the strict or at least high probability sense - acquires central importance. In other writings I have studied the problem of justifying inductive inferences, i.e., Hume's problem. I tried to show that there is a kind of higher-order justification that doesn't lead into a circle or infinite regress: an epistemic *optimality justification*. An optimality justification does not attempt to demonstrate that a given epistemic method or system is strictly or probabilistically reliable, in the sense of leading to the truth in all or most cases. It pursues a more modest epistemic goal, namely to demonstrate that a given method (or system) is *epistemically optimal* among all competing methods (of a given kind, e.g., induction or deduction) that are *cognitively accessible* to the given epistemic agent. Every optimality justification is relative to a given epistemic goal. In the case of induction, this goal is *predictive success*. I have proved that a certain method of *meta-induction* is predictively optimal in the long run among all prediction methods that are accessible to the forecaster, even in possible worlds in which the success rates of the competing prediction methods are permanently changing (Schurz (2008b), Schurz (2019); Thorn and Schurz (2020), Schurz and Thorn (2016)). The universal optimality result provides us with a weak a priori justification of meta-induction that can stop the justification regress for the problem of justifying induction.

In this paper I will apply the method of optimality justification to the domain of *logic*. More precisely, the paper is devoted to the problem of finding a non-circular justification for a system of logic. Thereby I will focus on the system of *classical* logic and, moreover, on the justification of classical *propositional* logic. However, as I will show in the end of the paper, similar methods can be applied to systems of non-classical logic.

Thus the primary question of this paper will be: how can we justify the rules of classical logic? The question becomes especially pressing insofar in present time a variety of *non-classical* alternatives to classical logics have been developed. When we prove the validity of the rules of classical logic, we have to assume these rules in our semantic, or metalogic. Thus, all direct demonstrations of the validity of logical rules are inherently *circular* (cf. Schurz (2018b), sec. 15.3). But how is a *non-circular* justification of a logical system possible? Is the threatening situation of an epistemic circle or infinite regress unavoidable? The situation seems hopeless. Yet, in this paper I will suggest a positive solution to the problem that will consist in an *optimality-based* justification, in relation to the epistemic goal of representation power. I will try to show that classical logic is representationally optimal in the sense that every non-classical logical system can be *translated* into classical logic.

2 The significance of non-circular justifications for contemporary philosophy of logic

To a certain extent, the basic logical operators can be justified in a Kantian 'transcendental' sense, as a presupposition of the possibility of cognition at all, by the following reasoning. (i) The possibility of describing any *manifold* presupposes the operation of *conjunction*, by which we can represent something that has several components. (ii) The possibility of expressing that a certain description is false presupposes the operation of *negation*. (iii) Finally the possibility of expressing that something can be described in several ways requires the operation of *disjunction*. (iv) If we enrich these operations by the idea of infinity, we obtain the universal quantifier (as an infinite conjunction) and existential quantifier (as an infinite disjunction). These 'quasi-Kantian' considerations give us a reason why every rational language will need these kinds of logical operations. However, this reasoning does not determine the precise logical rules or meaning of these logical operations.

The so-called *classical* logic is characterized by an additional semantic principle, the principle of *two-valuedness*: every statement p that is expressed in a semantically complete (non-indexical) way is *either true or false* (neither 'neither true nor false', which is the principle of excluded middle, nor 'both true and false', which is the principle of non-contradiction). Or more ontologically: every state of affairs either obtains, or does not obtain. It has to be emphasized that the principle of two-valuedness is meant in a purely *non-epistemic* and ontological way. It has nothing to do with our ability to *find* out or to *know* whether p or not-p is true, but merely with the truth-value itself. Since this truth value is understood in a correspondence-theoretic way, this means that ultimately the principle of two-valuedness expresses the *determinateness* of reality: if "p" is a semantically complete sentence, then either p or not p must obtain. In other words, the properties of reality are objectively determined.

I can see only two possible reasons for a philosopher to doubt the principle of two-valuedness: (i) either the philosopher rejects the idea of an objective reality and thus the correspondence-theoretic notion of truth, or (ii) the philosopher accepts the idea of an objective reality but doubts the determinateness of reality. If we look into the history of non-classical logic, we usually find one of these two possible reasons for the erection of a particular system of non-classical logic. Łukasiewicz' three-valued logic (Łukasiewicz (1920)) was based on the second reason (ii): it started from the thesis that there are sentences whose truth-value is objectively undetermined, because the corresponding states of affairs are neither 'being' nor 'non-being'. One reason for this view comes from the theory of *vagueness*. A more compelling reason comes from *quantum physics*: according to Heisenberg's *uncertainty relation* it is impossible that both the position s and the momentum p (or the corresponding wavelength $\lambda \sim 1/p$) of a quantum-mechanical (wave-particle) system can simultaneously be sharply realized; it rather holds that $\Delta s \cdot \Delta p \leq h/2\pi$.

Brouwer's *intuitionistic* logic was motivated by the first reason (i): in this logic the correspondence-theoretic notion of truth is replaced by the mathematical notion of verification, with the consequence that for the intuitionistic negation the classical law of double negation is no longer valid.

Łukasiewicz' three-valued logics has been mathematically generalized to *many-valued* logics with arbitrarily many 'truth-values', abstracting from any philosophical interpretation of these truth-values (Gottwald (1989), Rautenberg (1979), Malinowski (1993)). A second source of non-classical logics has been relevance logics (Anderson and Belnap (1975)), in particular their development into para-consistent logics by Priest (1979). Later, Priest (2006) gave independent philosophical motivations for the introduction of paraconsistent sentences, i.e. sentences that are both true and false, that are discussed rather controversially (cf. Williamson (2014), Williamson (2017)). A similar controversy has taken place about quantum logics, as developed by Birkhoff and von Neumann (1936) and philosophically supported by Putnam (1979). Quantum logic gives up the classical law of distributivity, $A \wedge (B \vee C) \rightarrow (A \wedge B) \vee (A \wedge C)$. However, this step is not 'enforced' by the facts of quantum physics. Critics of quantum logics have argued that the uncertainty relation is better explicated by classical logic, because in quantum logic the superposition of two exclusive states p, q is expressed as a disjunction $p \vee q$, which is a mistake according to the critics of quantum logics (Popper (1968), Dummett (1976), Stachel (1986)). In any case, the representation of the facts of quantum physics is also possible within classical logic.

In contemporary *philosophy of logic* the so-called '*anti-exceptionalists*' have argued that logics are not 'exceptional' compared to the empirical sciences. More precisely, (a) it is false that the laws of logic have an exceptional *apriori* status, as it was traditionally assumed in philosophy; rather (b) they have to be revised, corrected and abductively supported by empirical facts, similar as this has been the case for physical geometries (cf. Bueno (2010); Hjortland (2017), 632). *I agree with (a) but not with (b)*. In other words, I think that logic is not apriori, but yet exceptional. Logical systems are not apriori, because different possible logical systems can be constructed and reasonably applied. However, the second thesis seems to be untenable because, in contrast to geometry, every attempt to 'revise' or 'test' a system of logic is beset by the *problem of circularity*. The justification of different systems of geometry (say Euclidean versus non-Euclidean) can be based on independent logico-mathematical description systems that do *not* presuppose a particular logic. However, it is impossible to describe certain facts that are supposed to 'support' or 'test' a given logic without already assuming a certain logic in the description of these facts - for the reason that the logical operations are needed in every description of something, as explained above (or similar arguments see Rescher (1977), 240f.; Woods (2019), Sereni and Sforza-Fogliani (2017)). In particular, when the rules of non-classical logic are justified by their (non-standard) semantic principles, these principles are described in the so-called *meta-language* which uses itself a logic, the so-called *metalogic*.

Interestingly, even for non-classical logic it is rather common (though not ubiquitous) to use the classical logic in the meta-language. This seems to speak for a certain preference of classical logic. However, this is merely an indicator but not a systematic argument. One could argue that it is a just human 'convenience' to stick to classical logic in the meta-language. What is much more remarkable is that it is possible at all to justify the rules of a non-classical logic by a semantical framework that is classical. How can that be? I hope that I will be able to explain in this paper why this can be.

Is there any way to justify the laws of a logic in a way that is non-circular and, thus, foundation-theoretically acceptable? If by a "justification" one means a *demonstration* of the validity of these laws that the answer is *negative*. This insight seems to constitute a *threat* of foundation-theoretic epistemology, because it suggests that there are no objective criteria for the choice of logical systems. Since the laws of classical logic and non-classical logics are in mutual opposition, it seems that already at the most fundamental level of cognition, namely the level of logic, we are exposed to the dangers of relativism and incommensurability, that have been proclaimed by Kuhn (1962) for paradigms of natural sciences.

In this paper I try to show that optimality justifications can offer an escape. However, optimality justifications are always relative to a presupposed epistemic *goal*. What could the epistemic goal of logic be? In the controversies in philosophical logic, one often discusses goals such as intuitive naturality, usefulness in mathematics, computational simplicity or agreement with assumed metaphysical positions. The problem of goals of this sort is that they are much too subjective and context-dependent in order to enable a robust objective optimality justification. To give an example, both Priest (2006), Williamson (2017) and Bueno (2010) are anti-exceptionalists, but they draw opposite inferences about the preferred logic: while Williamson prefers classical logic, Priest argues for paraconsistent logic and Bueno for pluralism and context-sensitivity. For a robust epistemic justification, we need a much more general epistemic goal, indeed a most general one, since logic is the *most* general level of description.

I propose the *power of linguistic representation* to be this goal. The leading idea of the optimality justification proposed in this paper will be that the representational power of a logic L is at least as high than that of another logic L' if L' can be *translated* into L. It is a well-known fact that some non-classical logics L can be translated into classical logic L_2 ("2" for two-valued). If this is the case, then everything that is expressible in L' is also expressible in L_2, thus L_2 is representationally at least as powerful as L'. If we could prove this for all non-classical logics, at least for *all* non-classical logics of minimal plausibility, we would have a universal optimality argument for classical logic.

Let us explain the translation approach at hand of the translation of Łukasiewicz' three-valued (propositional) logic L_3 into the classical logic L_2. Assume the proponent of L_3 claims that the sentence "there is exactly one electron" is neither true nor false but objectively undetermined. Then nevertheless, the statement that this sentence is objectively undetermined is again two-valued, either true or false, but certainly not undetermined. More generally speaking, even in the three-valued logical framework the statements asserting that a certain statement is true, false or undetermined, are strictly two-valued. If we manage to translate all statements of the three-valued framework into combinations of two-valued statements of this sort, we have found a translation function. In the next section I will carry out this idea in a logically precise way.

3 Translating three-valued (or multi-valued) logic into classical logic

In what follows the indexed letter L_i varies over systems of propositional logic, and $\mathcal{L}_i$ designates the language of such a logic. L_2 denotes the classical (bivalent) propositional logic (consisting of its logical axioms, theorems and its valid inferences, the latter being denoted as $\vDash_{L_2}$). The language $\mathcal{L}_2$ contains $\neg, \wedge, \vee$ as primitive propositional connectives, moreover the material implication $\rightarrow$ and equivalence $\leftrightarrow$ being defined in the usual way. Languages are identified with the set of their well-formed formulas. I use

- $p_1, p_2, ..., q, r, ...$ as (propositional) variables,

- $A, B, ..., S, ...$ as schematic letters for arbitrary formulas (i.e., sentences), and

- $\Gamma, \Delta, ...$ for arbitrary sets of formulas.

L_3 is Lukasiewicz' three-valued logic (Łukasiewicz (1920)) with the truth-values true (t), false (f) and undetermined (u). The language $\mathcal{L}_3$ has four basic truth-functional connectives $\neg, \wedge, \vee$ and $\rightarrow$, where the three-valued conditional $\rightarrow$ is not being definable in terms of the other three connectives. As usual one assumes a linear ordering among the truth-values of a (finite) multi-valued logic; in our case the ordering is $f < u < t$; or represented as ranks: $-1, 0, +1$.

Based on this ordering, Lukasiewicz three-valued truth-tables for the four connectives are as follows: the truth value of $\neg p$ is the inverse of p's truth value, that of $p \wedge q$ is the minimum and that of $p \vee q$ the maximum of the truth values of p and q. Finally, $p \to q$'s truth value equals true, undecided or false, respectively, if the rank difference between q's truth value and p's truth value is not smaller than $0/-1/-2$.

p	$\neg p$		p	q	$p \wedge q$	$p \vee q$	$p \to q$
t	f		t	t	t	t	t
u	u		t	u	u	t	u
f	t		t	f	f	t	f
			u	t	u	t	t
			u	u	u	u	t
			u	f	f	u	u
			f	t	f	t	t
			f	u	f	u	t
			f	f	f	f	t

The notion of logical truth and validity in multi-valued logics is defined analogously as in bivalent logics. We let

- $\mathcal{P}$ be the denumerable set of propositional variables, and

- $val_3 : \mathcal{P} \to \{t, u, f\}$ range over trivalent truth-valuations over the (propositional) variables that are recursively extended to arbitrary complex formulas of $\mathcal{L}_3$ by way of the above truth tables. Then an $\mathcal{L}_3$-formula A is *logically true* in L_3, in short $\models_3 A$ iff $val_3(A) = t$ for all (possible) trivalent valuations, and A *follows* from a formula set Γ in L_3, in short $\Gamma \models_3 A$, iff all trivalent valuations making all formulas in Γ true make A true.

It is well known that some typical theorems and meta-theorems of classical L_2 are not among the theorems of L_3:

Some theorems of L_3: $p \to (q \to p), (\neg q \to \neg p) \leftrightarrow (p \to q),$
$(p \vee q) \leftrightarrow (p \to q) \to q$.
Some non-theorems of L_3: $p \vee \neg p, \neg(p \wedge \neg p), (p \vee q) \leftrightarrow (\neg p \to q)$.
Deduction theorem ($\Gamma, A \models B$ iff $\Gamma \models A \to B$) fails for L_3.

Our translation is based on the strategy to expand the classical two-valued language $\mathcal{L}_2$ by three operators T, U and F that express the truth values of being true, undetermined and false in three-valued logic. If S is a sentence of the three-valued logic, the sentences $T(S), F(S)$ and $U(S)$ are nevertheless two-valued, obeying the following truth table:

p	$T(p)$	$U(p)$	$F(p)$
t	t	f	f
u	f	t	f
f	f	f	t

We don't need to introduce these operators in $\mathcal{L}_3$, because they are definable in L_3 - $T(S)$ by $S \wedge \neg U(S)$, $F(S)$ by $\neg S \wedge \neg U(S)$, and $U(S)$ by $(S \vee \neg S) \rightarrow (S \wedge \neg S)$. As can be easily checked, the truth-functions of these formulas coincide with the above truth tables.

By adding the Łukasiewicz-operators T, U, F to the classical language $\mathcal{L}_2$ we obtain the extended classical language $\mathcal{L}_{2.Luk}$ whose formulas are still evaluated bivalently and whose basic logical laws are still the classical laws of L_2. Within L_2 the operators T, U and F figure as *intensional* (non-bivalently-truth-functional) operators, similar as the operators of modal logic. Based on the truth tables of these three operators, every semantic rule of three-valued logic can be translated into a set of corresponding axioms formulated in the expanded language of classical logic as follows:

- For negation: $T(\neg A) \leftrightarrow F(A), U(\neg A) \leftrightarrow U(A), F(\neg A) \leftrightarrow T(A)$

- For conjunction: $T(A \wedge B) \leftrightarrow T(A) \wedge T(B)$,
  $F(A \wedge B) \leftrightarrow F(A) \vee F(B)$,
  $U(A \wedge B) \leftrightarrow (U(A) \wedge \neg F(B)) \vee (U(B) \wedge \neg F(A))$

- For disjunction: $T(A \vee B) \leftrightarrow T(A) \vee T(B)$,
  $F(A \vee B) \leftrightarrow F(A) \wedge F(B)$,
  $U(A \vee B) \leftrightarrow (U(A) \wedge \neg T(B)) \vee (U(B) \wedge \neg T(A))$

- For implication:
  $T(A \rightarrow B) \leftrightarrow F(A) \vee (U(A) \wedge \neg F(B)) \vee (T(A) \wedge T(B))$,
  $U(A \rightarrow B) \leftrightarrow (T(A) \wedge U(B)) \vee (U(A) \wedge F(B))$,
  $F(A \rightarrow B) \leftrightarrow T(A) \wedge F(B)$

- Finally we add the trivalent truth-value axiom:

$$T(S) \,\dot\vee\, U(S) \,\dot\vee\, F(S)$$

("$\dot\vee$" for exclusive disjunction).

The set of these axiom schemata forms the axiom system Ax_{Luk} of Łukasiewicz' logic in the expanded language of classical logic $\mathcal{L}_{2.Luk}$.

Our translation of $\mathcal{L}_3$-statements into $\mathcal{L}_2$-statements is based on the truth view of assertion: asserting a sentence S means to assert that S is true. Thus our translation functions "$trans_{3\to 2}$" (from $\mathcal{L}_3$ into $\mathcal{L}_2$) is this:

$$\text{For all } S \in \mathcal{L}_3; trans_{3\to 2}(S) = T(S).$$

Note that the translation based on the assertion view is not recursive but *holistic*: it translates every complex $\mathcal{L}_3$-sentences at once into $\mathcal{L}_{2.Luk}$. By applying the axioms in Ax_{Luk} we transform every sentence of $\mathcal{L}_{2.Luk}$ into a truthfunctional combination of modalized variables (Tp_i, Up_i or Fp_i) and thus find out what the translation means for the truth-value of the modalized variables of the sentence.

Some examples of translations and their equivalent transformations: ("=" for identity, "$\leftrightarrow$" for material L_2-equivalence given Ax_{Luk}):

$trans(p) = T(p)$,
$trans(\neg p) = T(\neg p) \leftrightarrow F(p)$,
$trans(p \vee \neg p) = T(p \vee \neg p) \leftrightarrow Tp \vee Fp$,
$trans(p \wedge \neg p) = T(p \wedge \neg p) \leftrightarrow Tp \wedge Fp$,
$trans(p \to q) = T(p \to q) \leftrightarrow Fp \vee (Up \wedge (Uq \vee Tq)) \vee (Tp \wedge Tq)$, etc.

Note that we use the same logical symbols for the two-valued and the three-valued operators (e.g. both "$\to$" for two-valued and three-valued implication), but this is not a problem, because whenever we translate an $\mathcal{L}_3$-formula into $\mathcal{L}_2$, the three-valued logical operators are *hedged* in the scope of the intensional operators T, U and F.

We now show that the translation $trans_{3\to 2}$ preserves meaning and L_3-logical truth (or validity) in a precise sense. For this purpose, we have to introduce *some terminology*. In what follows, O_i ranges over the three trivalent truth-value operators, T, U and F. $\mathcal{P}(S) = p_1, ..., p_{n(S)}$ denotes the set of variables occurring in sentence S. We speak of the "p_i" as "unmodalized variables" and of the statements "$O_i p_j$" as the "modalized" variables. For $\mathcal{P}$ a set of unmodalized variables, $O\mathcal{P} = \bigcup_{p \in \mathcal{P}}\{Tp, Up, Fp\}$ denotes the corresponding set of modalized variables. If $\mathcal{P}$ is the (denumerable) set of unboxed variables common to $\mathcal{L}_3$ and $\mathcal{L}_{2.Luk}$, then following from the intensional nature of the operators O_i, truth-valuations over $\mathcal{L}_{2.Luk}$ are defined over the set of elementary formulas $\mathcal{P} \cup O\mathcal{P}$. Let $Val_3(\mathcal{P})$ be the set of all trivalent valuations over $\mathcal{P}$ and $Val_3(\mathcal{L}_3)$ be the set of (recursively extended) trivalent valuations over sentences of $\mathcal{L}_3$. Moreover, let $Val_{2.Luk}(O\mathcal{P})$ be the set of all bivalent valuations over $O\mathcal{P}$ satisfying the axiom $T(S) \dot\vee U(S) \dot\vee F(S)$ and $Val_{2.Luk}(\mathcal{L}_{2.Luk})$ be the set of all (recursively extended) bivalent truth-valuations over formulas of the expanded language satisfying the axioms of Ax_{Luk}. Then we can prove:

Theorem 1.

(1.1) Every three-valued valuation function val_3 over $\mathcal{P}$ corresponds exactly to a two-valued valuation function

$$val_2 =_{def} f(val_3)$$

over the modalized variables $O\mathcal{P} = \bigcup_{p \in \mathcal{P}} \{Tp, Up, Fp\}$ satisfying the axioms Ax_{Luk}, such that:

for every $\mathcal{L}_3$-formula A, $val_3 \models A$ iff $f(val_3) \models T(A)$.

(1.2) An $\mathcal{L}_3$-statement A is logically true in $\mathcal{L}_3$ iff $T(A)$ follows logically from Ax_{Luk} in L_2, and analogously for logical consequence. Thus:

$\models_{L_3} A$ iff $Ax_{Luk} \models_{L_2} T(A)$, and

$\Gamma \models_{L_3} A$ iff $Ax_{Luk} \cup T(\Gamma) \models_{L_2} T(A)$,

where $T(\Gamma) =_{def} \{T(B) : B \in \Gamma\}$.

Proof of the theorem: The proof is based on three lemmata:

Lemma 1: Every three-valued valuation val_3 over the unboxed variables in $\mathcal{P}$ corresponds to exactly one two-valued valuation val_2 over the corresponding boxed variables in $O\mathcal{P}$ satisfying the axiom $T(S) \dot\vee U(S) \dot\vee F(S)$.

Proof of lemma 1: We define $val_2 =_{def} f(val_3)$ as follows, for all $p \in \mathcal{P}$: $val_2(Tp) = t/val_2(Up) = t/val_2(Fp) = t$ exactly if $val_3(p) = t/u/f$. Then, the claim of lemma 1 is satisfied.

Lemma 2: For every $S \in \mathcal{L}_3$, Ax_{Luk} L_2-entails that $T(S)$ is equivalent with a distinguished disjunctive and negationless normal form $DN(S)$, each elementary disjunct being a conjunction of modalized variables $O_1p_1 \wedge ... \wedge O_np_n$ with one modalized variable for each unmodalized variable in $\mathcal{P}(S) = \{p_1, ..., p_n\}$. We call these conjunctions the *three-valued constituents*.

Proof of lemma 2: By successive application of the equivalence transformations corresponding to the axioms of Ax_{Luk}, one can drive the truth-value operators successively inside the formula until they stand immediately before the unmodalized variables; negations are eliminated thereby. By applying $\wedge - \vee$ distribution laws and expanding conjuncts in which certain modalized variables Op_i (for $p_i \in \mathcal{P}(S)$) are missing (via conjoining $Tp_i \vee Up_i \vee Fp_i$ to the conjunct), one can produce the required disjunction of three-valued constituents, abbreviated as $DN(S)$. The operations are analogous to those needed for producing distinguished normal forms in two-valued logics (we omit the inductive proof).

Lemma 3: Each elementary conjunct of $DN(S)$ corresponds to exactly one line in the three-valued truth-table of S that makes S true.

Proof of lemma 3: For every val_3 over $\mathcal{P}(S)$ making S true: val_3 verifies exactly one line in S's three-valued truth table of S. The corresponding two-valued valuation $val_2 = f(val_3)$ over $O\mathcal{P}(S)$ that verifies Ax_{Luk} makes exactly one disjunct of $DN(S)$ true (by lemma 1 and the truth table of bivalent conjunction). Thus val_2 makes $T(S)$ true, because $Ax_{Luk} \vDash_{L_2} S \leftrightarrow DN(S)$ holds, as proved in lemma 2. Vice versa, every val_2-function over $O\{p_1, ..., p_n\}$ satisfying Ax_{Luk} that makes $T(S)$ true makes exactly one disjunct of $DN(S)$ true, and this is the case exactly if the corresponding three-valued function $val_3 = f^{-1}(val_2)$ makes the corresponding line in the three-valued truth-table of S true.

Proof of theorem 1.1: For every $S \in \mathcal{L}_3$: $val_3 \vDash S$ iff some line in S's three-valued truth-table makes S true iff $val_2 = f(val_3) \vDash C_k(DN(S))$ for some elementary conjunct of $DN(S)$ (by lemma 3) iff $val_2 \vDash DN(S)$ iff $val_2 \vDash T(S)$ (by lemma 2).

Proof of theorem 1.2: $\vDash_{L_3} S$ iff $\forall val_3 \in Val_3(\mathcal{P}(S))$: $val_3 \vDash S$ iff $\forall val_2 \in Val_2(O\mathcal{P})$: $val_2 \vDash S$ (by theorem 1.1) iff $\vDash_2 S$. Analogously for inferences. Q.E.D.

4 Discussion and generalization of the proposed translation method

In the next three subsections we explain what we think has been achieved by the translation method.

4.1 Preservation of meaning: The translation $trans_{3\to 2}$ together with the axiom system Ax_{Luk} preserves the semantic meaning of the trivalent operators, which are part of $\mathcal{L}_{2.Luk}$ as well as of $\mathcal{L}_3$. The translation also preserves the meaning of the (propositional) variables 'as good as possible'. Of course, the meaning of "p" cannot be strictly the same in L_3 and L_2, because in L_3, p has three and in L_2, two truth values. However, the meaning of the more *fine-grained* propositions $T(p), O(p)$ and $U(p)$ is strictly the same in L_3 and in L_2. Asserting a sentence S in L_3 is expressed by asserting $T(S)$ in L_2; moreover by applying the Ax_{Luk}-equivalences (that are valid in $L_{2.Luk}$ as well as in L_3), the semantic composition of S in L_3 is fully reflected in L_2. We conclude that *every proposition that can be expressed in L_3 can be also expressed in L_2*.

4.2 Comparison with literature: By expanding the classical truth-functional logic by the intensional operators T, U, F we made it possible to preserve the meaning of the operators of the non-classical logic. This meaning-preservation distinguishes my account from the translation functions between logics studied in the previous literature. In the latter work, translations are not accompanied by expansions of the (classical) language, on the cost that these translation functions do not and cannot preserve the meaning and semantic composition of the translated statements; they only preserve the consequence operation. One example are the abstract 'translations functions' studied by Jerábek (2012). These translation functions map the formulas of the language $\mathcal{L}$ of a propositional logic L into formulas of a language $\mathcal{L}'$ of a logic L', such that if $A \vdash_L B$, then $f(A) \vdash_{L'} f(B)$ (where f need neither be injective nor surjective). Given an enumeration of all $\mathcal{L}$-formulas and the nth formula A_n of $\mathcal{L}$, Jerabek's translation of A_n into the language $\mathcal{L}'$ is, roughly speaking, defined as $X \vee (q_n \wedge Y)$, where X is the disjunction of translations of all premises with indices smaller than n that entail A_n, q_n is a new variable and Y is the conjunction of all translations of implications $C \to D$ with indices smaller than n such that $\{C, A_n\}$ entails D (ibid., 669). Jerábek (2012), theorem 2.6 proves that classical logic is 'translation-universal' in the sense that every finitary deductive system in countably many formulas can be conservatively translated into classical logic; moreover that many other but not all non-classical propositional logics are universal in this sense. The result is technically impressing, but obviously, Jerabek's 'translation' function neither preserves the meaning or semantic composition of formulas nor even their syntactic structure; the translation is constructed just for the purpose of preserving the consequence operation. Since the defender of a non-classical logic can argue that such a 'translation' does not preserve the meaning of the $\mathcal{L}_3$-statements, the translation is not useful for the purpose of an optimality justification in regard to representation power.

An example of a semantic 'translation of non-classical into classical logic is the bivalent reduction of multi-valued logic proposed by Suszko (1977). Given a standard multi-valued logic with a subset $Des \subset Val$ of designated truth values, Suszko proposed to translate the disjunction (or set) of the designated truth-values into the bivalent value "true" and the disjunction (or set) of the non-designated truth-values into "false". Suszko's translation is useful for many purposes (cf. Béziau (1999)). However, Suszko's translation does not preserve the semantic meaning of the propositional connectives; they become intensional under Suszko's bivalent semantics (Malinowski (1993), 79; Wansing and Shramko (2008)). For example, both p and $\neg p$ may have the truth-value false; thus the law of excluded middle, $p \vee \neg P$, is no longer valid in Suszko's bivalent semantics. Therefore, Suszko's bivalent semantics is not classical and, thus, does not yield a translation of L_3 into a classical logic.

4.3 Bridge axioms between L_3 and L_2: For every $S \in \mathcal{L}_3$, the truth value of $T(S)$ depends only on the truth values of its modalized but not of its unmodalized variables. So far, the bivalent truth-values of the modalized variables ($O_i p$) have not been related to the bivalent truth-value of the unmodalized variables. For the semantic coherence between L_3 and L_2 we require the following bridge axioms:

$$T(S) \rightarrow S \quad \text{and} \quad F(S) \rightarrow \neg S.$$

In words, a trivalently true (or false) sentence is also bivalently true (or false, respectively), while for undetermined sentences their bivalent truth value is *left open*. It is important that our translation does not prescribe whether a trivalently undetermined statement should be bivalently classified as true or false; this may depend on the particular context and content of p. This has to be so: the *converse implications* must not hold, since otherwise the translation would not be conservative and the translated three-valued logic would collapse into two valued logic. If we would accept the inverse bridge axioms we could infer $T(p \vee \neg p)$ and thus $\vDash_{L_3} p \vee \neg p$ from $\vDash_{L_2} p \vee \neg p$.

5 Generalizations

In general, the notion of validity in multi-valued logics is defined by assuming a subset $Des \subset Val$ of designated (truth-) values (Val being the set of all truth-values) and defining a formula A as valid in L_{Val} if all L_{Val}-valuations convey to A a designated value. The triple $\langle Val, Des, \{t_c : c \in C\}\rangle$ (with $\{t_c : c \in C\}$ being the set of truth-tables for a set of connectives C) is called a Val-valued logical matrix.

It is rather obvious that the translation strategy of sec. 3 applies to all many-valued logics that are representable by means of a matrix of finitely many truth values. Thus, if an n-valued logic L_n is based on a matrix $\langle Val_n, Des_k, \{t_c : c \in C\}\rangle$ with $|Val_n| = n$, $|Des_k| = k < n$ and $C = \{\neg, \wedge, \vee, \rightarrow\}$, then we introduce the n intensional operators $O_1, ..., O_n$ for the n truth values, the equivalence axioms for $\neg, \wedge, \vee$ and $\rightarrow$ describing the truth tables in terms of these n operators and the n-valent truth-value axiom $O_1(S)\dot{\vee}...\dot{\vee}O_n(S)$, and prove the translation theorem in the same way as above. Of course, with many truth values this translation can become rather cumbersome. It can be shown, however, that in many cases less intensional operators than truth values are needed to obtain an adequate translation function.

It is worth emphasizing that also paraconsistent logics can be characterized by means of finite truth value matrices (Priest (1979); (2013), sec. 3.6). The simplest paraconsistent logic, LP, contains the three truth values t, f and b for "both true and false". The designated values are t and b, which prevents the 'principle of explosion', i.e. that an inconsistent premise $p \wedge \neg p$ entails anything.

For logics that can only be algebraically represented by infinite matrices such as *intuitionistic logic*, the above translation strategy does not work. However, for all of these cases known to me there exist other methods of translation. It is well-known that intuitionistic logic is translatable into the modal logic $S4$, by interpreting the necessity operator as provable truth (Rautenberg (1979), 265). Also for quantum logic there exist translation functions into modal logic (cf. Dalla Chiara and Giuntini (2002), sec 5). All these fact support my *conjecture* that for every non-classical logic one can find a translation into an suitably expanded classical logic. My reason for this conjecture is the mentioned fact that all non-classical logics known to me use classical logic in their metalanguage, in which they describe the (correct and complete) semantics of their non-classical principles. Thus there should exist ways of translating their non-classical principles into classical logic, by introducing intensional operators into L_2 that correspond to the concepts used in the semantics of the non-classical logics. Elaborations of this conjecture are work for the future.

6 Epistemological conclusions and discussion of possible objections

If our argument is correct, it shows that every non-classical logic can be represented within classical logic, because everything expressible in the former can be expressed in the latter without loss of meaning, namely by expanding the classical language with appropriate operators and axioms for them. Since the basic laws and rules of classical logic are still valid in the expanded system, this argument gives us an *optimality justification* of classical logic: By using classical logic our conceptual representation system *can only gain but can never lose*, because if another logic has advantages for certain purposes, we can translate and thus embed it into classical logic. What is furthermore achieved by this result is *epistemic commensurability* and thus a refutation of logical relativism: different logical frameworks are not incommensurable (in the sense of Kuhn (1962)), because they are translatable into each other.

An apparent objection to my account points out that there is also the possibility of an inverse translation relation of classical logic into three-valued logic (or more generally, into a non-classical logic), by expanding the non-classical language with intensional operators for the bivalent truth-values and corresponding axioms. Indeed, it can be shown that such an inverse translation is often possible (because of space limitations we cannot demonstrate this here). Yet this 'objection' does not refute my optimality thesis, because *optimality does not entail dominance*. The two notions come from game theory. A method (here a logic) is dominant iff it has higher value than all other methods (in a given class of methods and in regard to the given goal, here linguistic representation power). It is optimal if it has maximal value, which does not exclude that there may be methods that have the same maximal value. Thus the defender of a three-valued logic may argue that his or her system is optimal, too, since (s)he can translate every bivalent system into his or her trivalent logic.

Yet the objection is not fully defeated by this move. For one may ask: if many logics are representationally optimal, what has been achieved by the proposed optimality justification of classical logic? There are three replies to this questions:

(1.) It is an open question how large the class of logics is that are provably representation-optimal, i.e. for which a universal translation theorem can be proved.

(2.) Even if the class of representationally optimal logics would be very large, optimality justifications would nevertheless be of high epistemological importance. These justifications show that different logics, that are incompatible on the level of their theorems, may on a deeper level be translated into each other. Therefore different logics are neither incommensurable nor do they put us into a situation of logical relativism. Instead we obtain the situation of a 'pluralism in harmony'. Every optimal logical system is foundation-theoretically justified and can be chosen as one's basic logic, because every other logical system can be embedded into it.

(3.) This is not all what can be said about epistemic preference of logics. There is an important ceteris paribus criterion for the choice of logical systems: among logics that are equally representationally optimal one should choose the simplest logic. One can plausibly argue that, at least in many respects, classical bivalent logic is the simplest logic. Moreover, translations of non-classical systems into this logic are most simple and straightforward. Moreover, one should not forget that in most contexts the principle of two-valuedness has enormous advantages. For example, the meta-logical properties of a logic should neither be undetermined nor paraconsistent. No logician wants to say that his or her preferred logic is neither correct nor incorrect, or is both consistent and inconsistent, etc. (cf. Batens (2014), §1). All this does not exclude, of course, that for specific application purposes a non-classical logic can have advantages.

References

[1] Achinstein, P. (1974). "Self-Supporting Inductive Arguments", in Swinburne (1974), *The Justification of Induction*. Oxford: Oxford University Press.

[2] Anderson, Alan R. and Belnap, N. D. (1975). *Entailment. The Logic of Relevance and Necessity*, Princeton Univ. Press, Princeton.

[3] Batens, D. (2014). "The Consistency of Peano Arithmetic. A Defeasible Perspective", in P. Allo and B. van Kerkhove (Hg.), *Modestly Radical or Radically Modest*, College Publications, London.

[4] Béziau, J.-Y. (1999): "A Sequent Calculus for Lukasiewicz's three-valued Logic based on Susko's Bivalent Semantics", *Bulletin of the Section of Logic* 28/2, pp. 89-97.

[5] Birkhoff, G., and von Neumann, J. (1936) "The Logic of Quantum Mechanics", *Annals of Mathematics* 37, pp. 823-843.

[6] Bueno, Otavio (2010). "Is Logic A Priori?", *The Harvard Review of Philosophy* 17, pp. 105–117.

[7] Dalla Chiara, M. L. and Giuntini, R. (2002). "Quantum Logics", in D. Gabbay and F. Guenthner (eds.), *Handbook of Philosophical Logic* Vol. 6, Kluwer, Dordrecht, 129-228.

[8] Dummett, M. (1976). "Is Logic Empirical?" in M. Dummett, *Truth and Other Enigmas*, Harvard University Press, 1978, pp. 269-289.

[9] Gottwald, Siegfried (1989). *Mehrwertige Logik*, Akademie-Verlag. Berlin.

[10] Hjortland, O. T. (2017). "Anti-Exceptionalism about Logic", *Philosophical Studies* 174, pp. 631–658.

[11] Jeřábek, E. (2012). "The Ubiquity of Conservative Translations", *Review of Symbolic Logic* 5/4, pp. 666-678.

[12] Kuhn, T.S. (1962). *The Structure of Scientific Revolutions, Die Struktur wissenschaftlicher Revolutionen.* Chicago: Univ. of Chicago Press (3rd edition 1996).

[13] Leitgeb, H. (2018). "HYPE: A System of Hyperintensional Logic". Appears in *Journal of Philosophical Logic* (doi.org/10.1007/s10992-018-9467-0).

[14] Lukasiewicz, J. (1920). "O logice trojwartosciowej". In: *Ruch Filozoficny* 5, pp. 170–171. Translated into English in: J. Łukasiewicz (1970), *Selected Works* (ed. by L. Borkowski). Amsterdam: North-Holland and Warsaw.

[15] Malinowski, G. (1993). *Many-Valued Logics*. Oxford: Clarendon Press.

[16] Popper, K. (1968). "Birkhoff and von Neumann's Interpretation of Quantum Mechanics", *Nature* 219, pp. 682-685.

[17] Priest, G. (1979). "Logic of Paradox"". *Journal of Philosophical Logic* 8, pp. 219-241.

[18] Priest, G. (2006). *In Contradiction* (2nd edition). Oxford: Oxford University Press.

[19] Priest, G. (2013). "Paraconsistent Logic". In: *Stanford Encyclopedia of Philosophy* (ed. by E, Zalta), http://plato.stanford.edu.

[20] Putnam, H. (1979). "The Logic of Quantum Physics", in H. Putnam. *Mathematics, Matter and Method: Philosophical Papers* Vol. 1, 2nd ed., Cambridge University Press, Cambridge, pp. 174–197.

[21] Rautenberg, W. (1979). *Klassische und nichtklassische Aussagenlogik*, Vieweg, Braunschweig.

[22] Reichenbach, H. (1949). *The Theory of Probability*. Berkeley: University of Cali-fornia Press.

[23] Rescher, N. (1977). *Methodological Pragmatism*. Oxford: B. Blackwell.

[24] Salmon, W. C. (1957). "Should We Attempt to Justify Induction?", *Philosophical Studies* 8/3, pp. 45-47.

[25] Schurz, G. (2008a). "Third-Person Internalism: A Critical Examination of Externalism and a Foundation-Oriented Alternative", *Acta Analytica* 23, pp. 9-28.

[26] Schurz, G. (2008b). "The Meta-Inductivist's Winning Strategy in the Predic-tion Game: A New Approach to Hume's Problem", *Philosophy of Science* 75, pp. 278-305.

[27] Schurz, G. (2018a). "Optimality Justifications: New Foundations for Foundation-Oriented Epistemology", *Synthese* 195, pp. 3877-3897.

[28] Schurz, G. (2018b). *Logik*. de Gruyter, Berlin und New York.

[29] Schurz, G. (2019). *Hume's Problem Solved: The Optimality of Meta-Induction*, MIT Press, Cambridge/Massachusetts.

[30] Schurz, G., and Thorn, P. (2016). "The Revenge of Ecological Rationality: Strategy-Selection by Meta-Induction", *Minds and Machines* 26(1), pp. 31-59.

[31] Sereni, A., and Sforza-Fogliani, M. P. (2017). "How to Water a Thousand Flowers. On the Logic of Logical Pluralism", *Inquiry*, https://doi.org/10.1080/0020174X.2017.1370064.

[32] Stachel, J. (1986). "Do Quanta Need a New Logic?" in R. G. Colodny, A. Coffa (Hg.), *From Quarks to Quasars: Philosophical Problems of Modern Physics*. University of Pittsburgh Press, Pittsburgh, pp. 229-347.

[33] Suszko, R. (1977). "The Fregean Axiom and Polish Mathematical Logic in the 1920's", *Studia Logica* 36, pp. 373-380.

[34] Thorn, P., und Schurz, G. (2020). "Meta-Inductive Prediction based on Attractiv-ity Weighting: An Empirical Performance Evaluation", *Journal of Mathematical Psychology* 89, pp. 13–30.

[35] Wansing, H., and Shramko, Y. (2008). "Suszko's Thesis, Inferential Many-Valuedness, and the Notion of a Logical System", *Studia Logica* 88, pp. 405-429.

[36] Williamson, T. (2014). "Logic, Metalogic, and Neutrality", *Erkenntnis* 79, pp. 211–231.

[37] Williamson, T. (2017). "Semantic Paradoxes and Abductive Methodology", in B. Armour-Garb (Hg.): *Reflections on the Liar*. Oxford University Press, Oxford, pp. 325–346.

[38] Woods, J. (2019). "Logical Partisanhood", *Philosophical Studies* 176, pp. 1203-1224.

ABSTRACT LOGIC WITH VOCABLES

Roderick Batchelor

Department of Philosophy, University of São Paulo, Brazil

Abstract

The paper proposes a new abstract notion of 'logical system', and sketches some basic developments of the general theory of such logical systems. Roughly speaking, a logical system in our sense consists of: ('abstract') sets of *models* and of *formulas*, with a binary relation of *verification* between them, plus (more distinctively) a set of *vocables* together with a *vocable-function* from formulas to sets of vocables. (In a typical concrete case, the 'vocables' of a formula will be the items of vocabulary, other than logical constants, occurring in the formula.) This notion seems to strike a quite good balance between simplicity and comprehensiveness on the one hand, and conceptual richness on the other. It is particularly noteworthy that it permits an elementary though non-trivial abstract study of *interpolation* and related ideas. – A final section of the paper briefly considers an extension of our basic notion of logical system by the addition of a *domain-function* permitting us to speak of the domain of a model.

1 Introduction

When one studies the semantical theory of various different logical systems, one is likely to be struck by the constant recurrence of patterns of definition of various notions in terms of simpler ones, as well as sometimes even proofs of various results from earlier ones; so much so that the eventual systematization of a corresponding 'abstract' scheme is almost inevitable. Thus various forms of such 'Abstract Logic' have appeared. The present paper proposes a new scheme of this kind.

Our scheme is meant to be very broadly applicable: in particular, applicable to both *classical* and *non-classical* (e.g. modal) logic, both *propositional* and *predicate logic*, both *first-order* and *higher-order* systems, both *finitary* and *infinitary* systems, both systems with '*functionally complete*' and systems with '*functionally incomplete*' logical languages (even languages lacking some of the classical truth-functional connectives). Thus, within the (rather varied) field of 'Abstract Logic', our formulation is along the lines of '*Abstract* Abstract Logic', like e.g. Tarski's original theory of abstract consequence relations or operations (Tarski 1930), rather than of 'More-or-less *Concrete* Abstract Logic', like the Polish theory of *structural* consequence relations (on which see e.g. Wójcicki 1988), or the 'Abstract Model Theory' tradition (see e.g. Barwise and Feferman 1985).

In Tarski's formulation there is a set of (abstract) 'formulas' and an (abstract) 'consequence relation' which is a relation between sets of formulas and formulas required to satisfy certain conditions ('axioms') like reflexivity, monotonicity, transitivity. (Actually Tarski has a consequence *operation* rather than *relation* – but it comes to the same.) Here we will go back one step in the analysis of the situation and will have a set of (abstract) '*models*' (in addition to the set of 'formulas') and a relation of *verification* between models and formulas; from which the consequence relation can be *defined* and its standard properties (reflexivity etc.) *derived* from the mere definition. (Our goal here is to develop an abstract analysis of recurrent aspects of *formal semantics*; so for us this is a natural step. *Tarski's* own goal by contrast was to develop an abstract analysis of recurrent aspects of *syntactic, axiomatico-deductive systems* ['formalized deductive disciplines']; so for *him* there was no corresponding natural step. [At most one can think of 'maximal consistent' sets of formulas; but obviously this is something that would come later than 'consequence' (i.e. deducibility) in the natural order of definitions in the syntactic context.])

This step is well known and can be found in many places in the literature of Abstract Logic. What is more distinctive about the approach taken here is that we have *also* (in addition to the ingredients already mentioned) a set of (abstract) '*vocables*' and a *vocable-function* L from formulas to sets of vocables. Thus e.g. in a typical concrete system of propositional logic, $L(\varphi)$ would be the set of propositional variables occurring in φ; in first-order predicate logic $L(\varphi)$ would be the set of predicate variables and free individual variables (and function variables and individual parameters if we have that) occurring in φ; and so on. (Or if we are not speaking of 'first-order predicate logic' in the sense where we have a single language with a fixed stock of predicate variables but rather considering the usual notion of a 'first-order language', then $L(\varphi)$ is the set of 'non-logical constants' and free individual variables occurring in φ [or simply the set of non-logical constants occurring in φ, if by 'formula' we understand sentence].) – With these additional ingredients we are able to make an abstract study of concepts and conditions which involve an interplay between 'semantical' notions like validity, satisfiability, consequence, etc., and 'syntactical' notions like the vocables which occur in a given formula or set of formulas – a paradigmatic example of which is of course *Interpolation*. – At the same time, by including no *other* ingredients we still have a framework that is quite simple and quite broadly applicable.

I will add that, although it is concrete logical systems of usual kind that I had primarily in mind as the paradigmatic concrete examples of my abstract 'logical systems', yet there is another very interesting application (concrete example) of these systems. Namely: the 'models' are now metaphysical *possible worlds*; the 'formulas' are *structured propositions* (states of affairs, situations); the 'vocables' are *simple non-logical entities*; 'verification' is converse of *true in*; and 'vocables-of' are *simple non-logical entities which are constituents of*. (See Batchelor 2013, esp. § 7.) Or again, instead of full-blown possible worlds one might have maximally specific possible states *relative to* a certain subject-matter, and a correspondingly limited notion of structured proposition.

Even in the more 'logical' cases there are also of course interesting variations on the more obvious concrete examples. For instance: 'models' might be something like maximal consistent sets of formulas, and 'verification' implication (or deducibility in a certain formal calculus); 'formulas' might be say propositional-logic formulas in *certain* variables only, and so on.

2 Basic notions

A *logical system* is a 5-tuple

$$(Mod, Fla, Voc, \vdash, L),$$

where:

Mod ('models') and *Fla* ('formulas') are arbitrary non-empty sets;

Voc ('vocables') is an arbitrary set (not necessarily non-empty);

$\vdash$ ('verification') is a binary relation between elements of *Mod* and elements of *Fla* (i.e. $\subseteq Mod \times Fla$);

L ('vocable-function') is a function from *Fla* to subsets of *Voc*.

Remark. We speak of the *set* of models; but we certainly mean to include e.g. classical predicate logic within the scope of the present notion of logical system, although its 'models' would not normally be said to 'form a set'. We leave to the reader the choice among alternative methods for resolution of this discrepancy. ⊣

As already indicated we call the elements of *Mod, Fla, Voc* respectively *models, formulas,* and *vocables*. We use the following variables:

Variables for models: $M, N, K, J, \ldots$

Variables for formulas: $\varphi, \psi, \theta, \chi, \ldots$

Variables for vocables: $p, q, r, s, \ldots$

Variables for sets of models: $\mathcal{M}, \mathcal{N}, \ldots$

Variables for sets of formulas: $\Gamma, \Delta, \Sigma, \Theta, \ldots$

Variables for sets of vocables: $\ell, \ell', \ldots$

– A *partial logical system* is like a logical system except that *Voc* and *L* are omitted.

We give now the following battery of definitions, which are relative to an arbitrary logical system (and are most of them applicable also to an arbitrary *partial* logical system): –

$M \vdash \Gamma =_{df} \forall \varphi \in \Gamma (M \vdash \varphi)$.

$Sat(\varphi) =_{df} \exists M (M \vdash \varphi)$.

$Sat(\Gamma) =_{df} \exists M (M \vdash \Gamma)$.

Remark. It is of course no part of our *definition* of logical system that models must be 'maximal' (at least w.r.t. the formulas of the system) and 'possible' ('internally coherent') (nor does it seem possible to *state* such requirements in the present abstract framework); but this will always be so in the intended applications; and the present definition of satisfiability already reflects this. If models were not 'maximal', a set of formulas could be intuitively 'satisfiable' but not verified (implied) by any model; and if models could be 'impossible', a set of formulas could be implied by a model without being intuitively 'satisfiable'. ⊣

$Unsat(\varphi) =_{df}$ not $Sat(\varphi)$.

$Unsat(\Gamma) =_{df}$ not $Sat(\Gamma)$.

$\varphi \vDash \psi =_{df} \forall M (M \vdash \varphi \Rightarrow M \vdash \psi)$.

$\Gamma \vDash \varphi =_{df} \forall M (M \vdash \Gamma \Rightarrow M \vdash \varphi)$.

$\Gamma \vDash \Delta =_{df} \forall M (M \vdash \Gamma \Rightarrow M \vdash \Delta)$.

$\varphi \simeq \psi =_{df} \forall M (M \vdash \varphi \Leftrightarrow M \vdash \psi)$.

$\varphi \simeq^{\Gamma} \psi$ ['φ is equivalent to ψ modulo Γ'] $=_{df} \forall M (M \vdash \Gamma \Rightarrow (M \vdash \varphi \Leftrightarrow M \vdash \psi))$.

$\Gamma \simeq \Delta =_{df} \forall M (M \vdash \Gamma \Leftrightarrow M \vdash \Delta)$.

$\vDash \varphi =_{df} \forall M (M \vdash \varphi)$.

$\vDash \Gamma =_{df} \forall M (M \vdash \Gamma)$.

$DisjVal(\Gamma)$ ['Γ is disjunctively valid'] $=_{df} \forall M \exists \varphi \in \Gamma (M \vdash \varphi)$.

$Cn(\varphi) =_{df} \{\psi : \varphi \vDash \psi\}$.

$Cn(\Gamma) =_{df} \{\varphi : \Gamma \vDash \varphi\}$.

Γ is closed $=_{df} \Gamma = Cn(\Gamma)$.

Γ is complete $=_{df} \forall\varphi(\Gamma \vDash \varphi \text{ or } Unsat(\Gamma \cup \{\varphi\}))$.

Γ is ℓ-complete $=_{df} \forall\varphi(L(\varphi) \subseteq \ell \Rightarrow (\Gamma \vDash \varphi \text{ or } Unsat(\Gamma \cup \{\varphi\})))$.

(Thus of course completeness tout court is *Voc*-completeness.)

Γ is coherent $=_{df}$ not $\exists\varphi(\Gamma \vDash \varphi \& Unsat(\Gamma \cup \{\varphi\}))$.

Γ is saturated $=_{df} Sat(\Gamma) \& \forall\Delta \supset \Gamma(Unsat(\Delta))$.

$Cons(\Gamma) =_{df} \forall\Gamma_0 \subseteq_{fin} \Gamma(Sat(\Gamma_0))$.

$MaxCons(\Gamma) =_{df} Cons(\Gamma) \& \forall\Delta \supset \Gamma(\text{not } Cons(\Delta))$.

$\Gamma \to \Delta =_{df} \forall M(M \vdash \Gamma \Rightarrow \exists\varphi \in \Delta(M \vdash \varphi))$.

φ settles $\psi =_{df} \varphi \vDash \psi$ or $Unsat\{\varphi, \psi\}$.

Γ settles $\varphi =_{df} \Gamma \vDash \varphi$ or $Unsat(\Gamma \cup \{\varphi\})$.

$Topic(\Gamma) =_{df} DisjVal(\Gamma) \& \forall\varphi, \psi \in \Gamma(\varphi \neq \psi \Rightarrow Unsat\{\varphi, \psi\})$.

$Th(M)$ ['the theory of M'] $=_{df} \{\varphi : M \vdash \varphi\}$.

$Th(\mathcal{M}) =_{df} \{\varphi : \forall M \in \mathcal{M}(M \vdash \varphi)\}$.

Γ is a theory $=_{df} \exists\mathcal{M}(\Gamma = Th(\mathcal{M}))$.

Γ is a simple theory $=_{df} \exists M(\Gamma = Th(M))$.

Remark. Obviously $M \vdash \varphi$ is always equivalent to $\varphi \in Th(M)$. Thus one might 'economize' by defining a logical system instead as a quadruple (*Fla*, *STh*, *Voc*, *L*) where *STh* is a non-empty set of *sets of formulas*, corresponding to the simple theories, and the rest is as before (and similarly for the *partial* case). Instead of $M \vdash \varphi$ for M in *Mod*, we would have now $\varphi \in \Gamma$ for Γ in *STh* ($\in$ being of course simply part of the ambient mathematical framework and not a constituent of the structure). (This kind of move is sometimes coupled with a requirement that $Fla \notin STh$, which however at least for our purposes would not be appropriate: e.g. in classical propositional logic with connectives say $\wedge$ and $\vee$ only, Fla *is* a simple theory. Incidentally, the same example shows the inadequacy [at least w.r.t. our purposes here] of the definition of $Sat(\Gamma)$ as $\exists \varphi (\Gamma \nvDash \varphi)$, usual in the theory of 'abstract consequence relations'. However, if one goes from structures $(Fla, \vDash)$ to structures $(Fla, \vDash, Sat)$, to which one may or may not add Voc and L, and of course with 'axioms' *both* for $\vDash$ and for Sat, one can do a good deal – though by no means *all* – of what we do here with our [partial or full] abstract 'logical systems'.) – This economy seems to me however somewhat artificial: the 'specifics' of the development of the basic semantical theory of a typical logical system stop precisely after the definitions of *model* and *verification* (and formulas, vocables, L); the definition of simple theory comes later and follows a fixed 'abstractable' pattern, like the definitions of satisfiability, validity, consequence, etc. ⊣

$\ell\text{-}Th(\mathcal{M})$ ['the ℓ-theory of $\mathcal{M}$'] $=_{df} \{\varphi : L(\varphi) \subseteq \ell \ \& \ \mathcal{M} \vDash \varphi\}$.

Γ is an ℓ-theory $=_{df} \exists \mathcal{M}(\Gamma = \ell\text{-}Th(\mathcal{M}))$.

$M \equiv^{\Gamma} N =_{df} \forall \varphi \in \Gamma (M \vDash \varphi \Leftrightarrow N \vDash \varphi)$.

$M \equiv^{\ell} N =_{df} M \equiv^{\{\varphi : L(\varphi) \subseteq \ell\}} N$.

$M \equiv N =_{df} M \equiv^{Fla} N$.

$Mod(\varphi)$ ['the models of φ'] $=_{df} \{M : M \vDash \varphi\}$.

$Mod(\Gamma) =_{df} \{M : M \vDash \Gamma\}$.

Γ axiomatizes $\mathcal{M} =_{df} Mod(\Gamma) = \mathcal{M}$.

$\mathcal{M}$ is axiomatizable $=_{df} \exists \Gamma (\Gamma$ axiomatizes $\mathcal{M})$.

$\mathcal{M}$ is finitely axiomatizable $=_{df} \exists$ finite $\Gamma (\Gamma$ axiomatizes $\mathcal{M})$.

$\mathcal{M}$ is axiomatizable by single formula $=_{df} \exists \varphi (\{\varphi\}$ axiomatizes $\mathcal{M})$.

Γ is redundant $=_{df} \exists \Delta \subset \Gamma (\Delta \simeq \Gamma)$.

Otherwise we say that Γ is *non-redundant*.

Γ is independent $=_{df} \forall \Delta \subseteq \Gamma \; \exists M (M \vDash \Delta \; \& \; \forall \varphi \in \Gamma - \Delta (M \nvDash \varphi))$.

Γ is essentially finite $=_{df} \exists$ finite $\Delta (\Delta \simeq \Gamma)$.

(Note that this is not always equivalent to $\exists$ finite $\Delta \subseteq \Gamma (\Delta \simeq \Gamma)$. E.g. in a suitable form of infinitary classical propositional logic, a denumerable set of propositional variables would be essentially finite, being equivalent to the conjunction of such variables, but would not be equivalent to any finite subset of itself.)

All the above definitions, with the exceptions only of ℓ-completeness and ℓ-theory and $\equiv^\ell$, make sense for arbitrary *partial* logical systems; the following on the other hand involve also vocables: –

$L(\Gamma) =_{df} \{p : \exists \varphi \in \Gamma (p \in L(\varphi))\}$.

p is essential to $\varphi =_{df} \forall \psi \simeq \varphi (p \in L(\psi))$.

p is inessential to $\varphi =_{df} \exists \psi \simeq \varphi (p \notin L(\psi))$.

p is inessential to φ modulo $\Gamma =_{df} \exists \psi \simeq^\Gamma \varphi (p \notin L(\psi))$.

$Ess(\varphi) =_{df} \{p : p \text{ is essential to } \varphi\}$.

$Ess(\Gamma) =_{df} \{p : \forall \Delta \simeq \Gamma (p \in L(\Delta))\}$.

$Can(\varphi)$ ['φ is canonical'] $=_{df} L(\varphi) = Ess(\varphi)$.

$Can(\Gamma) =_{df} L(\Gamma) = Ess(\Gamma)$.

The following two definitions give tentative abstract versions of the ideas of explicit definability and implicit definition.

p is explicitly definable w.r.t. $\Gamma =_{df} \forall \varphi$: p is inessential to φ modulo Γ.

Remark. This condition seems to be as close as we can get to the usual idea of explicit definability in our abstract framework. But note that it allows also cases of 'context-sensitive' definability: as long as p is always eliminable (modulo Γ), the condition is satisfied – the 'elimination' might be very different for different formulas. ⊣

Γ implicitly defines $p =_{df} \forall M \, \forall N[(M \vDash \Gamma \, \& \, N \vDash \Gamma \, \& \, \forall \varphi(p \notin L(\varphi) \Rightarrow (M \vDash \varphi \Leftrightarrow N \vDash \varphi))) \Rightarrow \forall \varphi(M \vDash \varphi \Leftrightarrow N \vDash \varphi)]$.

– Although our definition of logical system has no special conditions ('axioms'), from the above definitions it is possible to derive many properties of the notions of consequence ($\vDash$), satisfiability, validity, etc. – including e.g. all the usual 'axioms' for an 'abstract consequence relation' or 'abstract consequence operation' (but not Compactness of course – Tarski included that in his original papers [in keeping with his Abstract *Syntax* motivation] but it is no longer included nowadays), and hence also of course all the theorems which can be derived therefrom. – We list here some illustrative examples (all of which follow easily from the definitions above): –

$Sat(\Gamma \cup \Delta) \Rightarrow Sat(\Gamma)$.

$\Gamma \vDash \varphi \Rightarrow \Gamma \cup \Delta \vDash \varphi$.

$\varphi \in \Gamma \Rightarrow \Gamma \vDash \varphi$.

$\Gamma \vDash \Delta \, \& \, \Delta \vDash \Sigma \Rightarrow \Gamma \vDash \Sigma$.

$\Gamma \vDash \Delta \, \& \, Sat(\Gamma) \Rightarrow Sat(\Delta)$.

$Unsat(\Gamma) \Rightarrow \Gamma \vDash \Delta$.

$\simeq$ (between formulas, or between sets of formulas) is an equivalence relation.

$\vDash \Gamma \Rightarrow Sat(\Gamma)$.

$\vDash \Gamma \Rightarrow \Delta \vDash \Gamma$.

$\Gamma \subseteq Cn(\Gamma)$.

$Cn(\Gamma) = Cn(Cn(\Gamma))$.

$\Gamma \subseteq \Delta \Rightarrow Cn(\Gamma) \subseteq Cn(\Delta)$.

$Cons(\Gamma) \Rightarrow \exists \Delta \supseteq \Gamma (MaxCons(\Delta))$. (By Tukey's lemma.)

∀ finite Γ $\exists \Delta \subseteq \Gamma (\Delta \simeq \Gamma$ & Δ is non-redundant).

$Unsat(\Gamma) \Leftrightarrow \Gamma \to \varnothing$.

$DisjVal(\Gamma) \Leftrightarrow \varnothing \to \Gamma$.

$\Gamma \cup \{\varphi\} \to \Delta \cup \{\varphi\}$.

$\Gamma \to \Delta \Rightarrow \Gamma \cup \Sigma \to \Delta$.

$\Gamma \to \Delta \Rightarrow \Gamma \to \Delta \cup \Sigma$.

$\Gamma \cup \{\varphi\} \to \Delta$ & $\Gamma \to \Delta \cup \{\varphi\} \Rightarrow \Gamma \to \Delta$.

$\varphi \simeq \psi \Rightarrow Ess(\varphi) = Ess(\psi)$.

$\Gamma \simeq \Delta \Rightarrow Ess(\Gamma) = Ess(\Delta)$.

p is explicitly definable w.r.t. $\Gamma \Rightarrow \Gamma$ implicitly defines p.

$Cn(\Gamma) = Th(Mod(\Gamma))$.

Every simple theory is coherent.

Remarks. (1) Note however that in general a simple theory need not be *complete*. E.g. take classical propositional logic with connectives $\wedge$ and $\vee$ only, and the model M which gives T to p and q and F to the other variables. Then $Th(M) \simeq \{p, q\}$, and so e.g. neither $Th(M) \vDash r$ nor $Unsat(Th(M) \cup \{r\})$.

(2) Also, a non-simple theory may not be coherent: $Th(\varnothing) = Fla$ which of course may 'easily' not be coherent. However, this is the only such case: clearly $\forall \mathcal{M} \neq \varnothing$: $Th(\mathcal{M})$ is coherent. ⊣

– We have been considering so far a fixed arbitrary logical system; but it is also interesting of course to consider relations between different logical systems. An obvious one is the *sub-system* relation: (where A and B are logical systems)

A is a sub-system of B $=_{df}$ $Mod^A = Mod^B$, $Voc^A = Voc^B$, $Fla^A \subseteq Fla^B$, and $\vdash^A$, L^A are the restrictions of $\vdash^B$, L^B w.r.t. Fla^A.

(Alternatively, one might allow also *Mod* and/or *Voc* to shrink.)

Paradigmatic examples of sub-systems are 'functionally incomplete' systems w.r.t. a 'functionally complete' one – e.g. classical propositional logic with only →, or only ∧, →, or only ¬, ↔, etc., w.r.t. full classical propositional logic.

(Another, quite different notion of a 'part' of a system is: $Fla^A = Fla^B$, $Voc^A = Voc^B$, $L^A = L^B$; and Mod^A, Mod^B, $\vdash^A$, $\vdash^B$ are s.t. the defined relation $\vDash^A$ is subset of the defined relation $\vDash^B$ [whence in particular A-valid formulas are subset of B-valid formulas]. It is in this sense that e.g. S4 modal logic is a [proper] 'part' of S5 modal logic, or intuitionistic propositional logic [in connectives say ¬, ∧, ∨, →] is a [proper] part of classical propositional logic.)

3 Conditions on logical systems

We list here various interesting conditions which a logical system may satisfy (but which not *all* logical systems *do* satisfy).

Sat-Compactness: $Unsat(\Gamma) \Rightarrow \exists \Gamma_0 \subseteq_{fin} \Gamma(Unsat(\Gamma_0))$.

Sat-(κ-Compactness): $Unsat(\Gamma) \Rightarrow \exists \Gamma_0 \subseteq \Gamma(card(\Gamma_0) < \kappa \ \& \ Unsat(\Gamma_0))$.

$\vDash$-*Compactness*: $\Gamma \vDash \varphi \Rightarrow \exists \Gamma_0 \subseteq_{fin} \Gamma(\Gamma_0 \vDash \varphi)$.

$\vDash$-*(κ-Compactness)*: $\Gamma \vDash \varphi \Rightarrow \exists \Gamma_0 \subseteq \Gamma(card(\Gamma_0) < \kappa \ \& \ \Gamma_0 \vDash \varphi)$.

Formula Finiteness: $\forall \varphi(L(\varphi)$ is finite$)$.

Formula κ-Boundedness: $\forall \varphi(card(L(\varphi)) < \kappa)$.

In general, to a condition involving finiteness there always corresponds a more general condition-scheme where, in effect, $< \aleph_0$ is replaced by $< \kappa$. We will not always explicitly state here such corresponding general schemes.

Unrestricted Canonical Form for Formulas/Sets-of-Formulas:

$\forall \varphi \, \exists \psi (Can(\psi) \, \& \, \psi \simeq \varphi)$.

$\forall \Gamma \, \exists \Delta (Can(\Delta) \, \& \, \Delta \simeq \Gamma)$.

Semantically Restricted Canonical Form for Formulas/Sets-of-Formulas:

$\forall \varphi (\nvDash \varphi \, \& \, Sat(\varphi) \Rightarrow \exists \psi (Can(\psi) \, \& \, \psi \simeq \varphi))$.

$\forall \Gamma (\nvDash \Gamma \, \& \, Sat(\Gamma) \Rightarrow \exists \Delta (Can(\Delta) \, \& \, \Delta \simeq \Gamma))$.

Syntactically Restricted Canonical Form for Formulas/Sets-of-Formulas:

$\forall \varphi (Ess(\varphi) \neq \varnothing \Rightarrow \exists \psi (Can(\psi) \, \& \, \psi \simeq \varphi))$.

$\forall \Gamma (Ess(\Gamma) \neq \varnothing \Rightarrow \exists \Delta (Can(\Delta) \, \& \, \Delta \simeq \Gamma))$.

Semantic Rigidity of Closed Formulas: $L(\varphi) = \varnothing \Rightarrow (\vDash \varphi \text{ or } Unsat(\varphi))$.

Halldén Property for Joint Unsatisfiability:

$L(\Gamma) \cap L(\Delta) = \varnothing \Rightarrow [Unsat(\Gamma \cup \Delta) \Rightarrow (Unsat(\Gamma) \text{ or } Unsat(\Delta))]$.

Halldén Property for Disjunctive Validity:

$L(\Gamma) \cap L(\Delta) = \varnothing \Rightarrow [DisjVal(\Gamma \cup \Delta) \Rightarrow (DisjVal(\Gamma) \text{ or } DisjVal(\Delta))]$.

Halldén Property for Implication:

$L(\Gamma) \cap L(\Delta) = \varnothing \Rightarrow [\Gamma \vDash \Delta \Rightarrow (Unsat(\Gamma) \text{ or } \vDash \Delta)]$.

Halldén Property for Equivalence:

$L(\Gamma) \cap L(\Delta) = \varnothing \Rightarrow [\Gamma \simeq \Delta \Rightarrow (\vDash \Gamma \text{ or } Unsat(\Gamma))]$.

Remark. All these four conditions have straightforward correlates for *formulas*: in the first case we use $Unsat\{\varphi, \psi\}$ etc.; in the second $DisjVal\{\varphi, \psi\}$, $\vDash \varphi$, $\vDash \psi$; in the third $\varphi \vDash \psi$ etc.; and in the fourth $\varphi \simeq \psi$ etc. ⊣

Existence of Verum: $\exists \varphi (L(\varphi) = \varnothing \, \& \, \vDash \varphi)$.

Existence of Falsum: $\exists \varphi (L(\varphi) = \emptyset \ \& \ Unsat(\varphi))$.

Closure under Negation: $\forall \varphi \, \exists \psi \, \forall M (M \vdash \psi \Leftrightarrow \text{not } (M \vdash \varphi))$.

A stronger version of this condition adds to the matrix-statement: & $L(\psi) = L(\varphi)$. – If we put ψ neg $\varphi =_{df} \forall M(M \vdash \psi \Leftrightarrow \text{not } (M \vdash \varphi))$, then the condition above can be formulated more briefly as $\forall \varphi \, \exists \psi (\psi$ neg $\varphi)$. And similarly for conjunction etc. below, and Verum and Falsum above.

Closure under Binary Conjunction:

$\forall \varphi \, \forall \psi \, \exists \theta \, \forall M (M \vdash \theta \Leftrightarrow (M \vdash \varphi \ \& \ M \vdash \psi))$.

Again in a stronger version we add: & $L(\theta) = L(\varphi) \cup L(\psi)$. And similarly in other such cases.

Similar conditions can be formulated of course for closure under any other given finitary truth-function.

Closure under (All) Finitary Truth-Functions:

$\forall n \geq 1 \, \forall n$-ary truth-function $f \, \forall \varphi_1 \ldots \forall \varphi_n \, \exists \theta \, \forall M (M \vdash \theta \Leftrightarrow f(M \vdash \varphi_1, \ldots, M \vdash \varphi_n))$.

Remark. Closure under All Finitary Truth-Functions is of course equivalent to: Closure under Negation plus Closure under Binary Conjunction (or Closure under Negation plus Closure under Binary Disjunction, etc.). ⊣

Closure under Arbitrary Conjunctions: $\forall \Gamma \neq \emptyset \, \exists \varphi \, \forall M (M \vdash \varphi \Leftrightarrow M \vdash \Gamma)$.

Alternatively the condition may be strengthened by omitting '$\neq \emptyset$': the effect of this is adding the requirement of existence of some valid formula.

Closure under Conjunctions Bounded by Cardinal κ:

$\forall \Gamma \neq \emptyset$: $card(\Gamma) < \kappa \Rightarrow \exists \varphi \, \forall M (M \vdash \varphi \Leftrightarrow M \vdash \Gamma)$.

Again alternatively we may drop '$\neq \emptyset$', thus requiring existence of valid formula.

There are of course similar conditions for disjunctions etc.

'*Every vocable occurs in some formula*': $\forall p\ \exists \varphi (p \in L(\varphi))$.

This will hold of course in all 'normal' cases of logical systems. – There are also the similar conditions for arbitrary *finite* set of vocables and for completely arbitrary set of vocables. – Or again there is the condition that: for any set of vocables with two or more elements, if it is $L(\varphi)$ for some φ, then for any set of vocables strictly in-between $\emptyset$ and $L(\varphi)$ there is some formula ψ s.t. $L(\psi)$ is *that* set of vocables.

Indefinite Strengthening of Satisfiable Formulas:

$\forall Sat(\varphi)\ \exists Sat(\psi)(\psi \vDash \varphi\ \&\ \varphi \nvDash \psi)$.

Indefinite Weakening of Non-Valid Formulas:

$\forall \nvDash \varphi\ \exists \nvDash \psi (\varphi \vDash \psi\ \&\ \psi \nvDash \varphi)$.

Existence of Relative Verum w.r.t. Single Vocables:

$\forall p\ \exists \varphi (\vDash \varphi\ \&\ L(\varphi) = p)$.

Existence of Relative Verum w.r.t. Finite Sets ($\neq \emptyset$) of Vocables:

$\forall$ finite $\ell \neq \emptyset\ \exists \varphi (\vDash \varphi\ \&\ L(\varphi) = \ell)$.

Existence of Relative Verum w.r.t. Arbitrary Sets ($\neq \emptyset$) of Vocables:

$\forall \ell \neq \emptyset\ \exists \varphi (\vDash \varphi\ \&\ L(\varphi) = \ell)$.

There are also of course the similar conditions for Existence of Relative Falsum.

Identity of Equivalent Models: $M \equiv N \Rightarrow M = N$.

The following scheme represents 9 $(= 3 \times 3)$ conditions:

Axiomatizability [Finite Axiomatizability, Axiomatizability by Single Formula] of All Model-Classes [All Finite Model-Classes, All Models]:

$\forall \mathcal{M}$ [$\forall$ finite $\mathcal{M}$, $\forall M$] $\exists \Gamma$ [$\exists$ finite Γ, $\exists \varphi$] (Γ [$\{\varphi\}$] axiomatizes $\mathcal{M}$ [M]).

4 Interpolation etc.

We continue our list of conditions with a battery of conditions on various forms of interpolation and related ideas.

Unrestricted Formula Separation:

$Unsat\{\varphi, \psi\} \Rightarrow \exists \theta (L(\theta) \subseteq L(\varphi) \cap L(\psi)$ & $\varphi \vDash \theta$ & $Unsat\{\psi, \theta\})$.

Such θ may be said to be a 'separator' of φ and ψ. Note that $Unsat\{\psi, \theta\}$ is tantamount to $\psi \vDash \neg \theta$, which however is of course unsuitable to our abstract context where presence of negation is not assumed.

Unrestricted Formula Interpolation:

$\varphi \vDash \psi \Rightarrow \exists \theta (L(\theta) \subseteq L(\varphi) \cap L(\psi)$ & $\varphi \vDash \theta \vDash \psi)$.

Such θ is said to be an 'interpolant' for φ and ψ.

Unrestricted Formula Retrospection:

$DisjVal\{\varphi, \psi\} \Rightarrow \exists \theta (L(\theta) \subseteq L(\varphi) \cap L(\psi)$ & $\theta \vDash \varphi$ & $DisjVal\{\theta, \psi\})$.

Such θ is a 'retrospector' for ϕ and ψ: the disjunctive validity of $\{\varphi, \psi\}$ 'remounts' to the facts that θ (which is in the common vocables of φ and ψ) implies φ and that $\neg \theta$ implies ψ. Again, $\neg \theta \vDash \psi$ is the more suggestive but non-abstract equivalent of $DisjVal\{\theta, \psi\}$.

Semantically Restricted Formula Separation:

$Unsat\{\varphi, \psi\}$ & $Sat(\varphi)$ & $Sat(\psi) \Rightarrow \exists \theta (L(\theta) \subseteq L(\varphi) \cap L(\psi)$ & $\varphi \vDash \theta$ & $Unsat\{\psi, \theta\})$.

Semantically Restricted Formula Interpolation:

$\varphi \vDash \psi$ & $Sat(\varphi)$ & $\nvDash \psi \Rightarrow \exists \theta (L(\theta) \subseteq L(\varphi) \cap L(\psi)$ & $\varphi \vDash \theta \vDash \psi)$.

Semantically Restricted Formula Retrospection:

$DisjVal\{\varphi,\psi\}$ & $\nvDash \varphi$ & $\nvDash \psi \Rightarrow \exists \theta(L(\theta) \subseteq L(\varphi) \cap L(\psi)$ & $\theta \vDash \varphi$ & $DisjVal\{\theta,\psi\})$.

Syntactically Restricted Formula Separation:

$Unsat\{\varphi,\psi\}$ & $L(\varphi) \cap L(\psi) \neq \emptyset \Rightarrow \exists \theta$ etc.

Syntactically Restricted Formula Interpolation:

$\varphi \vDash \psi$ & $L(\varphi) \cap L(\psi) \neq \emptyset \Rightarrow \exists \theta$ etc.

Syntactically Restricted Formula Retrospection:

$DisjVal\{\varphi,\psi\}$ & $L(\varphi) \cap L(\psi) \neq \emptyset \Rightarrow \exists \theta$ etc.

Unrestricted Set Separation:

$Unsat(\Gamma \cup \Delta) \Rightarrow \exists \Theta(L(\Theta) \subseteq L(\Gamma) \cap L(\Delta)$ & $\Gamma \vDash \Theta$ & $Unsat(\Delta \cup \Theta))$.

Unrestricted Set Interpolation:

$\Gamma \vDash \Delta \Rightarrow \exists \Theta(L(\Theta) \subseteq L(\Gamma) \cap L(\Delta)$ & $\Gamma \vDash \Theta \vDash \Delta)$.

Unrestricted Set Retrospection:

$\forall M(M \vdash \Gamma$ or $M \vdash \Delta) \Rightarrow \exists \Theta(L(\Theta) \subseteq L(\Gamma) \cap L(\Delta)$ & $\Theta \vDash \Gamma$ & $\forall M(M \nvdash \Theta \Rightarrow M \vdash \Delta))$.

This seems to be the most natural set-version of the formula-retrospection condition. Where we had $\forall M(M \vdash \varphi$ or $M \vdash \psi)$ (i.e. $DisjVal\{\varphi,\psi\}$), we now have $\forall M(M \vdash \Gamma$ or $M \vdash \Delta)$; and so on. The clause $\forall M(M \vdash \Gamma$ or $M \vdash \Delta)$ is a kind of 'meta--disjunctive-validity'. But it comes to mind also to consider the 'ordinary' disjunctive validity of $\Gamma \cup \Delta$, and then the 'retrospector' Θ would be required to be such that, in vivid but non-abstract terms, $\Theta \vDash \vee(\Gamma)$ and $\neg \Theta \vDash \vee(\Delta)$ (i.e. $\neg \wedge(\Theta) \vDash \vee(\Delta)$). Translating this into abstract terms we get the condition:

$DisjVal(\Gamma \cup \Delta) \Rightarrow \exists \Theta(L(\Theta) \subseteq L(\Gamma) \cap L(\Delta)$ & $\forall M(M \vdash \Theta \Rightarrow \exists \varphi \in \Gamma(M \vdash \varphi))$ & $\forall M(M \nvdash \Theta \Rightarrow \exists \varphi \in \Delta(M \vdash \varphi)))$.

The *Restricted* versions of these Set-conditions are similar to the

Restricted versions of the corresponding Formula-conditions. Thus in the *Semantically* Restricted conditions, we have the qualifications:

For Separation: $Sat(\Gamma)$ & $Sat(\Delta)$.

For Interpolation: $Sat(\Gamma)$ & $\not\vDash \Delta$.

For Retrospection: $\not\vDash \Gamma$ & $\not\vDash \Delta$.

And for the modified Retrospection condition: not $DisjVal(\Gamma)$ & not $DisjVal(\Delta)$. As for the *Syntactically* Restricted conditions, the qualification is always the same, viz. $L(\Gamma) \cap L(\Delta) \neq \varnothing$.

There are also 'Set-Formula' versions of these conditions (and to be thorough also 'Formula-Set', though that is much less natural): it will suffice to give one illustrative formulation, in the case of Unrestricted Separation:

Unrestricted Set-Formula Separation:

$Unsat(\Gamma \cup \Delta) \Rightarrow \exists \theta (L(\theta) \subseteq L(\Gamma) \cap L(\Delta)$ & $\Gamma \vDash \theta$ & $Unsat(\Delta \cup \{\theta\}))$.

Again, there are also *Finite-Set* conditions where we use 'finite set' instead of 'set' or 'formula' at one or more places. E.g.:

Unrestricted Set-(Finite Set) Separation:

$Unsat(\Gamma \cup \Delta) \Rightarrow \exists$ finite $\Theta_0 (L(\Theta_0) \subseteq L(\Gamma) \cap L(\Delta)$ & $\Gamma \vDash \Theta_0$ & $Unsat(\Delta \cup \Theta_0))$.

Left-Uniform Unrestricted Formula Interpolation:

$\varphi \vDash \psi \Rightarrow \exists \theta (L(\theta) \subseteq L(\varphi) \cap L(\psi)$ & $\varphi \vDash \theta \vDash \psi$ & $\forall \varphi'(\varphi' \vDash \psi$ & $L(\varphi') = L(\varphi) \Rightarrow \varphi' \vDash \theta))$.

Right-Uniform Unrestricted Formula Interpolation:

$\varphi \vDash \psi \Rightarrow \exists \theta (L(\theta) \subseteq L(\varphi) \cap L(\psi)$ & $\varphi \vDash \theta \vDash \psi$ & $\forall \psi'(\varphi \vDash \psi'$ & $L(\psi') = L(\psi) \Rightarrow \theta \vDash \psi'))$.

There are also of course the corresponding *Restricted* conditions – Semantically Restricted ($\varphi \vDash \psi$ & $Sat(\varphi)$ & $\not\vDash \psi \Rightarrow \exists \theta$ etc.) and *Syn-*

tactically Restricted ($\varphi \vDash \psi$ & $L(\varphi) \cap L(\psi) \neq \emptyset \Rightarrow \exists \theta$ etc.) –; and the corresponding (Unrestricted or Restricted) *Set* Interpolation conditions.

Also, there are Uniform conditions of Separation and Retrospection. We state only the Unrestricted Formula conditions; the adaptations for *Set* conditions and *Restricted* conditions are as in other cases before.

Left-Uniform Unrestricted Formula Separation:

$Unsat\{\varphi, \psi\} \Rightarrow \exists\theta(L(\theta) \subseteq L(\varphi) \cap L(\psi)$ & $\varphi \vDash \theta$ & $Unsat\{\psi, \theta\}$ & $\forall\varphi'(Unsat\{\varphi', \psi\}$ & $L(\varphi') = L(\varphi) \Rightarrow \varphi' \vDash \theta))$.

Right-Uniform Unrestricted Formula Separation:

$Unsat\{\varphi, \psi\} \Rightarrow \exists\theta(L(\theta) \subseteq L(\varphi) \cap L(\psi)$ & $\varphi \vDash \theta$ & $Unsat\{\psi, \theta\}$ & $\forall\psi'(Unsat\{\varphi, \psi'\}$ & $L(\psi') = L(\psi) \Rightarrow Unsat\{\psi', \theta\}))$.

Left-Uniform Unrestricted Formula Retrospection:

$DisjVal\{\varphi, \psi\} \Rightarrow \exists\theta(L(\theta) \subseteq L(\varphi) \cap L(\psi)$ & $\theta \vDash \varphi$ & $DisjVal\{\theta, \psi\}$ & $\forall\varphi'(DisjVal\{\varphi', \psi\}$ & $L(\varphi') = L(\varphi) \Rightarrow \theta \vDash \varphi'))$.

Right-Uniform Unrestricted Formula Retrospection:

$DisjVal\{\varphi, \psi\} \Rightarrow \exists\theta(L(\theta) \subseteq L(\varphi) \cap L(\psi)$ & $\theta \vDash \varphi$ & $DisjVal\{\theta, \psi\}$ & $\forall\psi'(DisjVal\{\varphi, \psi'\}$ & $L(\psi') = L(\psi) \Rightarrow DisjVal\{\theta, \psi'\}))$.

We now state three conditions of 'Equivalential' Interpolation. Again we state only the *Unrestricted Formula* versions but there are also, in each case, the obvious modified *Restricted* and *Set* versions.

Unrestricted Formula Binary Equivalential Interpolation:

$\varphi \simeq \psi \Rightarrow \exists\theta(L(\theta) \subseteq L(\varphi) \cap L(\psi)$ & $\theta \simeq \varphi)$.

Unrestricted Formula Finite Equivalential Interpolation: ($\forall n \geq 2$:)

$\varphi_1 \simeq \varphi_2 \simeq \ldots \simeq \varphi_n \Rightarrow \exists\theta(L(\theta) \subseteq L(\varphi_1) \cap \ldots \cap L(\varphi_n)$ & $\theta \simeq \varphi_1)$.

Unrestricted Formula Generalized Equivalential Interpolation:

$\Gamma \neq \emptyset$ & $\forall \varphi, \psi \in \Gamma(\varphi \simeq \psi) \Rightarrow \exists\theta(L(\theta) \subseteq \bigcap\{L(\varphi) : \varphi \in \Gamma\}$ & $\theta \simeq \Gamma)$.

(In the *Set* version of this we take a set of equivalent *sets* of formulas.)

Unrestricted Right-wards Piecemeal Formula Interpolation:

$\varphi \vDash \psi \ \& \ p \in L(\varphi) - L(\psi) \Rightarrow \exists \theta (L(\theta) \subseteq L(\varphi) - \{p\} \ \& \ \varphi \vDash \theta \vDash \psi)$.

Unrestricted Left-wards Piecemeal Formula Interpolation:

$\varphi \vDash \psi \ \& \ p \in L(\psi) - L(\varphi) \Rightarrow \exists \theta (L(\theta) \subseteq L(\psi) - \{p\} \ \& \ \varphi \vDash \theta \vDash \psi)$.

There are also the versions for 'Restricted', 'Set', 'Separation', 'Retrospection', 'Equivalential Interpolation'. (But with 'Set' of course the Piecemeal condition [plus Formula Finiteness] will *not* imply the corresponding 'full' condition.)

Joint Satisfiability [or *Robinson Property*]:

[Γ is $L(\Gamma)$-complete & $\Gamma \subseteq \Delta$ & $\Gamma \subseteq \Sigma$ & $Sat(\Delta)$ & $Sat(\Sigma)$ & $(L(\Delta) - L(\Gamma)) \cap (L(\Sigma) - L(\Gamma)) = \varnothing] \Rightarrow Sat(\Delta \cup \Sigma)$.

Beth Property: Γ implicitly defines $p \Rightarrow p$ is explicitly definable w.r.t. Γ.

5 Illustrative theorems

We give here some theorems on logical systems. Our aim is only to *illustrate* the kind of results which it seems natural to try to prove in our Abstract Logic with Vocables; obviously a great deal more can be done along these lines.

Proposition. If S is a sub-system of S' and S' satisfies Compactness, then S satisfies Compactness.

Proof. Immediate from the definitions. □

Proposition. Same for Halldén Property (in all its versions). □

Proposition. (Closure under Negation) $\Rightarrow$ ($\vDash$-Compactness $\Leftrightarrow$ Sat-Compactness). □

Proposition. (Semantically Restricted Formula Interpolation) & (Existence of Verum) & (Existence of Falsum) ⇒ (Unrestricted Formula Interpolation). □

The similar Proposition for *Set* Interpolation also holds of course; and there are many other Propositions of this kind.

Proposition. [(Unrestricted Right-wards Piecemeal Formula Interpolation) or (Unrestricted Left-wards Piecemeal Formula Interpolation)] & [Formula Finiteness] ⇒ [Unrestricted Formula Interpolation]. □

An interesting class of Propositions consists of implications where the antecedent is Closure under Truth-Functions and the consequent is the equivalence of such things as the alternative forms of Halldén Property, or Interpolation/Separation/Retrospection, and so on.

We give now a series of Propositions concerning Equivalential Interpolation. We state the simplest Propositions, for *Unrestricted* and *Formula* conditions; but there are also the corresponding similar Propositions for *Restricted* and *Set* conditions.

Proposition. (Unrestricted Canonical Form for Formulas) ⇒ (Unrestricted Generalized [and so also a fortiori Unrestricted Finite and Unrestricted Binary] Equivalential Interpolation). □

Proposition. (Unrestricted Formula Interpolation) ⇒ (Unrestricted Formula Binary Equivalential Interpolation). □

Remark. Equivalential Interpolation is weaker than Interpolation, as can be seen from the case of constant-domain quantified S5 where the former holds but the latter does not hold. ⊣

Proposition. (Unrestricted Formula Binary Equivalential Interpolation) ⇒ (Unrestricted Formula Finite Equivalential Interpolation).

Proof. We can obtain equivalential interpolant for $\varphi_1 \simeq \ldots \simeq \varphi_n$ by successive applications of Binary Equivalential Interpolation. □

Proposition. (Unrestricted Formula Finite Equivalential Interpolation) ⇏ (Unrestricted Formula Generalized Equivalential Interpolation). (I.e. there are logical systems for which the former condition holds but the latter does not.)

Proof. A counterexample is provided by a logical system corresponding to a suitable form of infinitary classical propositional logic with 'few' (say denumerably many) variables, and with verum and falsum. Here Unrestricted Formula Interpolation holds: a simple argument for the case of finitary classical propositional logic in terms of disjunctive normal forms is generalizable to the infinitary case. Hence by the two preceding Propositions Unrestricted Formula Finite Equivalential Interpolation also holds. But Unrestricted Formula *Generalized* Equivalential Interpolation does *not* hold here: we can take e.g. the set $\Gamma :=$

{There are infinitely many truths in $Voc - \{p_1\}$,

There are infinitely many truths in $Voc - \{p_2\}$,

... ...},

where $Voc = \{p_1, p_2, p_3, \ldots\}$, and the set Γ has one formula (constructed in obvious way to have the import of the above verbal formulation) for each p_i. Each member of Γ is equivalent to 'There are infinitely many truths in Voc'. But $\bigcap\{L(\varphi) : \varphi \in \Gamma\} = \varnothing$, and obviously there is no formula θ with $L(\theta) = \varnothing$ and $\theta \simeq \Gamma$. (All formulas without vocables are here valid or unsatisfiable.) □

Remark. The same argument serves to show the corresponding Proposition with *Semantically* Restricted conditions, since (i) we have also Semantically Restricted Interpolation for the infinitary language without verum or falsum, by the same kind of argument from normal forms, and (ii) the elements of Γ are neither valid nor unsatisfiable. – As for the Proposition with *Syntactically* Restricted conditions: again Interpolation holds, by normal forms, but now we need to construct Γ in a somewhat different way, since in the Γ above the set of vocables common to all formulas in the set is empty, and so Syntactically Restricted Generalized Equivalential Interpolation is vacuously satisfied. But it is easy to modify the previous construction to circumvent this: we can take our new Γ as say, where $Voc^- := Voc - \{p_1\}$,

$\{p_1 \wedge$ There are infinitely many truths in $Voc^- - \{p_2\}$,

$p_1 \wedge$ There are infinitely many truths in $Voc^- - \{p_3\}$,

$\ldots \ldots \ldots\}$,

Now the set of common vocables is $\{p_1\}$, and $Ess(\Gamma) = \{p_1\}$, but of course there is no formula θ with $L(\theta) \subseteq \{p_1\}$ and $\theta \simeq \Gamma$. ⊣

Proposition. (Unrestricted Formula Binary Equivalential Interpolation) & (Formula Finiteness) ⇒ (Unrestricted Canonical Form for Formulas).

Proof. Assume the hypotheses, and let φ be a non-canonical formula with inessential vocables $p_1 \ldots p_n$ (i.e. $L(\varphi) - Ess(\varphi) = \{p_1, \ldots, p_n\}$; this set must be finite in view of Formula Finiteness). Since $p_1 \notin Ess(\varphi)$, there is formula $\psi \simeq \varphi$ with $p_1 \notin L(\psi)$. So by Equivalential Interpolation there is formula φ' equivalent to φ/ψ and with $L(\varphi') \subseteq L(\varphi) \cap L(\psi)$, whence $p_1 \notin L(\varphi')$ and $L(\varphi') - Ess(\varphi') \subseteq \{p_2, \ldots, p_n\}$. But then, if $p_2 \in L(\varphi')$ we repeat the procedure and take $\psi' \simeq \varphi'$ with $p_2 \notin L(\psi')$, and so on. After thus 'eliminating' all of $p_1 \ldots p_n$ we obtain a canonical equivalent of φ. □

(There are of course also the similar 'Restricted' Propositions.)

From this Proposition plus an earlier one, viz. the sixth displayed Proposition in this section, we get:

Proposition. (Formula Finiteness) ⇒ [(Unrestricted Formula Binary Equivalential Interpolation) ⇔ (Unrestricted Canonical Form for Formulas)]. □

Proposition. (Halldén Property for Implication, Formula Version) & (Existence of Relative Verum w.r.t. Single Vocables) & (Existence of Relative Falsum w.r.t. Single Vocables) ⇒ [(Semantically Restricted Formula Interpolation) ⇔ (Syntactically Restricted Formula Interpolation)].

Proof. Assume the hypotheses. Now *first*, suppose further (Semantically Restricted Formula Interpolation), and the hypotheses of (Syntactically Restricted Formula Interpolation), viz. $\varphi \vDash \psi$ and $L(\varphi) \cap L(\psi) \neq \emptyset$. Case (i): φ satisfiable and ψ not valid. Here the result (existence of interpolant) follows by (Semantically Restricted Formula Interpolation).

Case (ii): φ unsatisfiable or ψ valid. Here, for $p \in L(\varphi) \cap L(\psi)$, we can take the interpolant θ as a relative falsum or (as the case may be) relative verum w.r.t. p.

Secondly, suppose further (Syntactically Restricted Formula Interpolation), and the hypotheses of (Semantically Restricted Formula Interpolation), viz. $\varphi \vDash \psi$ and $Sat(\varphi)$ and $\nvDash \psi$. Recall that the assumed Halldén Property says:

$$L(\varphi) \cap L(\psi) = \varnothing \Rightarrow [\varphi \vDash \psi \Rightarrow Unsat(\varphi) \text{ or } \vDash \psi],$$

which reformulated by elementary transformations gives

$$\varphi \vDash \psi \Rightarrow [Sat(\varphi) \& \nvDash \psi \Rightarrow L(\varphi) \cap L(\psi) \neq \varnothing].$$

So using our hypotheses we get $L(\varphi) \cap L(\psi) \neq \varnothing$, whence by (Syntactically Restricted Formula Interpolation) there follows the existence of interpolant θ. □

Note that the Existence of Relative Verum/Falsum is used only in the *first* part of the above proof; and so we have also:

Proposition. (Halldén Property for Implication, Formula Version) $\Rightarrow$ [(Syntactically Restricted Formula Interpolation) $\Rightarrow$ (Semantically Restricted Formula Interpolation)]. □

Proposition. (Sat-Compactness) & (Semantically Restricted Finite-Set Separation) $\Rightarrow$ (Semantically Restricted Set Separation).

Proof. Suppose the antecedent and the hypotheses of the consequent, viz. $Unsat(\Gamma \cup \Delta)$ and $Sat(\Gamma)$ and $Sat(\Delta)$. From this by Sat-Compactness it follows that there are $\Gamma_0 \subseteq_{fin} \Gamma$ and $\Delta_0 \subseteq_{fin} \Delta$ with $Unsat(\Gamma_0 \cup \Delta_0)$. By (Semantically Restricted Finite-Set Separation) there is finite Θ_0 with $L(\Theta_0) \subseteq L(\Gamma_0) \cap L(\Delta_0)$ and $\Gamma_0 \vDash \Theta_0$ and $Unsat(\Delta_0 \cup \Theta_0)$; and such Θ_0 is then a fortiori a separator for Γ, Δ. □

Remark. The Θ being finite the consequent of the Proposition can be strengthened to (Semantically Restricted Set-[Finite Set] Separation). Various other more or less obvious modifications of this Proposition are also possible. ⊣

Proposition. (Semantically Restricted Set Separation) ⇒ (Robinson Property).

Proof. Assume the antecedent and suppose for contradiction the hypotheses of (Robinson Property) hold but not its conclusion – i.e. that:

Γ is $L(\Gamma)$-complete,

$\Gamma \subseteq \Delta$ and $\Gamma \subseteq \Sigma$ and $Sat(\Delta)$ and $Sat(\Sigma)$,

$(L(\Delta) - L(\Gamma)) \cap (L(\Sigma) - L(\Gamma)) = \varnothing$;

but $Unsat(\Delta \cup \Sigma)$.

Then by (Semantically Restricted Set Separation) there is Θ with $L(\Theta) \subseteq L(\Delta) \cap L(\Sigma)$, whence $L(\Theta) \subseteq L(\Gamma)$, such that $\Delta \vDash \Theta$ and $Unsat(\Sigma \cup \Theta)$. Since $Sat(\Delta)$ and $Sat(\Sigma)$, we must have $Sat(\Delta \cup \Theta)$, whence a fortiori $Sat(\Gamma \cup \Theta)$, and $\Sigma \nvDash \Theta$, whence a fortiori $\Gamma \nvDash \Theta$. So there is $\theta \in \Theta$ with $\Gamma \nvDash \theta$ and $Sat(\Gamma \cup \{\theta\})$. But since $L(\theta) \subseteq L(\Theta) \subseteq L(\Gamma)$, this contradicts the hypothesis that Γ is $L(\Gamma)$-complete. □

The following Propositions give a sort of partial abstract characterization of what goes on in various 'disjunctive normal forms'.

Proposition. $Topic(\Gamma)$ & $\forall \psi \in \Gamma(\psi$ settles $\varphi)$ & $\Delta = \{\psi \in \Gamma : \psi \vDash \varphi\} \Rightarrow \forall M(M \vdash \varphi \Leftrightarrow \exists \psi \in \Delta(M \vdash \psi))$. □

Proposition. $Topic(\Gamma)$ & $\forall \psi \in \Gamma(\psi$ settles $\varphi)$ & θ disj $\{\psi \in \Gamma : \psi \vDash \varphi\} \Rightarrow \varphi \simeq \theta$. □

Remark. In concrete cases 'topics' will be derived from certain 'parameters', such as e.g.: a finite ℓ, in classical propositional logic or S5 modal logic; a finite ℓ and a maximum number of nested quantifiers, in (classical or modal) first-order predicate logic (by Hintikka's 'distributive normal forms'). ⊣

6 Adding domains

An *extended logical system* is a logical system in the sense previously defined together with a *domain-function* dom which assigns, to each model $M \in Mod$, a non-empty set $dom(M)$ ('the domain of M'). – We will use the letters D, E, etc. as variables for domains. (There is also of course the notion of extended *partial* logical system, without Voc and L. Indeed nearly all the conditions on extended logical systems given below do not involve Voc and L.)

(These extended logical systems might be extended further to 'multi-sorted' systems with multiple domain-functions allowed. This would be useful to give a closer approximation not only to classical multi-sorted systems but also to quantified modal logic with possible world semantics. But here we will content ourselves with the simpler, 'uni-sorted' notion.)

$D\text{-}Sat(\varphi) =_{df} \exists M(dom(M) = D \ \& \ M \vDash \varphi)$.

$D\text{-}Unsat(\varphi) =_{df}$ Not $D\text{-}Sat(\varphi)$.

$\varphi \vDash^D \psi =_{df} \forall M(dom(M) = D \Rightarrow (M \vDash \varphi \Rightarrow M \vDash \psi))$.

$\vDash^D \varphi =_{df} \forall M(dom(M) = D \Rightarrow M \vDash \varphi)$.

Similarly for $D\text{-}Sat(\Gamma)$, $D\text{-}Unsat(\Gamma)$, $\Gamma \vDash^D \varphi$, $DisjVal^D(\Gamma)$ (i.e. $DisjVal^D(\Gamma) =_{df} \forall M(dom(M) = D \Rightarrow \exists \varphi \in \Gamma(M \vDash \varphi)))$.

$\kappa\text{-}Sat(\varphi) =_{df} \exists D(card(D) = \kappa \ \& \ D\text{-}Sat(\varphi))$.

$\kappa\text{-}Unsat(\varphi) =_{df}$ Not $\kappa\text{-}Sat(\varphi)$.

$\varphi \vDash^\kappa \psi =_{df} \forall D(card(D) = \kappa \Rightarrow \varphi \vDash^D \psi)$.

$\vDash^\kappa \varphi =_{df} \forall D(card(D) = \kappa \Rightarrow \vDash^D \varphi)$.

There are also the similar notions with '$\leq \kappa$' or '$< \kappa$' etc. in lieu of 'κ' – e.g.:

$\leq\kappa\text{-}Sat(\varphi) =_{df} \exists D(card(D) \leq \kappa \ \& \ D\text{-}Sat(\varphi))$.

Proto-Isomorphism Property:

$card(D) = card(E) \Rightarrow [(D\text{-}Sat(\Gamma) \Leftrightarrow E\text{-}Sat(\Gamma))\ \&\ (\Gamma \models^D \varphi \Leftrightarrow \Gamma \models^E \varphi)]$.

Remark. Any self-respecting logical system has this property. So there is something to be said for including satisfaction of this property as a condition in the *definition* of '(extended) logical system'. (As the 'Isomorphism Property' is included in the definition of 'abstract logics' in Lindström and the subsequent 'abstract model theory' tradition.) ⊣

Formula Inflation: $\kappa\text{-}Sat(\varphi)\ \&\ \kappa < \lambda \Rightarrow \lambda\text{-}Sat(\varphi)$.

Formula Deflation: $\models^\kappa \varphi\ \&\ \lambda < \kappa \Rightarrow \models^\lambda \varphi$.

Similarly for 'Set Inflation' and 'Set Deflation', with Γ instead of φ.

'Overspill' Property: $\forall n \geq 1\, \exists m \geq n(m\text{-}Sat(\varphi)) \Rightarrow \aleph_0\text{-}Sat(\varphi)$.

(Again there is of course the similar property with Γ instead of φ.)

Strict Löwenheim Property (for Satisfiability): $Sat(\varphi) \Rightarrow \aleph_0\text{-}Sat(\varphi)$.

Weak Löwenheim Property: $Sat(\varphi) \Rightarrow \leq\!\aleph_0\text{-}Sat(\varphi)$.

Strict Löwenheim-Skolem Property: $Sat(\Gamma) \Rightarrow \aleph_0\text{-}Sat(\Gamma)$.

Weak Löwenheim-Skolem Property: $Sat(\Gamma) \Rightarrow \leq\!\aleph_0\text{-}Sat(\Gamma)$.

Downward Löwenheim-Skolem: $\kappa > \aleph_0\ \&\ \kappa\text{-}Sat(\Gamma) \Rightarrow \aleph_0\text{-}Sat(\Gamma)$.

Upward Löwenheim-Skolem: $\aleph_0\text{-}Sat(\Gamma)\ \&\ \kappa > \aleph_0 \Rightarrow \kappa\text{-}Sat(\Gamma)$.

– There are also the similar conditions but with some other cardinal κ instead of $\aleph_0$; also the conditions which say that *such* condition holds for *some* cardinal κ. E.g.:

Generalized Strict Löwenheim-Skolem Property (or: *Existence of Universally Representative Model-Size*): $\exists \kappa\, \forall \Gamma\colon Sat(\Gamma) \Rightarrow \kappa\text{-}Sat(\Gamma)$.

Generalized Weak Löwenheim-Skolem Property: Same only with $\leq\!\kappa\text{-}Sat$ in lieu of $\kappa\text{-}Sat$.

If this last condition holds for a given system, the *least* such cardinal

κ is sometimes called the *Löwenheim number* of the system (e.g. in Bell & Slomson 1974, p. 85). (As will be seen below, this is an exceptional case in that actually the condition holds for *all* logical systems, given our requirement that Fla is a 'set' [as opposed to 'proper class' or whatever like Mod might be].)

The various Löwenheim and Löwenheim-Skolem conditions above are all *Satisfiability* conditions, and there are cognates for *Validity* and *Implication* – e.g.:

Strict Löwenheim Property for Validity: $\models^{\aleph_0} \varphi \Rightarrow \models \varphi$.

Weak Löwenheim-Skolem Property for Implication: $\Gamma \models^{\leq \aleph_0} \varphi \Rightarrow \Gamma \models \varphi$.

There is a batch of conditions concerning expressibility of size of models (which we won't bother to name individually):

$\forall \kappa \, \exists \varphi \, \forall M (M \vdash \varphi \Leftrightarrow card(dom(M)) = \kappa)$.

(Since Fla is a set, this can only hold if 'few' cardinalities are model-sizes.)

$\forall \kappa \, \exists \Gamma \, \forall M (M \vdash \Gamma \Leftrightarrow card(dom(M)) = \kappa)$.

$\forall n \, \exists \varphi$ etc. (or: $\exists \Gamma$ etc.).

Similarly with $\leq \kappa$ or $\leq n$, or $\geq \kappa$ or $\geq n$, instead of $= \kappa$ or $= n$.

$\exists \varphi \, \forall M (M \vdash \varphi \Leftrightarrow card(dom(M)) \geq \aleph_0)$.

– And so on.

And there are various other conditions to do with size of models. To give just one more example:

$\exists \varphi \, \forall D (D\text{-}Sat(\varphi) \Leftrightarrow card(D) \geq \aleph_0)$.

– Also 'local' versions of some 'global' properties: –

Local Compactness: $D\text{-}Unsat(\Gamma) \Rightarrow \exists \Gamma_0 \subseteq_{fin} \Gamma (D\text{-}Unsat(\Gamma_0))$.

Local Interpolation: $\varphi \models^D \psi \Rightarrow \exists \theta (L(\theta) \subseteq L(\varphi) \cap L(\psi)$ & $\varphi \models^D \theta$ & $\theta \models^D \psi)$.

All the numerous variations on Interpolation property which we saw above in the 'global' context can again be made in the 'local' case.

Again, there is 'Local Canonical Forms', 'Local Halldén Property', etc. etc.

– Here are a few 'Illustrative Theorems':

Proposition. (Closure under Negation) $\Rightarrow$ [(Strict Löwenheim Property for Satisfiability) $\Leftrightarrow$ (Strict Löwenheim Property for Validity)]. □

Proposition. (Strict Löwenheim Property for Satisfiability) & (Local Compactness) & (Proto-Isomorphism Property) & (Closure under Binary Conjunction) $\Rightarrow$ (Strict Löwenheim-Skolem Property for Satisfiability). □

Proposition. [Local Interpolation] & [Generalized Strict Löwenheim Property for Implication] & [Proto-Isomorphism Property] $\Rightarrow$ [(Global) Interpolation]. □

Proposition. [Finite-Set (Global) Interpolation] & [Proto-Isomorphism Property] & [Expressibility by single formula without vocables of each model-size] $\Rightarrow$ [Local Interpolation].

Proof. Assume hypotheses and that $\varphi \models^D \psi$. Let κ be $card(D)$, and let δ be formula without vocables 'saying' that there are exactly κ things (i.e. $\forall M(M \vdash \delta \Leftrightarrow card(dom(M)) = \kappa)$). Since every model based on D which verifies φ verifies ψ, we have by the Proto-Isomorphism Property that: every model with domain of cardinality κ which verifies φ verifies ψ. Thus $\{\varphi, \delta\} \models \psi$. Now let θ be an interpolant for this (global) implication. Thus $\{\varphi, \delta\} \models \theta$, whence $\varphi \models^D \theta$; and $\theta \models \psi$, whence a fortiori $\theta \models^D \psi$; also of course $L(\theta) \subseteq L(\varphi) \cap L(\psi)$, since by assumption $L(\delta) = \varnothing$. □

Proposition (cp. Bell and Slomson 1974 pp. 85–86 Thm. 4.3). Every logical system satisfies the Generalized Weak Löwenheim-Skolem Property – or equivalently, has Löwenheim number.

Proof. Since Fla is supposed to be a set, also

$$SAT := \{\Gamma \subseteq Fla : Sat(\Gamma)\}$$

is a set. (Note that $SAT \neq \emptyset$, since $Sat(\emptyset)$ and so $\emptyset \in SAT$!) But for each such satisfiable set Γ there is the least cardinal λ s.t. Γ is λ-satisfiable – let us call this cardinal $\mu(\Gamma)$. Then

$$\{\mu(\Gamma) : \Gamma \in SAT\}$$

is a set. And clearly its supremum (the least cardinal $\geq$ every cardinal in the set) is the Löwenheim number for the given system. $\square$

Remark. By similar arguments, we can show also other results of this kind – e.g. that every logical system satisfies the Generalized Weak Löwenheim Property for Validity, i.e. $\exists \kappa \; \forall \varphi (\vDash^{\leq \kappa} \varphi \Rightarrow \vDash \varphi)$. $\dashv$

References

[1] Jon Barwise and Solomon Feferman (eds.) (1985) *Model-Theoretic Logics*, Berlin, Springer.

[2] Roderick Batchelor (2013) 'Complexes and their constituents', *Theoria* (Lund), vol. 79, pp. 326–352.

[3] J. L. Bell and A. B. Slomson (1974) *Models and Ultraproducts: An Introduction*, 3rd printing, Amsterdam, North-Holland.

[4] Alfred Tarski (1930) 'On some fundamental concepts of metamathematics', incl. in his *Logic, Semantics, Metamathematics: Papers from 1923 to 1938*, 2nd ed., Oxford U. P., 1983, pp. 30–37.

[5] Ryszard Wójcicki (1988) *Theory of Logical Calculi: Basic Theory of Consequence Operations*, Dordrecht, Kluwer.

An abstract definition of normative system

Juliano Maranhão

Law School, University of São Paulo, Brazil

Abstract

In this paper we provide a formal and abstract definition of normative system as a class of sequences of normative sets closed by input/output operators, a definition of normative theories and of normative coherence.

1 Introduction

There are two features of logic that Edelcio Gonçalves de Souza finds most attractive and therefore thoroughly cultivates in his writings.

The first is precision. De Souza appreciates particularly the contribution of formal methods and logical analysis to describe philosophical problems more precisely. And that is key to foresee and advance solutions or make relevant distinctions, which provide clarification to philosophical inquiry.

The second is abstraction. De Souza believes that core philosophical problems are abstract in nature and therefore can only be sufficiently explored by abstract methods. And logic is particularly suited for such a conceptual task. No wonder that one of de Souza's main interests lies on "abstract logic", which is the study of abstract properties of logical systems.

Edelcio's interest on normative reasoning is located exactly in this tradition of inquiry. If there is a form of reasoning which is normative, what are its abstract and distinctive properties? What does it mean to say that a rule or obligation is derived or implied from other rules? And if normative knowledge is systematized in some theory about rules, what is exactly the object of such theory and how its propositions are derived?

In a recent paper, which I have co-authored with de Souza [5], we have made an effort to describe legal interpretation as a dynamic of changing concepts employed in rules and modifying rules themselves, aiming at coherent descriptions of obligations and permissions derived from those concepts and rules.

In the process of building a representation of reasoning with concepts and the content of rules, many philosophical questions have appeared, which bothered both of us but in particular de Souza, in his quest for precision. For instance, what does a theory about rules describe? Are the rules systematized or is it the theory that systematize its propositions about the rules. Is this theory a description of the rules, or a description of the obligations, which are inferred from rules. So what is the relation between rules, the logic that grounds normative reasoning and the obligations which are derived? In one word what is a normative system?

In this paper I shall make an attempt to reply those difficult questions, which we have somehow detoured in that paper. In what follows I shall provide a definition of normative systems and of theories about normative systems. Since there are different kinds of rules that may be combined in normative reasoning (constitutive, regulative, value rules, technical rules, consequential rules, preferences, etc.) and such rules may be subject to different principles of reasoning, we shall talk about architectures of reasoning upon which normative systems are constructed. I shall conceive a normative system as a combination of sets of rules with a combination of underlying logics. A theory about a normative system is a description of the obligations, prohibitions and permissions which are the outcome of such systems, in given contexts.

2 Normative sets

A very influential concept of normative system in the legal domain was provided by Alchourrón and Bulygin [1]. They have defined a normative system simply as the logical closure of a set of sentences from which at least a deontic sentence is derived. A deontic sentence was conceived as a sentence involving deontic modalities as obligations, permissions or prohibitions. It is clear that Alchourrón and Bulygin had in mind so called "regulative rules", that is positively rules enacted by authorities which demand or authorize a future behaviour from individuals. Provided that, according to Alchourrón and Bulygin, in line with Hume's guillotine, deontic modalities (representing prescritive statemens) are only derivable from sets of sentences containing deontic modalities (among other sentences) the normative set which is the base of the system necessarily contains at least a deontic sentence.

But their definition of normative system is abstract enough and quite general, since it has no commitment with the kinds of rules and sentence that may pertain to the basic set of sentences from which the obligations and permissions are derived. The normative system may contain different kinds of rules, regulative, constitutive and also may contain descriptive sentences. What is important is that the system delivers at least a deontic statement. Given any set of sentences A, then $Cn(A)$ is a normative system if there is a sentence x, such that $x \in Cn(A)$ and x has a deontic modality in at least one subformula.

Such abstract approach to the definition of the normative system has two limitations.

The first limitation regards the distinction between basic and implied (or derived rules). Since the normative system is defined by a characteristic of the set of consequences, it makes no difference whether a rule belongs to the set of basic rules or whether it is the content of an implied obligation or permission. This limitation raises both a philosophical and a practical concern.

The philosophical concern relates to the problem about the "ontological status" of derived rules [8]. If rules are positively enacted by acts of will, is it possible to say that derived rules exist with the same status of enacted rules, even though they were not explicitly willed, and the legislator may not even be aware of the derivation?

The practical concern relates to the fact that normative systems with equivalent sets of consequences in terms of implied obrigations and permissions, but based on different formulations of rules, may react differently to a normative change [3].

Hence, from the synchronic perspective, i.e. considering the normative system at a particular moment of time, one may assume that two normative systems are the same if they derive the same set of obligations/permissions, even if they have different formulations. That is, from that perspective, the formulation of the base of explicit rules is irrelevant. However, from the dyachronic perspective, that is, considering the normative system's change at a moment to a second moment where a new rule is promulgated or derogated, the formulation of the base of explicit rules becomes relevant, given that the revision of different sets of explicit rules with the same derived obligations/permissions may lead to different outcomes.

The second limitation of Alchourrón and Bulygin's abstract definition is that it is concerned only with prescritive or positively enacted rules. It consciously excludes other kinds of rules and particularly rules regarding axiological considerations about positively enacted rules.

It is not the case here of arguing whether these limitations are virtues or vices. I shall provide a definition which is able to overcome such limitations by distinguishing explicit from derived rules and also by encompassing different sorts of rules that one may find relevant to legal reasoning. For those who believe that these capacities are vices that distorts the "real understanding" of what a normative system is, we have to say that the general definition here provided may be reduced to a model which is exactly Alchourrón and Bulygin's.

I shall use the term *"normative set"* to refer to sets of different kinds of rules. These rules might be (i) *conceptual rules*; (ii) *consequential rules*; (iii) *deontological rules* (iv) *axiological rules*; (v) *technical rules*; (vi) any other sort of rules that one may believe has a different logical behaviour with respect to the others. The only formal requirement is that such rules may be reduced to conditionals which are pairs of propositions. Such pairs of propositions may be indexed to some valuation, which would be simply a function taking each pair to a particular interval of real numbers.

Conceptual rules state that the entities described by certain factors (relevant features) *count as* (are to be classified as) instances of the ascribed concept. Conceptual rules have the form (a, c) where a is the triggering factor (or conjunction of factors) and c is the ascribed concept. For instance, a conceptual rule stating that text message stored in a mobile phone (sms) counts as "data" can be represented as (sms, dat). Conceptual rules may be indexed to degrees of confidence on a meaning ascription. For instance, one may have a degree of confidence of .6 that text messaging counts as a communication $(sms, com)_{.6}$.

Consequential rules specify the extent to which the presence of a factor affects the impact of actions on values. That is what are the consequences of an action and how they impact the promotion or demotion of relevant values.

Consequential rules may be indexed to degrees of influence that the presence of a certain context (or factor) has on the impact of an action on a value. For an example consider the rule according to which the impact that the action of searching and seizuring property items has on privacy is increased if the item is a mobile phone $(mob, Priv^{acc})_{.8}$.

We distinguish two kinds of rules establishing obligations or permissions, deontological and axiological ones. Such rules lead to deontic conclusions which may be in conflict. *Deontological rules* link the (deontological) prohibition or permission of a given action to the presence of certain antecedent conditions. For instance, we represent as $(\neg sord, \neg acc)$ the rule prohibiting police officers from accessing personal documents without a search & seizure order. *Axiological rules* are partitioned into two sets: those linking the prohibition of an action to a value demoted by that action; and those linking the permission of an action to a value promoted by that action. We represent axiological rules in the form $(V^x, x)_i$, where V is the value demoted or promoted by action x and i is the weight of the value. For instance, access to a mobile phone by police officers demotes privacy, which is a reason for prohibiting it $(Priv^{acc}, \neg acc)_{.4})$, while it promotes public safety, which is a reason to permit it $((Saf^{acc}, acc)_{.6})$.

There are also *technical rules*, which describes as duties those means which are necessary means to reach some normative goal. For instance, that there is a duty to provide enough beds in hospitals, as a necessary means to fulfill the goal of providing universal health. Such rules may also be indexed according to the degree of necessity that one may assign to such means. One may also conceptualize different sorts of enunciates that may be relevant to normative reasoning, where one may argue that its logical behaviour differs from the others we have mentioned here. For each kind of rule one may conceptualize, one may propose a different box of conditionals that may be be combined in a chain of reasoning.

3 Normative systems

Reasoning with each kind of rule (conceptual, consequential, deontological, axiological, or technical) has different logical properties and therefore requires a different output operator in an architecture of i/o logics (see [7], for an introduction).

Let L be a standard propositional language with propositional variables and logical connectives: $\neg, \wedge, \vee, \rightarrow, \bot, \top$. Let $Val = \{V_1^x, V_2^x, ... V_1^y, V_2^y, ...\}$ be a set of values. We say that $N \subseteq G \times G$, where $G \in \{L, Val\}$ is a *normative set* and that each $r \in N$ is a *rule*. For any $A \subseteq G$, $N(G)$ is the image of N under G, that is $N(G) = \{x : (a, x) \in N,$ for some $a \in G\}$. We write simply $N(a)$ to abbreviate $N(\{a\})$. To state that x is the output of input a to normative set N, we may write $x \in out_i(N, a)$, or $(a, x) \in out_i(N)$. For any normative set N we define

$body(N) = \{a : (a,x) \in N\}$.

Therefore, normative sets contain pairs of propositions or pairs linking a proposition or value to another proposition or value. Cl denotes the classical consequence operator, which will not be applied to values. As a consequence, we shall employ basic (out_2) and basic-reusable (out_4) operators [4] as well as a *weakened output*, defined below.

Definition 1. Let N be a normative set, $A \subseteq L$ and $\mathcal{V}$ the set of all maximal consistent sets v in classical propositional logic. Then:
(i) *simple minded*: $out_1(N, A) = Cl(N(Cl(A)))$
(ii) *weakened*: $out_{1-}(N, A) = N(Cl(A))$
(iii) *basic*: $out_2(N, A) = \bigcap\{out_1(N, v) : A \subseteq v, \text{ for } v \in \mathcal{V} \text{ or } v = L\}$
(iv) *weakened basic*: $out_{2-}(N, A) = \bigcap\{out_{1-}(N, v) : A \subseteq v, \text{ for } v \in \mathcal{V} \text{ or } v = L\}$
(v) *basic reusable*: $out_4(N, A) = \bigcap\{out_1(N, v) : A \subseteq v \text{ and } out_1(N, v) \subseteq v, \text{ for } v \in \mathcal{V} \text{ or } v = L\}$

We shall also consider sets $P \subseteq (L \times L)$ of explicit permissions and corresponding output operators $perm_i(P, N)$ defined as $(a, x) \in perm_i(P, N)$ iff $(a, x) \in out_i(N \cup Q)$, for some singleton or empty $Q \subseteq P$.

One may combine normative sets N_1 and N_2 and output operators out_i, out_j, by making the output of a normative set (possibly joined with the input set) the input of the output operation on the other normative set, that is $out_{i,j}(N_1, N_2, A) = out_i(N_1, out_j(N_2, A) \cup I)$, where $I \in \{A, \emptyset\}$. We call *sequence* a chain of combinations of normative sets.

Definition 2. (*Normative System*) Let $A, I \subseteq L$. Let $N_1, ..., N_n, N$ be normative sets and $r \in \{0, 1\}$. Then $(N_1^{out_i, r_1}, ..., N_n^{out_m, r_n})$ is a *sequence of normative sets* if, and only if, for all N_j, $1 \leq j \leq n$, it holds that $out_{k,l}(N_j, N_{j+1}, ..., N_n, A) = out_k(N, out_l(N_{j+1}, ..., N_n, A) \cup I)$, where $N_j \subseteq N$ and $I = A$, if $r_i = 1$, or $I = \emptyset$, if $r_i = 0$. A *normative system* is a class of sequences of normative sets.

For instance, the arquitecture presented at [6], where the set of conceptual rules (C) contributes to the determination of which deontological rules and which axiological rules are triggered. In that architecture, the set of conceptual rules is governed by a basic reusable output operator and the set of deontological rules could be governed by a basic output operator. Their combination is given by the identities:

$$out_{2,4}(O_d, C, A) = out_2(O_d, out_4(C, A) \cup A)$$
$$perm_{2,4}(P_d, C, A) = perm_2(P_d, out_4(C, A) \cup A).$$

From now on, we may write O/P_d or O/P_v for referring to both obligation and permission rules, $O_{d/v}$ and $P_{d/v}$ for both deontological and axiological rules and $O/P_{d/v}$ to include all modalities.

I am not going to define a particular operator for a value assessment from which one may derive axiological obligations, prohibitions or permissions. We may assume that one set of axiological rules (P_v) links each value to the permission of the action that promotes it, and the other (O_v) links each value to the prohibitions of the action that demotes it. Both are governed by an axiological output operator $out_\succ$. We may also assume that the axiological outputs are first modulated by an analysis of the influence that factors have on the impact of an action on a velue, that is the derivation of the axiological boxes may be preceded by a modulation box M of consequential rules. An example of a combination of such sets, proposed by [6], is given by:

$$out_{\succ,2^-,4}(O/P_v, M, C, A) = out_\succ(O/P_v, out_{2^-}(M, out_4(C, A) \cup A))$$

Below we have some examples of structures that may be specified to form normative systems:

$$\langle O/P_d, C \rangle = \{(O_d^{2,0}, C^{4,1}), (P_d^{2,0}, C^{4,1})\}$$
$$\langle O/P_v, M, C \rangle = \{(O_v^{\succ,0}, M^{2^-,0}, C^{4,1}), (P_v^{\succ,0}, M^{2^-,0}, C^{4,1})\}$$
$$\langle O/P_{d/v}, M, C \rangle = \langle O/P_d, C \rangle \cup \langle O/P_v, M, C \rangle$$

A particular normative system is specified by indicating the rules of each normative set in the corresponding structure.

Definition 3. (*Argument*) A sequence $(X_1, ..., X_n)$ is an *argument* for (a, x) based on the normative sequence $(N_1, ..., N_n)$ of normative sets N_i if, and only if: (i) $X_j \subseteq N_j$, for $1 \leq j \leq n$; (ii) for every $out_{k,l}(N_m, N_{m+1})$ and $out_{i,j}(X_m, X_{m+1})$, it holds that $k = i$ and $l = j$; (iii) $(a, x) \in out_{i_1,...,i_n}(X_1, ..., X_n)$; (iv) if $X'_j \subset X_j$, then $(a, x) \notin out_{i_1,...,i_n}(X_1, ..., X'_j, ..., X_n)$.

We denote by $Args_{(N_1,...,N_n)}(a, x)$ the set of arguments for (a, x) based on $(N_1, ..., N_n)$. If $(a, x) \in out_{i_1,...,i_n}(X_1, ..., X_n)$ we say that (a, x) is entailed by $X_1, ..., X_n$.

Correspondingly, an argument is based on a normative system if it is based on a normative sequence in the system, and a rule (a, x) is entailed by a normative systems if it is entailed by a sequence in the system. If the argument $(X_1, ..., X_n)$ supporting the derivation (a, x) is such that its top element X_1 is a non-empty set of deontological (axiological) rules, we say that (a, x) is an entailed or derived deontological (axiological) rule, or that the normative system implies the rule (a, x).

Hence, in this norm-based semantic there is a clear difference between a rule pertaining to a normative set, or particularly, to a set of deontological rules $((a, x) \in O_d)$ and a rule being derived from a normative system $((a, x) \in out(O_d))$.

4 Normative theories

The implications of a normative system for particular real or hypothetical fact situations are described by normative propositions, i.e., statements that certain obligations and permissions would hold given certain factors. Normative propositions, while being descriptive of a given normative system, may be also evaluative, reflecting the evaluations that are embedded in the normative systems itself, i.e., the ascription of indexes to each pair of rules in the corresponding normative sets. For instance, these indexes may represent, intensities of influence (through consequential rules) or the ascription of weights of values (through axiological rules) or the ascriptions of degrees of confidence to conceptual rules, or degrees of necessity for technical rules. Let us call such ascription of particular indexes λ-evaluations. These λ-evaluations contribute to determine the axiological obligations/permissions delivered by the system.

Each normative proposition describes an entailed deontological or axiological rule, with the exception of negative permissive propositions, which describe the non-derivability of such a rule. Thus, following Alchourron [2], we distinguish a negative sense of permission $\mathbb{P}^-_{d/v}(x/b)$, as the absence of the prohibition of x given factor b, from a positive sense of permission, as an entailed by a deontological or axiological permission $\mathbb{P}^+_{d/v}(x/b)$. We express normative propositions as dyadic *formulae*, which link possible inputs to deontological or axiological deontic qualifications.

Definition 4. Let $NS = \langle O/P_{d/v}, M, C \rangle$ be a normative system and $b, x \in L$, then:
$NS \models \mathbb{O}_d(x/b)$ iff $x \in out_{2,4}(O_d, C, b)$
$NS \models \mathbb{P}^-_d(x/b)$ iff $\neg x \notin out_{2,4}(O_d, C, b)$

$NS \models \mathbb{P}^+{}_d(x/b)$ iff $x \in perm_{2,4}(O_d, P_d, C, b)$
$NS \models \mathbb{O}_v(x/b)$ iff $x \in out_{\succ,2-,4}(O_v, M, C, b)$
$NS \models \mathbb{P}^-{}_v(x/b)$ iff $\neg x \notin out_{\succ,2-,4}(O_v, M, C, b)$
$NS \models \mathbb{P}^+{}_v(x/b)$ iff $x \in out_{\succ,2-,4}(P_v, M, C, b)$

A normative theory about a normative system NS is the $Th_{NS} = \{\alpha : NS \models \alpha\}$ of normative proposition. Such a theory may be inconsistent, incoherent or unstable, as defined below:

Definition 5. (*Consistency, Coherence and Stability*) For any given $b, x \in L$, a normative theory is:
b-inconsistent iff $\bot \in out_2(O/P_d, b)$
b-incoherent iff $\bot \in out_{2,4}(O/P_d, C, b)$
b-properly incoherent iff it is consistent but incoherent
b-unstable iff it is inconsistent, incoherent, or if it is the case that
$\{\mathbb{O}_v(\neg x/b), \mathbb{P}_d(x/b)\} \subseteq Th_{NS}$ or $\{\mathbb{O}_d(\neg x/b), \mathbb{P}_v(x/b)\} \subseteq Th_{NS}$
b-properly λ-unstable iff it is consistent, coherent but b-unstable.

In other words, inconsistency captures cases in which deontological rules directly deliver incompatible conclusions, proper incoherence the case in which the conflict of deontological rules is triggered by a conceptual classification, and proper instability the case in which deontological rules are in conflict with axiological rules expressing the values of the system. We also say that a normative theory is strongly stable, relatively to an input and an output, if the corresponding deontological normative proposition is matched by an axiological proposition, and that it is weakly stable, if the deontological propositions are not conflicted by an axiological proposition.

Definition 6. (*Weak and Strong Stability*) Consider Th^λ_{NS} and $b, x \in L$. Then, Th_{NS} is:
- *strongly stable*: if $\mathbb{O}_d(x/b) \in Th_{NS}$ then $\mathbb{O}_v(x/b) \in Th_{NS}$ and if $\mathbb{P}_d(x/b) \in Th_{NS}$ then $\mathbb{P}_v(x/b) \in Th_{NS}$
- *weakly stable*: if $\mathbb{O}_d(\neg x/b) \in Th_{NS}$ then $\mathbb{P}_v(x/b) \notin Th_{NS}$ and if $\mathbb{P}_d(x/b) \in Th_{NS}$ then $\mathbb{O}_v(\neg x/b) \notin Th_{NS}$

We say that a theory is consistent if it is b-consistent for every b such that $b \in Cl(a)$ and $a \in body(O_d \cup P_d)$. It is coherent (stable) if it is b-coherent (b-stable) for every b such that $b \in Cl(a)$ and $a \in body(C) \cup body(O_d \cup P_d)$.

5 Final remarks

Deontological rules in the regulation are meant to serve the values aimed at by the regulation. The alignment is successfully achieved when the circumstances under which rules prohibit (permit) an action correspond to the circumstances under which the action would be detrimental (favourable) to the relevant values. However, a mismatch is also possible: what is deontologically prohibited may be axiologically required (having a positive impact on the relevant values) and what it deontologically permitted may be axiologically prohibited.

The general framework we have proposed here may be compatible with any legal theory regarding the relations between law and morality. It is actually a framework to build any system one may like, with any sort of rules or conditionals which one may consider relevant in normative reasoning.

The final outcome of the normative system, here conceived as any class of sequences of normative sets built with output operators, will be determined by the λ-evaluation and the criteria a particular theory may assume to solve incoherences and instabilities when there is a mismatch between axiology and deontology.

Alchourrón and Bulygin's conception may also be represented in this general framework simply as a sequence where the top elements, or the last box in the chain of derivations is a set of (obligatory or permission) rules and which is closed under logical consequence. That is, since Alchourrón & Bulygin also admit that the base of elements of normative system may include sentences other than rules (deontic sentences), their conception may be reduced, in this framework to any any structure of the form $(N_1^{out_i, r_1}, ..., N_n^{out_m, r_n})$, where N_1 is a deontological box (of obligations or permissions), which is closed, that is $N_1 = out_{i,...,m}(N_1, ..., N_n)$.

References

[1] Carlos E Alchourrón and Eugenio Bulygin. *Normative Systems*. Springer, 1971.

[2] C.E. Alchourrón. Logic of norms and logic of normative propositions. *Logique et analyse*, 12:242–68, 1969.

[3] Risto Hilpinen. On Normative Change. In Morscher and Strazinger, editor, *Ethics: Foundations, Problems and Applications (Proceedings of the 5th international Wittgenstein Symposium)*. Wien, 1981.

[4] David Makinson and Leendert van der Torre. Input/output logics. *Journal of Philosophical Logic*, 29:383–408, 2000.

[5] Juliano Maranhão and Edelcio G. de Souza. Contraction of Combined Normative Sets. In *Deontic Logic and Normative Systems: 14th International Conference, DEON 2018*, pages 247–261. Springer, 2018.

[6] Juliano Maranhão and Giovanni Sartor. Value assessment and revision in legal interpretation. In *Proceedings of the 17th International Conference on Artificial Intelligence and Law, ICAIL 2019*, pages 219–223. Association for Computing Machinery, Inc, jun 2019.

[7] Xavier Parent and Leendert van der Torre. Input/output logic. In J. Horty, D. Gabbay, X. Parent, R. van der Meyden, and Leon van der Torre, editors, *Handbook of Deontic Logic and Normative Systems, Volume 1*, pages 499–544. College Publications, 2013.

[8] Alf Ross. Imperatives and Logic. *Philosophy of Science*, 11:30–46, 1944.

SUPPES PREDICATE FOR CLASSES OF STRUCTURES AND THE NOTION OF TRANSPORTABILITY[1]

Newton C. A. da Costa
Decio Krause

Department of Philosophy, Federal University of Santa Catarina, Brazil

Abstract

Patrick Suppes' maxim "to axiomatize a theory is to define a set-theoretical predicate" is usually taking as entailing that the formula that defines the predicate needs to be transportable in the sense of Bourbaki. We argue that this holds for *theories*, where we need to cope with *all* structures (the models) satisfying the predicate. For instance, in axiomatizing the theory of groups, we need to grasp all groups. But we may be interested in catching not all structures of a species, but just some of them. In this case, the formula that defines the predicate doesn't need to be transportable. The study of this question has lead us to a careful consideration of Bourbaki's definition of transportability, usually not found in the literature. In this paper we discuss this topic with examples, recall the notion of transportable formulas and show that we can have significant set-theoretical predicates for classes of structures defined by non transportable formulas as well.

1 Introduction

A real revolution in the discussion of scientific theories arose in the 1950s having Patrick Suppes as one and perhaps the most important responsible. The 'revolution' was directed to the logical empiricist view (started in the 1920s) that a scientific theory would be seen as a formal calculus to which an interpretation is ascribed via what Carnap termed *correspondence rules* (other philosophers used other names for the very same thing). The axiomatics would be, in principle, within classical first order logic, but later they have acknowledged that modal operators could also be used; Federick Suppe's (not Suppes) article presents us

---

[1]Dedicated to Edelcio Gonçalves de Souza, who always had demonstrated interest in these matters.

three steps in the development of the empiricists' ideas, the last one involving modal logics (see his article in [18]).

The precise nature of the correspondence rules (CR) is not clear at all. They look as informal associations (Carnap recalls that N. R. Campbell called this set of rules a 'Dictionary' [7, Chap.24]) connecting theoretical terms (of the language) with observable terms. These two concepts, roughly speaking, mean the following. Observable terms are those terms either directly perceived by the senses (such as 'hard' and 'hot') or those that can be measured by "a simple apparatus" [7, Chap.23], such as the temperature of a certain body. Carnap's own example of a CR is the following: "The temperature (measured by a thermometer and, therefore, an observable in the wider sense explained earlier) of a gas is proportional to the mean kinetic energy of the molecules" (loc. cit.). That is, the connection is given by an informal (almost *ad hoc*) association. Theoretical terms, as the name indicates, are 'theoretical' and cannot be availated as above; a typical example is the kinetic energy in the above example. Despite Carnap had written his book after the raising of modern model theory [8], he doesn't consider *formal semantics*, where the association of language-terms with 'the reality' is given by formal (mathematical) rules, and this 'reality' is taken as a set-theoretical structure. This brings important distinctions we shall made later.

Several criticisms were posed to this view, some of them by Suppes himself. It is enough to remember here that in providing this kind of approach to a huge theory such as general relativity, which requires several 'step theories' such as tensorial calculus, Riemannian geometry, partial differential equations, real analysis and so on, would turn the axiomatics something rather difficult to follow, for all these step theories would be in need of being axiomatized too.

Suppes started by considering all these step theories as done in advance, presupposing them as already given by (informal) set theory and directed the efforts to the interested theory itself. This seems simple today, but, as we have said, constituted a real advance in the axiomatic approach to scientific theories, a program that was suggested by David Hilbert in the sixth of his celebrated 23 Problems of Mathematics [14]. The new resulting program was named the *semantic approach* to theories, in contrast to the 'syntactical' approach of the logical empiricists. Suppes refers to these two views as constituting the *extrinsic* approach (his semantic view) and the *intrinsic* one [19].

Why informal set theory? Precisely because he didn't wish to discuss the foundational issues; if we need tensors, they can be defined set-theoretically. Do we need partial differential equations? Set theory does the job. Do we need to be more specific about proofs? We proceed as the mathematicians usually do. But of course if it is necessary to make this base explicit, we can choose an axiomatic set theory that gives the desired results, say (for most physical theories) the ZFC first-order system [8, pp.592-3]. The important thing is that we don't need any more to be occupied with the details of these step theories, but just assume them, as for instance physicists do in most cases.[2]

Let us give a more detailed simple example, to be used also later. This is the case of semi-groups. As it is known from any basic book on algebra, a semi-group is a (non-empty) set endowed with a binary operation which is associative. We can write this in terms of structures (see more on this below) of the form

$$\mathcal{S} = \langle M, \star \rangle, \tag{1}$$

where $M \neq \emptyset$ and $\star$ a function from $M \times M$ to M, that is, an element of $\mathcal{P}(M \times M \times M)$. As said before, the only axiom being this one (the quantifiers range over M):

$$\forall x \forall y \forall z (x \star (y \star z) = (x \star y) \star z). \tag{2}$$

Examples of semi-groups, or models of this axiomatics, abound. The set of real numbers endowed with addition of these numbers, the set of natural numbers with multiplication, the set of all $n \times n$ matrices with multiplication of matrices, etc. Suppes' account was to show that this move can be summed up by a certain formula of the language of set theory,[3] namely, the predicate (if we need to say, a formula with just one free variable) $\mathsf{P}(X)$ defined as follows:

$$\mathsf{P}(X) := \exists M \exists \star (X = \langle M, \star \rangle \wedge \star \in \mathcal{P}(M \times M \times M) \wedge (2)). \tag{3}$$

Important to realize that those who analise the definition claim for instance that the formula is in the free variable X, as we see, and doesn't

---

[2] Notice that in informal set theory we can do practically everything we wish in mathematical terms.

[3] Notice that the informal set theory has not a well defined language, but if necessary we can reason as if the only specific predicate is membership, $\in$. All other symbols are either logical symbols or defined ones.

depend on any specific property of the set M, referring to only to the way it enters in the formula by means of the whole expression [9]; so, it is *transportable* in the sense of Bourbaki (see below). This means that there cannot be imposed restrictions whatever on the principal sets occurring in the formula (in the case, in the set M), for instance, the requirement that M must be different from another non-empty set N (see below). But, could we use the predicate below as the axiom, as in almost all standard books? That is,

$$\mathsf{P}(X) := \exists M \exists \star (X = \langle M, \star \rangle \wedge M \neq \emptyset \wedge \star \in \mathcal{P}(M \times M \times M) \wedge (2)). \quad (4)$$

Since this is important for that what follows, and since neither Bourbaki nor those who mention his definitions are clear, we shall try to make this claim precise.[4] In fact, at §4 of his *Algebra I* book [4, p.30], Bourbaki introduces the definition of group this way:

> "Definition 1 – A set with an associative law of composition, possessing an identity element and under which every element is invertible, is called a *group*."

The adaptation to semi-groups is immediate, just by requiring that the operation (law of composition) be only associative. We see that the empty set is not excluded of being a group (or a semi-group), and there is no justification for that. We guess this has to be with his notion of transportable formula, although as we shall see soon, the adding of something like $M \neq \emptyset$ does not violates the transportability condition.

But sometimes we would be interested in collecting not all semi-groups, but just some of them. Thus, we must add some restriction to the above predicate, say by avoiding that some specific structure (or structures) enter in the range of the predicate. In this case, some care is to be taken into account.

Thus, we see that if we wish to axiomatize all models of a certain kind, the formula must of course be transportable, which is the case when we are (in Suppes' sense) axiomatizing a certain *theory*, as the theory of semi-groups, for in this case we are interested in collecting (as models of the predicate) all semi-groups. But we can also use set-theoretical predicates for collecting a certain classes of models, which doesn't require the formula to be transportable. Let us go to the details.

---

[4]Really, people in general simply adopt Bourbaki's definitions taking into account that they are clear. In our opinion, they are not.

2 Mathematics: a world of structures

Nicholas (or 'Nicolas') Bourbaki is the pseudonym of a group of (mainly) French mathematicians who, in the 1930s, intended to rescue French mathematics to actuality, that is, to keep it in the level and dealing with the methods (the axiomatic method) that were being developed mainly in Germany (the book by van der Waerden, *Modern Algebra* [22], was taken as a paradigm).[5] It is known that France lost many of its more important scientists during the first world war, so that the mathematics taught in the universities during the post-war times were still the 'old' mathematics of the XIX century, and did not cover the most recent subjects with new methods (the axiomatic method), such as abstract algebra. The group, initially formed by Jean Dieudonné, Henri Cartan, André Weil, Claude Chevalley and Jean Delsarte, started in 1934 a project termed *Éléments de Mathématique* and, according to Dieudonné, planed to finish it in three years. Dieudonné recalls that this was a plan of young and ill informed students and that they never had planed such a thing if they were more informed [13].[6] Important to notice that the members of the group change from time to time, so that the group is still alive today. The initial objective was yet not achieved at all (which gives an idea of the wide task they ascribed to themselves). A very nice historical account about the group and its realizations can be seen in [16]; further considerations in [9], [17].

The idea was to see all mathematical theories as formed by structures of a kind.[7] These structures were to be build from some fundamental structures, termed *mother structures*, which are the *algebraic*, *ordering* and *topological* structures, Bourbaki also acknowledges that this expressed a stage in the development of mathematics, so that further developments could suggest other 'mother' structures. According to Corry [9], since Bourbaki reputed ordering structures as fundamental, his axiomatics for sets initially took the notion of ordered pair as

---

[5]In explaining (p.ix) the purpose of his book, van der Waerden said that he wished "to introduce the reader into the whole world of [algebraic] concepts", by considering the "recent expansion" in this field, "due to the 'abstract', 'formal', or 'axiomatic' school".

[6]Meaning that they realized later the huge work they intended to cover in so few years.

[7]Important to notice that Bourbaki didn't deal with all fields of the mathematics of the day, for instance leaving number theory and geometry aside. There is no apparent reason for that.

primitive [6], what was modified in later versions, when he turned to the usual way of taking the unordered pair as given by a specific axion, from which the ordered pair results by definition. All other mathematical structures would arise from suitable 'combinations' of these mother structures. So, the field of the real numbers was characterized by being a complete ordered field, these three words indicating the (topological) completeness, ordering, and an algebraic structure [5, p.264]. A semigroup is an algebraic structure, so are the monoids, groups, rings, etc.

We clearly see the formalistic purely syntactical approach of Bourbaki. Mathematics is got by writing symbols in the paper according to the rules stated in the Theory of Sets. If something was not written yet, it does not belong to the field of mathematics. So, his notion of truth is quite peculiar, in a certain sense constructive: something is *true* if we have a proof of it, even an indirect proof, which shows that his mathematics is 'classic', that is, the excluded middle law, so as *reductio at absurdum*, among other 'classical' procedures, hold. In the same vein, something is false if there is a proof for its negation. Thus, something for which there is no proof neither by the affirmation nor to its negation, is neither true nor false. In other words, the *mathematics* is classic, by the *metamathematics* is not, being constructive. Bourbaki doesn't think of semantics as we are accustomed nowadays.

2.1 Species of structures

In Chapter 4 of the book on set theory, Bourbaki develops his 'theory of structures', which interests us here. In [11], the authors proposed a modification and adaptation of Bourbaki's notions, grounding them in a set theory with atoms. Here we shall follow Bourbaki, but without the technical details and subtleties. As said before, most of the content of present day mathematics, according to Bourbaki, fall under the notion of structure.[8] But, what is a structure? Bourbaki speaks of *species of structures* and of *structures of a certain species*; intuitively speaking, using a terminology which departs from him, a (mathematical) structure

---

[8]Interesting to mention that the theory of categories [16] was never mentioned by Bourbaki [9] (another omission are the Gödel's incompleteness theorems [17]), although the concept of category arose with one of its members, Samuel Eilenberg (together with Saunders MacLane), and further developed for instance by another of members, Alexander Grothendieck. Categories are 'big enough' to be treated as usual sets, so it would be necessary to expand the logical (ZF) basis to cope with them, something that perhaps Bourbaki was not being up to do.

is precisely that we are spectating it to be since our logical courses: a set, or a collection of sets, endowed with relations and operations not only among their members (which would characterize *order-1* structures [15]) but also among collections of elements of these sets (the elements of the basic sets are called *individuals*), relations among them, etc. So, we may have relations whose relata are also relations, that is, the structures may by of *order-n*, $n > 1$.[9]

To give an idea of how things proceed, let us start with a finite collection of sets $E_1, E_2, \ldots, E_n$ which will be the *principal* sets and a finite collection $A_1, A_2, \ldots, A_m$ of *auxiliary* sets (in our examples, we shall use just three principal sets, F, G, and H). The auxiliary sets must not contain any references to the principal sets. For instance, the vector space structure comprises a principal set V of vectors and an auxiliary set K of scalars, while semi-groups have just one principal set M and no auxiliary sets. Using the set-theoretical operations of taking the power set and the Cartesian products, we can obtain a sequence of new sets $\mathcal{P}(F)$, $F \times H$, $\mathcal{P}(G) \times H$, ... These sets are constructed from a very subtle schema S he terms an *echelon construction schema* we will not recall here (but see [5, chap.4], [11], [9]). The last set in the sequence is the echelon of scheme S. For instance, to get a binary operation over a set F, we build a sequence of sets with a certain schema S (omitted here), for instance (there is not just one schema, so different sequences can be obtained) F, $F \times F$, $F \times F \times F$, and finally we get the echelon of scheme S, $\mathcal{P}(F \times F \times F)$. The binary operation will be an element $\mathfrak{s} \in \mathcal{P}(F \times F \times F)$ (before, in our given example of semi-groups, we have termed $\star$ such an $\mathfrak{s}$), to which we impose the restrictions we wish, written in the form of postulates, say that $\mathfrak{s}$ must be associative, that is [5, p.60],

$$\mathfrak{s}(\mathfrak{s}(a,b), c) = \mathfrak{s}(a, \mathfrak{s}(b, c)), \tag{5}$$

which in our above terminology means $\star(\star(a,b), c) = \star(a, \star(b, c))$, or simply $(a \star b) \star c = a \star (b \star c)$.

Another important concept is that of *canonical extensions of mappings*. Given an echelon construction schema S and two collections of

---

[9]We use the terminology 'order-n' instead of 'first-order', 'second-order' etc. to avoid confusion with the order of the languages. Really, we can define order-n structures ($n > 1$) in first-order languages, say in first-order ZF; for instance, well-ordering structures are not order-1, as it is easy to see (the postulate requires the reference to 'all non-empty subsets', which is not a sentence of first-order.

sets $E_1, \ldots, E_n$ and $E'_1, \ldots, E'_n$, let us consider mappings (functions) $f_i : E_i \to E'_i$. Bourbaki defines extensions of these mappings from the sets in an echelon based in S constructed over the E_i to the corresponding sets in the echelon also based in S but now on the E'_i, until getting a mapping from the echelon of scheme S based on the E_i to the echelon of scheme S based on E'_i (the reader must be attentive with the terminology). This last mapping is the *canonical extension* of the f_i, written $\langle f_1, \ldots, f_n \rangle^S$. If the f_i are injective (surjective, bijective), then $\langle f_1, \ldots, f_n \rangle^S$ will be injective (surjective, bijective) [5, p.261].

A *species of structure* Σ is defined this way. We take a collection of principal base sets $x_1, \ldots, x_n$, a collection of auxiliary bases sets $A_1, \ldots, A_m$ and a specific echelon construction schema $S(x_1, \ldots, x_n, A_1, \ldots, A_m)$. An element $\mathfrak{s} \in S(x_1, \ldots, x_n, A_1, \ldots, A_m)$ is the typification of Σ. The typification is written by Bourbaki as a formula $T(x_1, \ldots, x_n, \mathfrak{s})$. Let now $R(x_1, \ldots, x_n, \mathfrak{s})$ be a transportable formula (see below) with respect to the given typification, with the x_i as the principal sets and the A_j as the auxiliary sets. This formula will be the *axiom* of the species of structures with typification T. If we select some particular sets $E_1, \ldots, E_n, U$ so that both $T(E_1, \ldots, E_n, U)$ and $R(E_1, \ldots, E_n, U)$ hold, then U is said to be *a structure of species* Σ. His first example is that of the species of structures of ordered sets, where from a set A, we get (by a suitable echelon construction schema S) the set $\mathcal{P}(A \times A)$ and the typification $\mathfrak{s} \in \mathcal{P}(A \times A)$ (a binary relation on A), with the axiom $\mathfrak{s} \circ \mathfrak{s} = \mathfrak{s}$ (reflexivity) and $\mathfrak{s} \cap \mathfrak{s}^{-1} = \Delta_A$ (transitivity), being Δ_A the diagonal of A (informally, the set $\Delta_A = \{(x,x) : x \in A\}$). Other examples can be found in [5, pp.263ff].

Thus, the restrictions imposed to $\mathfrak{s}$ constitute the axioms of the species of structure (in the case of semigroups, the restriction is that $\mathfrak{s}$ must be associative). As we see, all of this relies on the notion of transportability, for the axiom (the conjunction of the formulas we standardly use) must be transportable. So, the predicate (4) defines the species of structure of semi-groups, the semi-groups (the structures that satisfy the predicate) being the structures of that species.

2.2 Transportable formulas

The notion of transportable formulas is important for our account, so that it deserves a particular subsection. The definition is marked by Bourbaki with the symbol '¶', which means 'difficult exercise'. So, let us go slow, even without providing all the details. Important to remark that it does not constitute a simple and easy definition. Interesting enough that people who mention Bourbaki's account do not discuss it in full and, in our opinion, neglect important aspects of it. This is why we shall give some attention to it.

Think again of a semi-group. Remember that we (by hypothesis) intend to develop the *theory* of semi-groups, which requires an axiomatization of *all* of them (by the way, this is one of the main advantages of the axiomatic method).[10] So, we need to provide a definition that does not exclude any semi-group from the list, which requires that our definition should not refer to any particularity of the domain M that could leave some semi-group out, being not covered by the definition. So, we cannot characterize semi-groups by saying things like 'a semi-group is a set M distinct from the set of the natural numbers so that blah-blah-blah', for in doing this we would be eliminating important semi-groups, such as $\langle \mathbb{N}, + \rangle$. In other words, the formula which characterizes the species of structure must be *transportable*, or invariant by substitutions of the principal set(s), as we shall see soon.

Bourbaki calls (in our terminology) formulas 'relations'. Suppose we have an echelon construction schema S for $n + m$ terms (sets), where there are n principal sets $x_1, \ldots, x_n$ and m auxiliary sets $A_1, \ldots, A_m$, which, as before, is written $S(x_1, \ldots, x_n, A_1, \ldots, A_m)$; let us abbreviate by $S(x_i, A_j)$. As we have seen, an element $\mathfrak{s} \in S(x_i, A_j)$ characterizes a *typification* of $\mathfrak{s}$. Notice that to typify something is just to select if from a certain set constructed by set-theoretical operations from base sets (principal and auxiliary). So, as seen before, $\star \in \mathcal{P}(M \times M \times M)$ is a typification of a binary operation on the base set M. Another example is useful. Let us consider vector spaces again, with V as principal set and K as the auxiliary set. We form the Cartesian product $K \times V$ and chose an element $\cdot \in K \times V$. This element can be written as

$$\cdot = \{\langle k, \alpha \rangle : k \in K \land \alpha \in V\}. \tag{6}$$

If we write $\langle k, \alpha \rangle$ as $k \cdot \alpha$ of simply $k\alpha$ for short, we see that the typification characterizes the operation of multiplication of vectors by

---

[10]Bourbaki emphasizes this. He says that the main task of axiomatization is that enable us to study non-categorical theories (*multivalent* in his terminology). As he says, "The study of multivalent theories is the most striking feature which distinguishes modern mathematics from classical mathematics" [5, p.385].

scalars.

The typification could involve several choices say $s_1 \in S_1(x_i, A_j)$, ..., $\mathfrak{s}_p \in S_p(x_i, A_j)$ if we have also several echelon construction schemes $S_1, \ldots, S_p$. This of course defines a formula, which we write, as above, adapting Bourbaki's notation, $T(x_1, \ldots, x_n, \mathfrak{s}_1, \ldots, \mathfrak{s}_p)$. Now comes the ¶ part.

Let $R(x_1, \ldots, x_n, \mathfrak{s}_1, \ldots, \mathfrak{s}_p)$ be a formula, and let $y_1, \ldots, y_n, f_1, \ldots, f_n$ be variables other than the x_i and the $\mathfrak{s}_j$. The f_i are bijections from x_i onto y_i, and Id_j are the identity functions of the auxiliary sets A_j. Once we have the canonical extension

$$\langle f_1, \ldots, f_n, Id_1, \ldots, Id_m \rangle^S, \tag{7}$$

we can get the $\mathfrak{s}'_j$ by applying these extensions to the $\mathfrak{s}_j$, namely,

$$\mathfrak{s}'_j = \langle f_1, \ldots, f_n, Id_1, \ldots, Id_m \rangle^{S_j}(\mathfrak{s}_j), \tag{8}$$

so that the formula $R(x_1, \ldots, x_n, \mathfrak{s}_1, \ldots, \mathfrak{s}_n)$, by bijective mappings, gives $R(y_1, \ldots, y_n, \mathfrak{s}'_1, \ldots, \mathfrak{s}'_m)$. Then, the formula R is *transportable* if these two formulas are equivalent, that is, iff we can prove in the system that

$$R(x_1, \ldots, x_n, \mathfrak{s}_1, \ldots, \mathfrak{s}_n) \leftrightarrow R(y_1, \ldots, y_n, \mathfrak{s}'_1, \ldots, \mathfrak{s}'_m). \tag{9}$$

The notation is in fact far-fetched. So, let us try to translate the definition to a language closer to that we use today. Let $R(x_1, \ldots, x_n, \mathfrak{s})$ a formula (we take just one $\mathfrak{s}$) and S be an echelon construction schema. If $f_i : x_i \to y_i$ $(i = 1, \ldots, n)$ are bijections, any canonical extension $\langle f_1, \ldots, f_n \rangle^S$ is also a bijection. So, we get that

$$\langle f_1 \ldots, f_n \rangle^S \Big(S(x_1, \ldots, x_n, \mathfrak{s}) \Big) = S(y_1, \ldots, y_n, \mathfrak{s}'), \tag{10}$$

being $\mathfrak{s}' = \langle f_1, \ldots, f_n \rangle^S(\mathfrak{s})$. Then, if $R(y_1, \ldots, y_n, \mathfrak{s}')$ also holds, the formula $R(x_1, \ldots, x_n, \mathfrak{s})$ is transportable. Let us remark that (9) is speaking in syntactical terms, that is, in *proof*. The equivalence must be shown on syntactical grounds.[11]

---

[11] We emphasize once more the purely syntactical aspect of Bourbaki's approach.

But nowadays we are accustomed with semantics, so some authors prefer to express the idea on semantical groundings [11]. In this case, we can grasp the concept by considering two isomorphic structures $\mathfrak{A}$ and $\mathfrak{B}$. Let α be a sentence of the language appropriated for both structures.[12] In the present day 'semantical' language, α is transportable if and only if

$$\mathfrak{A} \models \alpha \text{ iff } \mathfrak{B} \models \alpha, \tag{11}$$

that is, if and only if α is preserved under isomorphisms [11]. Let us take an example. Think of the Peano's axioms for arithmetics (within set theory). The axioms can be written, in a standard language, as follows, where the quantifiers range over the set of natural numbers $\mathbb{N}$, except for the third one, where quantification over subsets of $\mathbb{N}$ is also allowed, and where 0 means 'zero' and n' stands for the sucessor of n:

1. $\forall n (0 \neq n')$

2. $\forall n \forall m (n' = m' \to n = m)$

3. $\forall A (A \subseteq \mathbb{N} \to (0 \in A \land \forall n (n \in A \to n' \in A) \to A = \mathbb{N}))$

Thus, $\mathcal{N} = \langle \mathbb{N}, 0, ' \rangle$ is a model of these axioms, the *standard* model. According to Bourbaki, the axioms must be transportable. Let us prove that using the semantic approach, by considering the first axiom. It is easy to show that it is transportable. Really, let us consider another set $\mathbb{N}_1$ such that $\mathcal{N}_1 = \langle \mathbb{N}_1, 0_1, " \rangle$ is also a model of the above axioms (that is, it is also a structure of *that* species). Thus, let $f : \mathbb{N} \to \mathbb{N}_1$ be a bijection such that $f(0) = 0_1$ and $f(n') = (f(n))"$, the sucessor of $f(n)$ in the second structure. If $m \in \mathbb{N}_1$, let $n = f^{-1}(m) \in \mathbb{N}$, so that since $n' \neq 0$, then $f(n') \neq f(0) = 0_1$. Thus $m" \neq 0_1$. In other words, $\mathcal{N} \models \forall n (0 \neq n')$ entails $\mathcal{N}_1 \models \forall m (0_1 \neq m")$. The converse is also easy to prove. With more patience, we can prove that the other two axioms are also transportable.

Notice that this 'semantic' account is an *interpretation* of Bourbaki's notions and, although we agree that most mathematicians will take it for granted, it can't t be shown to be equivalent to the original approach,

---

[12] We leave the formal definition of 'appropriate' out, keeping only with its intuitive aspect. But see [11], [15].

for there is no way of comparison between then: one is syntactical, the other is semantical, and we know that set theory is not a complete theory (when syntax agrees with semantics). So, we must take care.

We also remark that neither axiom poses a restriction on the principal set (namely, $\mathbb{N}$). The restriction of being different of 0 is ascribed to the sucessor of n, and this does not violate the definition of transportability.[13] But let us take the following formula $\mathfrak{s}(\mathfrak{s}(0)) = \{\{\emptyset\}\}$ (Zermelo's 'two'). Notice that now we have something different, namely, the presence of the set $\{\{\emptyset\}\}$ which is not part of the formal language. And, of course, taking another definition of 'two', we could have $\mathfrak{s}(\mathfrak{s}(0))$ being associated to another set, say $\{\emptyset, \{\emptyset\}\}$ (von Neuman's 'two'). Thus, $\mathfrak{s}(\mathfrak{s}(0)) = \{\{\emptyset\}\}$ is not transportable.

Bourbaki gives us the following not so clear example, as fas as we know, never discussed elsewhere:

> "For example, if $n = p = 2$ and if the typification $(\ldots)$ is '$\mathfrak{s}_1 \in x_1$ and $\mathfrak{s}_2 \in x_1$', [then] the relation $\mathfrak{s}_1 = \mathfrak{s}_2$ is transportable. On the other hand, the relation $x_1 = x_2$ is not transportable." [5, p.262].

Our explanation is as follows. The typification takes elements of a same set x_1, hence we need no more than this set in our echelon. The (only) bijection will be some $f : x_1 \to y_1$, being y_1 a set whatever. Hence the canonic extension $\langle f \rangle^S$ is f itself. Then, the formula (1) $\mathfrak{s}_1 = \mathfrak{s}_2$ conduces to (2) $f(\mathfrak{s}_1) = f(\mathfrak{s}_2)$ by the bijection. Obviously, if (1) holds, so does (2). For the second case, we have two sets x_1 and x_2, and two bijections $f_1 : x_1 \to y_1$ and $f_2 : x_2 \to y_2$. But $x_1 = x_2$ doesn't entail that the set x_1 (or x_2, since they are equal) is lead by the two bijections f_1 and f_2 in the same set, so that not necessarily y_1 and y_2 are equal.

This last remark and the example may suggest something already mentioned earlier, namely, why Bourbaki didn't made the exigence that the domain of a group (and this applies also to semi-groups) needs to be not empty. Apparently, this could be due to the fact that the formula $x \neq \emptyset$ seems to be not transportable, for the negation of a transportable formula is also transportable and we could just take $x_2 = \emptyset$ above. But this is false. The emptyset has specific properties; let us see. Suppose

---

[13] Really, the formula is a particular case of Bourbaki's own example shown in the quotation below, namely, that (the negation of) $\mathfrak{s}_1 = \mathfrak{s}_2$ is transportable, just taking $\mathfrak{s}_1$ as n' and $\mathfrak{s}_2$ being 0, both in $\mathbb{N}$.

we have a set M (to go along with our example) for defining the species of structures of semi-groups; should we use (1) or (4) as the axiom? It is indifferent, and this is due to the restriction. Really, suppose we have again another N (which plays the role of y_1 in the definition), and let $f : M \to N$ be a bijection. Since $M \neq \emptyset$, we conclude that $N \neq \emptyset$, so the restriction doesn't impede the transportability of the formula (as we shall see with more details below, this will be not the case with other non-empty sets). The difference in using (1) or (4) is that, as we have remarked, with the first we enable the emptyset to be a semi-group, something that is avoided in the second case.

Of course the above reasoning is grounded on the following immediate theorem:

METATHEOREM 1 *A formula α is transportable if and only if all subformulas of α are transportable.*
Proof: If α has some subformula β that is not transportable, then β will be not invariant under isomorphisms, so not will be α. The converse is trivial. ∎

So, from the perspective of Bourbaki, we cannot select *some* structures to be the models of some set-theoretical predicate; we must consider *all* structures that satisfy the predicate. As we shall see, this is precisely the case of Suppes' set-theoretical predicates when used to axiomatize theories. But, as anticipated, sometimes we are interested in selecting some particular model or some class of models. Next, let us consider this.

3 Selecting classes of models

According to one of most widespread characterization of the semantic approach to scientific theories, a theory is specified by a family of structures, the *models* of the theory [21, p.77]. Models of most scientific theories are, as said already, set-theoretical structures. In first-order logic, models (of first-order languages or theories) are *order-1* structures, or structures comprising sets as their domain(s) and the relations (and operations) are defined among the elements of the domain(s). No relation (operations are particular cases of relations, so we shall speak of relations only) can have as arguments other relations or sets of such elements, as the case of topological spaces illustrates.[14] With respect of such languages, we have an important theorem which goes as follows. Given a certain collection of order-1 structures, there exists a necessary and sufficient condition for axiomatizing such a collection, that is, a condition that says that there can exist a theory (set of postulates) whose models are exactly the chosen structures, namely, the collection must be closed by elementary equivalence and by ultraproducts [8, Thm.4.1.12, p.220]. The right definitions are not important here, but just the fact that there is such a criterium for first-order languages. Concerning higher-order languages or classes of structures, there is no a similar theorem; given classes of such structures, we need to study them case by case. The importance of this fact is that most structures that model postulates of a scientific theory are higher-order structures, or order-n structures with $n > 1$. A typical example is that of classical particle mechanics, which can be summarized as follows (this is one of the most simple examples we have, so it is explored to exhaustion by several authors too).

According to Suppes, going to the characterization of a theory in terms of structures, "A system of classical (particle) mechanics is a mathematical structure of the following sort ...", and then specifies a basic finite domain P of entities, the *particles*, a set T of instants of time (usually an interval of the real number line), and some other elements which are not of our interest here (but see [20, p.320]). All of this is collected in a structure $\mathcal{P} = \langle P, T, \ldots \rangle$, subjected to suitable postulates.

The set-theoretical predicate would be something saying that some set X is a classical particle mechanics iff it obeys the predicate

$$\mathsf{CPM}(X) := \exists P \exists T \ldots (X = \langle P, T, \ldots \rangle \wedge P \neq \emptyset \wedge T = [a,b] \subseteq \mathbb{R} \wedge \ldots (12)$$

But there are two important restrictions in this definition: the principal set P must be finite and non empty, and the auxiliary set T must be an interval of real numbers (*ibidem*). So, as we have seen already, the set-theoretical predicate is a transportable formula (the other elements of the structure are typifications), and so it defines a specie of structures in the sense of Bourbaki, selecting a huge class of structures that satisfy it, the models of the predicate or, as Suppes suggests, the 'classical particle mechanics'.

---

[14] A topological space (in terms of structures) is an ordered pair $\mathcal{T} = \langle X, \tau \rangle$ where τ is a collection of subsets of X satisfying certain axioms [5, p.263], that is, τ comprises *sets* of elements of the domain.

But sometimes we may wish to consider just a small part of the whole class of models. Let us take a simple example. Suppose again that we have a set-theoretical predicate for semi-groups, but we wish, by some hidden reason, to avoid considering all semi-groups having the set of real numbers as domain, say $\mathcal{A} = \langle \mathbb{R}, + \rangle$, the semi-group of the real numbers with usual addition. How should we proceed? This is simple, one may say. Just take the predicate (4) and impose that the domain must be different from $\mathbb{R}$, that is, something like

$$\mathsf{P}(X) := \exists M \exists \star \, (X = \langle M, \star \rangle \land M \neq \mathbb{R} \land \star \in \mathcal{P}(M \times M \times M) \land (2)). \quad (13)$$

It is easy to see that the models of such a predicate are all semi-groups (the empty set included) *except* those that have $\mathbb{R}$ as the domain. But wait! The formula, as we have seen, is not transportable due to the imposed restriction. So, it doesn't axiomatize the theory of semi-groups, for in this case no semi-group should be leaved out.

A more relevant example should be the following. Suppose we wish to consider just those models characterized by a *representation theorem* our theory may admit. Let us say something on this point. Suppes emphasizes that sometimes a theory is so that there is a subclass of models with the following property: for every model of the theory, there is in this class a model which is isomorphic to the given model. In mathematics, it is simple to give examples, namely, every group is isomorphic to a group of permutations (Cayley's representation theorem) or Stone's representation theorem, which says that every Boolean algebra is isomorphic to a a certain field of sets. Thus, in a certain sense, it is enough to have this subclass in order to know all possible models of a theory (up to isomorphisms). (Just to comment, as Suppes recalls, in the case of empirical theories it may be quite difficult to find a representation theorem, or to prove it. But let us move on without discussing the details.)

The important fact is that we may be interested in some specific class of models, or then we wish at to disregard some specific model of the theory which is not interesting to us. In the case of the example, we need to impose that the models of the predicate will be precisely those we wish and not others. Although we shall not provide the details here, it is to be acknowledged that this can be done. The problem is that this step, which seems to be justified by the interest of the scientist, finds problems with the above definition of a set-theoretical predicate or Bourbaki's species of structures. Really, as we have seen, the formula which stand for the predicate that defines the theory must be transportable, that is, we cannot impose arbitrary restrictions on the principal sets, which need to be instantiated by any sets whatever (with possible exceptions such as the empty set, as seen earlier).

So, we see that set-theoretical predicates can be used both to define theories and also to select classes of structures, yet that sometimes things may go to not useful results, as in the case when we take the predicate $P(x) := x = \emptyset$ [11], which apparently does not define any theory whatsoever, yet selects a structure having the emptyset as its domain (and vacuous typifications, of course).

Two things need to be enlighten: the first is that, as we have said, in considering such (set-theoretical) predicates for defining theories or classes of structures, all the step theories are being presupposed. Secondly, remember that (in general) we are working within a set theory such as the ZF system, and we could be interested in finding a set-theoretical predicate for ZF itself. In this case, as it results from Gödel's incompleteness theorems, being consistent, ZF does not admit ZFC-sets as models, that is, models that are sets of ZF. For doing that, we need to strengthen ZF with additional postulates, say by assuming the existence of *universes*,[15] or going to another stronger theory. But this (apparently) is not necessary for most theories in the empirical domain.

The fact is that there is no *one* solution to all problems. The use of set-theoretical predicates will depend on the set-theory being used and the needs of the scientist. Important is to be aware of the technique and of its importance. The details must be fulfilled in each particular case.

4 Conclusion

In this paper we have shown the dependence of Bourbaki's notion of species of structures to the concept of transportable formulas. Furthermore, we have enlighten that his definition of transportable formulas does not enable us to introduce arbitrary restrictions on the principal sets, for once some restriction is made, the formulas may not be invariant by isomorphisms. So, there is a strong difference between Bourbaki's

---

[15]Universes, initially introduced by Alexander Grothendieck (who was a member of the Bourbaki group) to cope with categories in set theoretical terms; see [3]. The existence of universes is equivalent to the existence of inaccessible cardinals.

species of structures and Suppes' set-theoretical predicates, which also characterize certain structures, the models of the predicate, but enabling us to introduce restrictions on the principal sets, thus allowing the selection of just the models we may be interested in. But, when the set-theoretical predicate is a transportable formula, Suppes' approach coincides with Bourbaki's.

Summing up, Bourbaki's approach and Suppes' account using transportable formulas are directed to the axiomatization of *theories*, where no model can be left out, while the use of set-theoretical predicates without such a restriction (of being a transportable formula) is more general, for it enables also to grasp just *some* relevant models of a certain class of models.

References

[1] Balzer, W., Moulines, C. U. and Sneed, J. D. (2012), *Una Arquitectónica para la Ciencia: El Programa Estructuralista*. Buenos Aires: Bernal – Universidad Nacional de Quilmes Editorial.

[2] Bourbaki, N. (1950), The Architecture of mathematics. *The American Mathematical Montly* 57 (4): 221-32.

[3] Bourbaki, N. (1972), *Appendice* to Artin, E., Grothendieck, A. et Verdier, J. L. (administrateurs), *Séminaire de Géométrie Algébrique du Bois-Marie 1963-1964*, Theorie des Topos et Cohmologie Etale des Schemas (SGA 4): Tome 1 – Théorie des Topos, Exposés I à IV. *Lecture Notes in Mathematics* 269, 270 and 305.

[4] Bourbaki, N. (1989), *Algebra I – Chapters 1-3*. Berlin & Heidelberg: Springer.

[5] Bourbaki, N. (2004), *Theory of Sets*. Berlin & Heidelberg; Springer-Verlag.

[6] Bourbaki, N. (2006), *Théorie des Ensembles*. Berlin-Heidelberg: Springer.

[7] Carnap, R. (1966), *Philosophical Foundations of Physics: An Introduction to the Philosophy of Science*. New York & London: Basic Books.

[8] Chang, C. C. and Keisler, H. J. (1977), *Model Theory*. 2nd.ed. Amsterdam: North-Holland.

[9] Corry, L. (1992), Nicolas Bourbaki and the concept of mathematical structure. *Synthese* 92 (3): 315-48.

[10] da Costa.N. C. A. (1994), Review of 'An Architectonic for Science: The Structuralist Program', by Wolfgang Balzer, C. Ulises Moulines and Joseph D. Sneed *The Journal of Symbolic Logic* 59 (2): 671-3.

[11] da Costa, N. C. A. and Chuaqui, R. (1988), On Suppes' set theoretical predicates. *Erkenntnis* 29: 95-112.

[12] da Costa, N. C. A. and Doria, F. A. (2008), *On the Foundations of Science*. Cadernos do Grupo de Altos Estudos, Vol. II. Programa de Engenharia de Produção da COPPE/UFRJ.

[13] Dieudonné, J. A. (1970), The work of Nicholas Bourbaki. *The American Mathematical Montly* 77 (2): 134-45.

[14] Hilbert, D. (1902), Mathematical problems, *Bull. American Mathematical Society* **8**, 437-79.

[15] Krause, D. and Arenhart, J. R. B. (2017), *The Logical Foundations of Scientific Theories: Languages, Structures, and Models*. London: Routledge.

[16] Mashaal, M. (2006), *Bourbaki: A Secret Society of Mathematicians*. American Mathematical Society.

[17] Mathias, A. R. D. (1992), The ignorance of Bourbaki. *The Mathematical Intelligencer* 14 (3): 4-13.

[18] Suppe, F. (ed) (1977), *The Structure of Scientific Theories*, 2a. ed. Illinois: Un. Illinois Press, 2a. ed.

[19] Suppes, P. (1967), What is a scientific theory? In Morgenbesser, S. (ed.), *Philosophy of Science Today*. New York: Basic Books, 55-67.

[20] Suppes, P. (2002), *Representation and Invariance of Scientific Structures*. Stanford: Center for the Study of Language and Information – CSLI Lecture Notes 130.

[21] van Fraassen, B. (1980), *The Scientific Image*. Oxford: Oxford Un. Press.

[22] van der Waerden, B. L. (1949/1950), *Modern Algebra*. Vols. I and II. New York: F. Ungar Pu. Co. Original German edition from 1930/1931.

On a reconstruction of the valuation concept[1]

Patrícia Del Nero Velasco

Center for Natural and Human Sciences, Federal University of ABC, Brazil

Abstract

The aim of this work is to show that the concept of valuations proposed by Da Costa can serve as a general framework for defining the usual notions of consequence, consistency and validity. I show that, developed in this way, these notions satisfy some intuitively expected properties, which is an interesting result, if we remember that the original purpose of the valuation theory was the construction of a correct and complete semantic to the paraconsistent calculus, and not to furnish a framework for defining such notions.

The results herein presented were first obtained in my PhD thesis, *On a reconstruction of the valuation concept*, written under the supervision of Professor Edelcio Gonçalves de Souza.[2]

1 Introduction

In 1963, N.C.A da Costa presented his Chair Thesis at the Department of Mathematics at the Federal University of Paraná. The work, "Formal Inconsistent Systems" (da Costa, 1963), synthesized previous inquiries from the author. The formal systems in questions are simply axiomatic systems for which it is possible to precisely define a deduction notion, understood merely as a relation between sentences and sets of sentences in a given formal language; examples of such axiomatic systems are the usual formalizations of classical logic.

---

[1] The present article is a result of my PhD thesis with the same title, obtained at PUC-SP in 2004, under the supervision of Professor Edelcio Gonçalves de Souza— a clear existence proof that academic life can also be permeated with generosity, affection, and good humor. *Saudações alviverdes!*

[2] The English version of this chapter was revised by Daniel Arvage Nagase; I'm most grateful for his precious contributions to its final version.

Recall that, from the classical logician perspective, inconsistent systems are those that can deduce a sentence as well as its negation; classically, this implies that any sentence could be deduced in an inconsistent system. In other words, in classical logic, inconsistent systems are trivial (everything is deductible). In contrast, da Costa presented in his work systems whose inconsistency did not imply trivialization. We can therefore say that paraconsistent logics were born in that work (cf. de Souza; da Costa; Maranhão, 2001).

Paraconsistent logics are a class of non-classical logics that reject the principle of non-contradiction. Another example of a class of non-classical logics is the class of multi-valued logics that reject the principle of the excluded middle. There is, however, a basic difference in how such logics came into being. Multi-valued logics were first created by considering a logic from the semantic point of view, introducing tables with more than two truth values, and one of the first problems concerning such logics was to find axiomatic systems that were sound and complete with respect to this new semantics.

The creation of paraconsistent logics followed the inverse direction. They were born as formal systems, and the problem was to find an adequate semantics for these systems. It was known that logical matrices with truth-functional connectives were not suitable for this end. The theory of valuations appeared in this context to furnish a sound and complete semantics for the paraconsistent calculus, abandoning the idea of truth-functional connectives and creating a powerful instrument for the demonstration of completeness of formal systems that obey certain basic properties (cf. da Costa and Loparic, 1984; da Silva, 2000; de Souza, 2001).

The present article considers a different path toward the valuation notion, a path that takes as its starting point an informal (but adequate) definition of logical consequence and related notions. This work can be understood as an attempt to answer the question: what is the simplest mathematical framework (i.e. the one that requires the least number of primitive concepts) in which the concepts of consequence, consistency and validity can be defined, fulfilling certain requirements of material adequacy?

I will argue that the theory of valuations from da Costa is precisely this simplest mathematical framework adequate for this task.

2 A intuitive notion of consequence

The notion of logical consequence is widely discussed by philosophers and logicians. Let us, at first, investigate the informal and usual definition of consequence.

We say that an argument is valid "if the conclusion follows necessarily from the premises", or, in other words, "if it is impossible for the premises to be true and the conclusion false". Such definition is not very enlightening, since it does not contain clearer concepts than those that we want to define, viz. the notion of logical consequence. What does "follows necessarily from" mean? In what consists the "impossibility"? Is it, for instance, a material or logical impossibility? It seems desirable for a good definition of logical consequence to use the least possible number of concepts, such as the definition proposed by G. Priest:

> So what is a valid inference? One, we saw, where the premises can't be true without the conclusion also being true. But what does that mean? In particular, what does the *can't* mean? (...) It is natural to understand the 'can't' relevant to the present case in this way, to say that the premisses can't be true without the conclusion being true is to say that in all situations in which all the premisses are true, so is the conclusion. (Priest, 2000, [12], p. 05)

Here, Priest mentions only "situations" and "truth". It is nevertheless possible to make a further generalization: we could replace the notions of "conclusion" and "premise" by the notion of "formula". This would result in a notion of consequence as a property of formulas, that is, as a relation between formulas and sets of formulas. Proceeding in this way, we would obtain the following definition:

Definition 1. *A formula is a* logical consequence *of a set of formulas if and only if there is no situation in which the formulas of this set are true and the formula in question is false.*

Even if we took the notion of formula as primitive (by not considering its internal structure), the notion of logical consequence so defined would, however, remain obscure, since this definition still depends on the definitions of situation and of formula true in a situation. It therefore leads us to the following questions: (1) what is situation? And (2) once a formula and a situation are given, when is the formula true in the situation? (By answering this second question, we will also answer the question of when a formula is false in a situation, since we can consider here truth and falsity as opposites.)

Still, let us bracket these questions for the moment and see whether any interesting properties follow from this informal (and typical) definition of logical consequence. The following three properties, called, respectively, "reflexivity", "monotonicity" and "transitivity", are a good example of such properties.

Property 1. *If a formula is an element of a set of formulas, then it is a consequence of the set.*

Property 2. *If a formula is a consequence of a set of formulas and this set is contained in another one, then the formula is also a consequence of this second set, i.e., if a formula is consequence of a set, it is also a consequence of any other set that contains the first set.*

Property 3. *If a formula is logical consequence of a set and the formulas of this set are consequence of another set, then the given formula is a consequence of this further set.*

To obtain further interesting results related to the notion of consequence, it will be necessary to introduce new concepts, which will be defined from previously mentioned concepts.

Definition 2. *A set of formulas is* **consistent** *if and only if there is a situation in which all the formulas of the set are true. Otherwise, the set is said to be* **inconsistent**.

As with the concept of consequence, it is possible to state some properties of this notion of consistency.

Property 4. *Every subset of a consistent set is consistent.*

Property 5. *Any formula is a consequence of an inconsistent set of formulas.*

Next, let us state the definition of validity:

Definition 3. *A formula is* **valid** *if and only if it is true in all situations, i.e., if there is no situation in which the given formula is false.*

Property 6. *If a formula is valid, then it is a consequence of every set of formulas.*

Even though there are other properties involving the notions of consequence, consistency and validity, we will limit our exposition to the ones stated above.

Now that we have seen the notions of consequence, consistency and validity, it is possible to establish a material adequacy criterion for each one of these.

In order to formalize the criterion of material adequacy mentioned above, let us fix a set FOR of formulas, denoting the elements of FOR by lower case Greek letters and the subsets of FOR by upper case Greeks letters. I will use the symbol $\models$ to indicate the relation of logical consequence, so that $\Gamma \models \alpha$ means that the formula α is a consequence of the set Γ of formulas; similarly, $\Gamma \models \Delta$ means that for each formula δ that belongs to the set Δ, $\Gamma \models \delta$.

Definition 4. *A relation $\models$ is **materially adequate** if and only if for every formula α and sets of formulas Γ, Δ, we have:*
  (i) if $\alpha \in \Gamma$, then $\Gamma \models \alpha$;
  (ii) if $\Gamma \models \alpha$ and $\Gamma \subseteq \Delta$, then $\Delta \models \alpha$;
  (iii) if $\Gamma \models \Delta$ and $\Delta \models \alpha$, then $\Gamma \models \alpha$.

The conditions in the above definition correspond to the previously mentioned properties 1-3, that is, respectively, to *reflexivity, monotonicity* and *(generalized) transitivity.*

The material adequacy criterion for definitions of consistency may be formalized this way:

Definition 5. *A definition of consistency is **materially adequate** if and only if:*
  (i) if $\Delta \subseteq \Gamma$ and Γ is consistent, then Δ is consistent, i.e., every subset of a consistent set is consistent;
  (ii) if α is a formula and Γ is an inconsistent set, then $\Gamma \models \alpha$, i.e., every formula is a consequence of an inconsistent set of formulas.

Finally, the material adequacy criterion for definitions of validity is:

Definition 6. *A definition of validity is **materially adequate** if and only if:*
  (i) if α is a valid formula and Γ is any set, then $\Gamma \models \alpha$, i.e., every valid formula is a consequence of every set of formulas.

Having formulated the material adequacy criteria for definitions of consequence, consistency and validity, I will now give formal definitions for these notions using the language of set theory.

3 Situation framework

However we understand the notion of situation, it is possible to formalize the definitions of consequence, consistency and validity by using the definition of a situational structure $\mathbb{S}$. Such a structure is composed of a non empty set FOR whose elements will be called formulas, an equally non-empty SIT whose elements will be called situations, a set VAL of truth-values ($VAL = \{\mathbf{V}, \mathbf{F}\}$) and a function f_S, which attributes to each formula in a given situation a truth-value. In the definition below, the symbol $\times$ denotes the usual Cartesian product of sets.

Definition 7. *A situational structure $\mathbb{S}$ is a 4-tuple*

$$\mathbb{S} = \langle FOR, SIT, VAL, f_S \rangle \text{ such that:}$$

$$f_S : FOR \times SIT \to VAL$$
$$(\alpha, s) \mapsto f_S(\alpha, s)$$

We thus say that a given formula α is true in a situation s if and only if $f_S(\alpha, s) = \mathbf{V}$. The situational structure $\mathbb{S}$ is, therefore, a possible interpretation of the language, and it answers the questions we had asked above: (1) a situation is an element of the set SIT and (2) a formula α is true in a situation s if and only if $f_S(\alpha, s) = \mathbf{V}$, that is, if the function f_S assigns to α the truth-value $\mathbf{V}$.

Now that we have defined situational structures, as well as specified when a formula is true in a situation, we can, finally, introduce the definitions of consequence, consistency and validity in a situational structure.

Definition 8. *A formula α is a* **logical consequence** *(in the structure $\mathbb{S}$) of a set Γ of formulas, in symbols, $\Gamma \models_{\mathbb{S}} \alpha$, if and only if $\forall s \in SIT$ such that $f_S(\gamma, s) = \mathbf{V}$ for all $\gamma \in \Gamma$, then $f_S(\alpha, s) = \mathbf{V}$.*

This definition states that a formula α is consequence of a set Γ (in the structure $\mathbb{S}$) if and only if for every situation in which all the formulas of Γ are true, α is also true. In other words, the formula α is logical consequence (in the structure $\mathbb{S}$) of a set Γ of formulas if and only if there is no situation in which the formulas of Γ are true and α is false.

The formal definition suggested above corresponds to the informal definition of logical consequence that we saw in the previous section. Still, before accepting this definition, it is necessary to show that it satisfies our material adequacy criterion.

Proposition 1 (Material adequacy for $\mathbb{S}$-consequence). $\models_{\mathbb{S}}$ *is materially adequate ($\models$ is materially adequate in $\mathbb{S}$), i.e., for all α and all Γ, Δ, we have:*
  (i) if $\alpha \in \Gamma$ then $\Gamma \models_{\mathbb{S}} \alpha$;
  (ii) if $\Gamma \models_{\mathbb{S}} \alpha$ and $\Gamma \subseteq \Delta$, then $\Delta \models_{\mathbb{S}} \alpha$;
  (iii) if $\Gamma \models_{\mathbb{S}} \Delta$ and $\Delta \models_{\mathbb{S}} \alpha$, then $\Gamma \models_{\mathbb{S}} \alpha$.

(I will not provide here a proof this and other results; the interested reader can find the proofs in my PhD thesis (Cf. Velasco, 2004).)

Next, let us examine the notions of consistency and validity in a situational structure.

Definition 9. *A set Γ of formulas is **consistent** (in the structure $\mathbb{S}$) if and only if there is at least some situation $s \in SIT$ such that $f_S(\gamma, s) = \mathbf{V}$ for all $\gamma \in \Gamma$. Otherwise, the set Γ is said to be **inconsistent**.*

It would be more accurate to call consistency in the structure $\mathbb{S}$, $\mathbb{S}$-consistency. However, for simplicity, I will employ *consistency* whenever there is no threat of ambiguity.

Proposition 2 (Material adequacy for $\mathbb{S}$-consistency). *This definition of consistency in $\mathbb{S}$ is materially adequate, i.e.:*
  (i) if $\Delta \subseteq \Gamma$ and Γ is consistent, then Δ is consistent;
  (ii) if α is any given formula and Γ is inconsistent, then $\Gamma \models_{\mathbb{S}} \alpha$.

Definition 10. *A formula α is **valid** (in the structure $\mathbb{S}$), or $\mathbb{S}$-**valid**, if and only if for every situation s, $f_S(\alpha, s) = \mathbf{V}$, i.e., if there is no situation that attributes the value $\mathbf{F}$ to α.*

Proposition 3 (Material adequacy for $\mathbb{S}$-validity). *This definition of validity in $\mathbb{S}$ is materially adequate, i.e.:*
  (i) if α is a valid formula and Γ a set, then $\Gamma \models_{\mathbb{S}} \alpha$.

Given the above definitions, it is natural to ask: is there a set of situations in which all the formulas of a particular set are true? This set can be defined in the following way. Start with $s \in SIT$ and $\Gamma \subseteq FOR$. We say that s is an $\mathbb{S}$-model of Γ if and only if $f_S(\gamma, s) = \mathbf{V}, \forall \gamma \in \Gamma$.

Using this, the set of the models of Γ in the structure $\mathbb{S}$, in symbols, $MOD_\mathbb{S}(\Gamma)$, can be defined in the following way: $MOD_\mathbb{S}(\Gamma) =_{def} \{s \in SIT : s \text{ is an } \mathbb{S}\text{-model of } \Gamma\}$.

This new notion allows another characterization for the above notions of consequence, consistency, and validity: a formula α is logic consequence of a set Γ in the structure $\mathbb{S}$ if and only if all models of Γ are models of $\{\alpha\}$, i.e., $\Gamma \models_\mathbb{S} \alpha$ if and only if $MOD_\mathbb{S}(\Gamma) \subseteq MOD_\mathbb{S}(\{\alpha\})$; a set Γ is $\mathbb{S}$-consistent if and only if there is at least one situation s such that $s \in MOD_\mathbb{S}(\Gamma)$, i.e., Γ is $\mathbb{S}$-consistent if and only if $MOD_\mathbb{S}(\Gamma) \neq \emptyset$; a formula α is $\mathbb{S}$-valid if and only if the set of models of $\{\alpha\}$ in the structure $\mathbb{S}$ coincides with the set of all models in the same structure $\mathbb{S}$, i.e., α is $\mathbb{S}$-valid if and only if $MOD_\mathbb{S}(\{\alpha\}) = MOD_\mathbb{S}$.

This new characterization does not mention the terms "truth" or "situation". It is therefore possible to define another logical framework by using concept of model as a primitive. Of course, if we opted for this new framework, we would need to show that the new definitions of consequence, consistency and validity were materially adequate. I will examine this possibility in the next section.

4 Model-based framework

I will denote the structures defined in this new framework—called *model-based*—by $\mathbb{M}$. Let FOR be a non-empty set of formulas from a previously fixed language and MOD a non-empty set whose elements are called models of the structure $\mathbb{M}$. I will construct a function f_M that attributes to each element of MOD a subset of FOR.

Definition 11. *A model-based structure $\mathbb{M}$ is a 3-tuple*

$$\mathbb{M} = \langle FOR, MOD, f_M \rangle \text{ such that:}$$

$$f_M : MOD \to \mathbb{P}(FOR)$$
$$m \mapsto f_M(m)$$

Now, we can define the set of the models of Γ in the structure $\mathbb{M}$:

Definition 12. *Let $\Gamma \subseteq FOR$. We define the set of the models of Γ in the structure $\mathbb{M}$, in symbols, $MOD_{\mathbb{M}}(\Gamma)$, in the following way:*

$$MOD_{\mathbb{M}}(\Gamma) =_{def} \{m \in MOD : \forall \gamma \in \Gamma, \gamma \in f_M(m)\}.$$

This allows us to define the notions of consequence, consistency and validity.

Definition 13. *A formula α is logical consequence (in a structure $\mathbb{M}$) of a set Γ of formulas, in symbols, $\Gamma \models_{\mathbb{M}} \alpha$, if and only if $MOD_{\mathbb{M}}(\Gamma) \subseteq MOD_{\mathbb{M}}(\{\alpha\})$, i.e., all models of Γ are also models of $\{\alpha\}$.*

Whenever there is no threat of ambiguity, I will simply write MOD (instead of $MOD_{\mathbb{M}}$).

The set of the models of Γ in a structure $\mathbb{M}$ is denoted by $MOD_{\mathbb{M}}(\Gamma)$ and is thus defined: $MOD_{\mathbb{M}}(\Gamma) =_{df} \{m \in MOD : \forall \gamma \in \Gamma, \gamma \in f_M(m)\}$. Therefore, a set Γ of formulas is said to be consistent in a model-based structure if and only if there is at least some model m such that $\gamma \in f_M(m)$, $\forall \gamma \in \Gamma$, i.e., $MOD_{\mathbb{M}}(\Gamma) \neq \emptyset$.

Definition 14. *A set Γ of formulas is consistent (is structure $\mathbb{M}$) if and only if $MOD_{\mathbb{M}}(\Gamma) \neq \emptyset$. Otherwise, the set Γ is said to be inconsistent.*

Definition 15. *A formula α is valid (in a structure $\mathbb{M}$), or $\mathbb{M}$-valid, if and only if $MOD_{\mathbb{M}}(\{\alpha\}) = MOD_{\mathbb{M}}$.*

The following results show the material adequacy of the above definitions of consequence, consistency and validity.

Proposition 4 (Material adequacy for $\mathbb{M}$-consequence). *$\models_{\mathbb{M}}$ is materially adequate, i.e.:*
(i) if $\alpha \in \Gamma$ then $\Gamma \models_{\mathbb{M}} \alpha$;
(ii) if $\Gamma \models_{\mathbb{M}} \alpha$ and $\Gamma \subseteq \Delta$, then $\Delta \models_{\mathbb{M}} \alpha$;
(iii) if $\Gamma \models_{\mathbb{M}} \Delta$ and $\Delta \models_{\mathbb{M}} \alpha$, then $\Gamma \models_{\mathbb{M}} \alpha$.

Proposition 5 (Material adequacy for $\mathbb{M}$-consistency). *Consistency in $\mathbb{M}$ is materially adequate, i.e.:*
(i) if $\Delta \subseteq \Gamma$ and Γ is consistent, then Δ is consistent;
(ii) if α is any formula and Γ is an inconsistent set, then $\Gamma \models_{\mathbb{M}} \alpha$.

Proposition 6 (Material adequacy for $\mathbb{M}$-validity). *Validity in $\mathbb{M}$ is materially adequate, i.e.:*
(i) if α is a valid formula and Γ is any set, then $\Gamma \models_{\mathbb{M}} \alpha$.

Now that I have demonstrated the material adequacy of the definitions of consequence, consistency and validity in a model-based framework, it is possible to give a new characterization for such notions using the definition of a *valuation*.

Let $\mathcal{K} = Im(f_M)$, *i.e.*, let $\mathcal{K}$ be the image of the function f_M. Elements of $\mathcal{K}$ are called **valuations**. Consequently, we can define the set of valuations for a given subset of formulas: let be $\Gamma \subseteq FOR$. We define **the set of valuations for** Γ, in symbols, $\mathcal{V}(\Gamma)$, in the following way: $\mathcal{V}(\Gamma) = V \in \{\mathcal{K} : \Gamma \subseteq V\}$.

Notice that each element of a valuation **is** a set of formulas. It is therefore possible to develop the theory without using a function that selects interesting formulas (i.e., the true ones), since the set of valuations is **defined**, directly, for a subset of formulas of the language. Thus, the family $\mathcal{K}$ of subset of formula of the language gives us the subset of the true interpretations, precisely because it is defined as the image of the function f_M.

Using this notion of valuation, it is possible to formulate—in the model framework—a new characterization of logical consequence, namely: a formula α is a consequence of a set Γ of formulas if and only if the set of the valuations for Γ is subset of the set of the valuations for $\{\alpha\}$, *i.e.*, $\Gamma \models_M \alpha$ is and only if $\mathcal{V}(\Gamma) \subseteq \mathcal{V}(\{\alpha\})$.

Similarly, a consistent set can be defined as one whose set of the valuations is different from the empty set, that is, Γ is M-consistent if and only if $\mathcal{V}(\Gamma) \neq \emptyset$. Therefore, we say that a formula is valid if and only if the set of valuations for it coincides with the set $\mathcal{K}$ of valuations, *i.e.*, α is M-valid if and only if $\mathcal{V}(\{\alpha\}) = \mathcal{K}$.

Such a characterization of the notion of logical consequence (and its correlates) suggests an even simpler framework: one that only involves the notion of formula.

5 Valuation framework

This new paradigm for the notion of logical consequence (and its correlates) was created from the theory of valuations, which originated in the work of professor Newton C. A. da Costa.[3]

I will denote by $\mathbb{V}$ a *valuation-based structure*, which consists only in a set FOR of formulas of a fixed language and a family $\mathcal{K}$ of subsets of

---

[3]Cf. Da Costa, N.C.A and Loparic, A., 1984, [4].

FOR, that is:

Definition 16. *A structure of valuations $\mathbb{V}$ is a pair $\mathbb{V} = \langle FOR, \mathcal{K} \rangle$.*

Thus, we can define the *set of valuations of a set of formulas.*

Definition 17. *Let $\mathcal{K} \subseteq \mathbb{P}(FOR)$. We define the set of valuations for a set Γ of formulas, in symbols, $\mathcal{V}_{\mathbb{V}}(\Gamma)$, in the following way:*

$$\mathcal{V}_{\mathbb{V}}(\Gamma) =_{def} \{V \in \mathcal{K} : \Gamma \subseteq V\}.$$

Let us now introduce the definitions of logical consequence, consistency and validity for the structure $\mathbb{V}$, as well as their corresponding properties.

Definition 18. *Let FOR be a non empty set, $\mathcal{K} \subseteq \mathbb{P}(FOR)$, $\alpha \in FOR$ and $\Gamma \subseteq FOR$. We define that α is logical consequence of Γ (in structure $\mathbb{V}$), in symbols, $\Gamma \models_{\mathbb{V}} \alpha$, if and only if $\mathcal{V}_{\mathbb{V}}(\Gamma) \subseteq \mathcal{V}_{\mathbb{V}}(\{\alpha\})$.*

Proposition 7 (Material adequacy for $\mathbb{V}$-consequence). *$\models_{\mathbb{V}}$ is materially adequate, i.e.:*
  (i) if $\alpha \in \Gamma$ then $\Gamma \models_{\mathbb{V}} \alpha$;
  (ii) if $\Gamma \models_{\mathbb{V}} \alpha$ and $\Gamma \subseteq \Delta$, then $\Delta \models_{\mathbb{V}} \alpha$;
  (iii) if $\Gamma \models_{\mathbb{V}} \Delta$ and $\Delta \models_{\mathbb{V}} \alpha$, then $\Gamma \models_{\mathbb{V}} \alpha$.

Definition 19. *Γ is consistent (in structure $\mathbb{V}$), or $\mathbb{V}$-consistent, if and only if $\mathcal{V}_{\mathbb{V}}(\Gamma) \neq \emptyset$. Otherwise, the set Γ is said inconsistent.*

Proposition 8 (Material adequacy for $\mathbb{V}$-consistency). *Consistency in $\mathbb{V}$ is materially adequate, i.e.:*
  (i) if $\Delta \subseteq \Gamma$ and Γ is consistent, then Δ is consistent;
  (ii) if α is any formula and Γ is an inconsistent set, then $\Gamma \models_{\mathbb{V}} \alpha$.

Definition 20. *Let $\Gamma \subseteq FOR$, $\alpha \in FOR$ and $\mathcal{K} \subseteq \mathbb{P}(FOR)$. We say that α is valid (in structure $\mathbb{V}$), or $\mathbb{V}$-valid, if and only if $\mathcal{V}_{\mathbb{V}}(\{\alpha\}) = \mathcal{K}$, that is, $\alpha \in \bigcap \mathcal{K}$.*

Proposition 9 (Material adequacy for $\mathbb{V}$-validity). *Validity in $\mathbb{V}$ is materially adequate, i.e.:*
  (i) if α is a valid formula and Γ a set, then $\Gamma \models_{\mathbb{V}} \alpha$.

6 Equivalence of the frameworks

In the preceding sections we built three logical frameworks that formalized the notions of consequence, consistency and validity in a materially adequate and also extremely general way, without specifying the internal structure of the concept of formula. Therefore, it is possible to ask: are these frameworks equivalent? The answer is affirmative, as the theorems below state.

Theorem 1 (1st Theorem of equivalence). *Let $\alpha \in FOR$ and $\Gamma \subseteq FOR$. Thus, we have:*

(a) let $\mathbb{S} = \langle FOR, SIT, VAL, f_S \rangle$ be a situational structure. There is then a model-based structure $\mathbb{M} = \langle FOR, MOD, f_M \rangle$ such that:
(i) $\Gamma \models_\mathbb{S} \alpha$ if and only if $\Gamma \models_\mathbb{M} \alpha$;
(ii) Γ is $\mathbb{S}$-consistent if and only if Γ is $\mathbb{M}$-consistent;
(iii) α is $\mathbb{S}$-valid if and only if α is $\mathbb{M}$-valid;
(b) let $\mathbb{M} = \langle FOR, MOD, f_M \rangle$ be a model-based structure. There is then a situational structure $\mathbb{S} = \langle FOR, SIT, VAL, f_S \rangle$ such that:
(i) $\Gamma \models_\mathbb{M} \alpha$ if and only if $\Gamma \models_\mathbb{S} \alpha$;
(ii) Γ is $\mathbb{M}$-consistent if and only if Γ is $\mathbb{S}$-consistent;
(iii) α is $\mathbb{M}$-valid if and only if α is $\mathbb{S}$-valid;

Theorem 2 (2nd Theorem of equivalence). *Let $\alpha \in FOR$ and $\Gamma \subseteq FOR$. Thus, we have:*

(a) let $\mathbb{M} = \langle FOR, MOD, f_M \rangle$ be a model-based structure. There is then is a valuation-based structure $\mathbb{V} = \langle FOR, \mathcal{K} \rangle$ such that:
(i) $\Gamma \models_\mathbb{M} \alpha$ if and only if $\Gamma \models_\mathbb{V} \alpha$;
(ii) Γ is $\mathbb{M}$-consistent if and only if Γ is $\mathbb{V}$-consistent;
(iii) α is $\mathbb{M}$-valid if and only if α is $\mathbb{V}$-valid;
(b) let $\mathbb{V} = \langle FOR, \mathcal{K} \rangle$ be a valuation-based structure. There is then a model-based structure $\mathbb{M} = \langle FOR, MOD, f_M \rangle$ such that:
(i) $\Gamma \models_\mathbb{V} \alpha$ if and only if $\Gamma \models_\mathbb{M} \alpha$;
(ii) Γ is $\mathbb{V}$-consistent if and only if Γ is $\mathbb{M}$-consistent;
(iii) α is $\mathbb{V}$-valid if and only if α is $\mathbb{M}$-valid;

7 Conclusion

The two theorems from the previous section show us the equivalence of the three studied frameworks: the situational, the model-based, and the valuation-based. Moreover, I have also shown that, given the two theorems, there is no epistemological loss in adopting the simplest framework, namely the valuation one. Therefore, even if it is less intuitive than the others, the valuation framework is shown to better satisfy the methodological precept known as *Ockham's Razor*—being thus the natural endpoint for our analysis of the notions of logical consequence, validity, and consistency. Indeed, the very organization of this work is an argument for this thesis: one can see the paring down of the concepts involved in the definitions above as successive applications of Ockham's Razor, applications that culminate in the construction of a valuation-based structure! This is surprising, given that the original motivation for the introduction of such structures by Da Costa was not to find a common core for these notions , but rather the construction of a semantics for the paraconsistent calculi.

References

[1] Arruda, A.I. e N.C.A. da Costa. "Une sémantique pour le calcul $\mathcal{C}_1^=$". *C. R. Acad. Sc. Paris*, 284, 1977, pp. 279-282.

[2] Da Costa, N.C.A. *Sistemas Formais Inconsistentes*. Tese de Cátedra. Universidade Federal do Paraná. Curitiba, 1963.

[3] Da Costa, N.C.A. e E.H. Alves. "Une semantique pour le calcul $\mathcal{C}_1$". *C. R. Acad. Sc. Paris*, 283, 1976, pp. 729-731.

[4] Da Costa, N.C.A. e A. Loparic. "Paraconsistency, paracompleteness and valuations". *Logique et Analyse*, 27 (106), 1984, pp. 119-31.

[5] Da Silva, J.A. *Sistemas formais e valorações: sobre um teorema geral de completude*. Dissertação de Mestrado. Programa de Estudos Pós-Graduados em Filosofia. Pontifícia Universidade Católica de São Paulo. São Paulo, 2000.

[6] De Souza, E.G. "Lindenbaumologia I: a teoria geral". *Cognitio: Revista de Filosofia*, 2. Departamento de Filosofia da PUC-SP. São Paulo: EDUC/Editora Angra, 2001, pp. 213-219.

[7] De Souza, E.G. e P.D.N. Velasco. "Lindenbaumologia II: Cálculos Lógicos Abstratos". *Cognitio: Revista de Filosofia*, 3. Departamento de Filosofia da PUC-SP. São Paulo: EDUC/Editora Angra, 2002, pp. 115-121.

[8] De Souza, E.G., Da Costa, N.C.A. e J.S.A. Maranhão. *Introdução à lógica paraconsistente: a hierarquia C_n*. Pré-publicação. São Paulo: Instituto de Estudos Avançados da Universidade de São Paulo, 2001.

[9] Kotas, J. e N.C.A. da Costa. "Some problems on logical matrices and valorizations". *Proceedings of the Third Brazilian Conference on Mathematical Logic.* Editado por A.I. Arruda, N.C.A. da Costa e A.M. Sette. Sociedade Brasileira de Lógica, 1980, pp. 131-146.

[10] Loparic, A. "Une étude sémantique de quelques calculs propositionnels". *Comptes Rendus de l'Académie des Sciences de Paris, tom. 284A*, 1977, pp. 835-838.

[11] Loparic, A. "The method of valuations in modal logic". *Mathematical Logic: Proceedings of the First Brazilian Conference.* Editado por A.I. Arruda, N.C.A. da Costa e R. Chuaqui. Marcel Dekker, 1978, pp. 141-157.

[12] Priest, G. *Logic: A Very Short Introduction.* New York: Oxford University Press, 2000.

[13] Tarski, A. *Logic, semantics, metamathematics.* Hackett Publishing Company, 1983.

[14] Tarski, A. "On extensions of incomplete systems of the sentential calculus". *Logic, semantics, metamathematics.* Hackett Publishing Company, 1983, pp. 393-400.

[15] Tarski, A. e J. Lukasiewicz. "Investigations into the sentential calculus". *Logic, semantics, metamathematics.* Hackett Publishing Company, 1983, pp. 38-59.

[16] Velasco, P. D. N. *Sobre uma reconstrução do conceito de valoração.* 2004. 187f. Tese (Doutorado em filosofia) - Pontifícia Universidade Católica de São Paulo, São Paulo, 2004.

Part 2

Categories, logics and arithmetic

Internal logic of the $H-B$ topos[1]

Vladimir L. Vasyukov

Institute of Philosophy, Russian Academy of Sciences, Russia

Abstract

Chris Mortensen in his book "Inconsistent Mathematics" introduced the notion of complement topos which internal logic is dual to the usual logic of standard topos. Since complement-classifier is indistinguishable (via categorial methods) from a standard subobject classifier then topos, in fact, always can be considered as $H-B$ topos in which we have both Heyting algebra of subobjects of any object and co-Heyting (Brouwerian) algebra too. A formulation of internal logic of such $H-B$ topos is proposed which is based on C.Rauszer's $H-B$ logic.

1 Introduction

For each topos one can define a language which would be employed as a convenient mean for yielding statements on objects and arrows of the topos in question or even for proving theorems about them (cf. [3, p.172]). Brief description of the language and zero-order topos logic, formulated in this language, according C.MacLarty's version [4, p.126] is as follows.

For any topos $\mathcal{C}$ the internal language is typed, with each $\mathcal{C}$-object as one type[2]. Terms and their types are defined inductively.

(LT1) Each $\mathcal{C}$-object A has a list of variables over A, $x_1, x_2, x_3, ...$ Every variable over A is a term of type A.

(LT2) For any arrow $f : A \to B$ and term s type A, fs is a term of type B. Every arrow $c : 1 \to A$ with domain type 1 is itself a term of type A (we will call it a constant of type A). Let ! means a constant for identity arrow on 1.

---
[1] This resrach is supported by RFBR grant 19-011-00799.
[2] MacLarty considers not types but sorts while we will speak of types for convenience, following the commonly accepted usual practice.

(LT3) For every term s_1 of type A and s_2 of type B there is a term $\langle s_1, s_2 \rangle$ of type $A \times B$.

(LT4) For every term s of type B and variable y of type A there is a term $(\lambda y)s$ of type B^A.

A variable x is *free* unless it is *bound* by a lambda operator (λx). We will write $(\lambda x.A)$ to indicate the type of the variable. A term with no free variables is *closed*.

Now let us a term s has type B and all its free variables are in the list $y_1, ..., y_k$, where the y's are variables over $A_1, ..., A_k$ respectively. Then s refers to a morphism $|s| : A_1 \times ... \times A_k \to B$, which we call the interpretation of s. Informally, any assignment of a value to each variable, giving y_i a value in A_i, determines a value for s. The morphism $|s|$ actually depends on the list of variables involved, therefore we should show the list in the notation.

Let us use $\bar{x}$ to abbreviate a list $x_1, ..., x_n$. Then $A_1, ..., A_n$ is the list of types of the variables in the same order. A variable can only appear in a list, but an object A will appear as many times as there are variables over A in the list. For a term s of type B and a list $\bar{x}$ including all the free variables of s we write $|s|_{\bar{x}} : A_1 \times ... \times A_k \to B$ for the interpretation of s relative to the list $\bar{x}$. We always assume that lists of variables include all those that are free in the terms we apply them to. If s has no free variables then $\bar{x}$ can be an empty list of variables, and of course the product of an empty list of types is $\top$.

Now we define inductively the interpretation relative to the lists:

(I1) For any list $\bar{x}$ and variables x_i in the list, $|x_i|_{\bar{x}}$ is the ith projection
$$A_1 \times ... \times A_n \to A_i$$

(I2) For any arrow $f : A \to B$, if s is a term of type A then $|fs|_{\bar{x}}$ is
$$A_1 \times ... \times A_n \xrightarrow{|s|_{\bar{x}}} A \xrightarrow{f} B$$

For any constant c, $|c|_{\bar{x}}$ is
$$A_1 \times ... \times A_n \to 1 \xrightarrow{c} B$$

(I3) For any terms s_1 of type A and s_2 of type B, $|\langle s_1, s_2 \rangle|_{\bar{x}}$ is the pair morphism to $A \times B$ induced by $|s_1|_{\bar{x}}$ and $|s_2|_{\bar{x}}$.

(I4) For any term s of type B, if the variable y over A is not in the list $\bar{x}$ then $|(\lambda y)s|_{\bar{x}}$ is the *transpose* of $|s|_{\bar{x},y} : A_1 \times ... \times A_n \times A \to B$, i.e. an arrow $\overline{|s|_{\bar{x},y}} : A_1 \times ... \times A_n \to B^A$. If the bound variable y is in the list $\bar{x}$ that is an irrelevant coincidence. Then we replace y in $(\lambda y)s$ by some variable over A neither in s nor in the list $\bar{x}$.

According to (LT2) for any term g of type B^A and s of type A there is a term $ev(\langle g, s \rangle)$ which we will abbreviate to $g(s)$. Also, we will use set builder notation for lambda abstraction over Ω writing $\{x.A : s\}$ instead of $(\lambda x.A)s$ when s has type Ω.

Formulas are the terms of type Ω. By (LT2) for any formulas φ and ψ there are formulas $\wedge(\varphi, \psi)$ and $\to (\varphi, \psi)$ which we will write as $\varphi \wedge \psi$ and $\varphi \to \psi$. There are also formulas $\top, \bot, \neg \varphi, \varphi \vee \psi$.

For any formula φ and variable y over any object A, we define the formula $(\forall y.A)\varphi$ as an abbreviation for $\forall_A \{y.A : \varphi\}$. Thus $(\forall y.A)\varphi$ says that $\{y.A : \varphi\}$ is all of A. Its interpretation over a list of variables $\bar{x}$ follows the definition. It is

$$A_1 \times ... \times A_n \xrightarrow{\overline{|\varphi|_{\bar{x},y}}} \Omega^A \xrightarrow{\forall_A} \Omega$$

if the variable y does not occur in the list $\bar{x}$. If y is in the list it is first replaced by some new variable. Note that y is bound in $(\forall y.A)\varphi$.

The existential quantification of formula φ at a variable x can be defined to be $(\forall w)((\forall x)(\varphi \to w) \to w)$, where w is a variable over Ω not free in φ.

As in [4, p.120], for any sub-objects $q : Q \rightarrowtail A$ and $r : R \rightarrowtail A$, we define $Q \implies R$ to be the sub-object classified by $\chi_q \to \chi_r$ and call $Q \implies R$ the (material) inplicate of R by Q (here $\chi_q \to \chi_r$ abbreviates $\to \circ \langle \chi_q, \chi_r \rangle$ and $\to : \Omega \times \Omega \to \Omega$ is the *material conditional* arrow).

The *extension* of φ over a list of variables $\bar{x}$ is the subobject $A_1 \times ... \times A_n$ classified by $|\varphi|_{\bar{x}}$. We will write $[\bar{x} : \varphi]$ for this extension intending 'all $\bar{x}$ such that φ'. For example, we have:

$$[\bar{x} : \top] = A_1 \times ... \times A_n$$
$$[\bar{x} : \varphi \wedge \psi] = [\bar{x} : \varphi] \cap [\bar{x} : \psi]$$
$$[\bar{x} : \varphi \vee \psi] = [\bar{x} : \varphi] \cup [\bar{x} : \psi]$$
$$[\bar{x} : \varphi \to \psi] = [\bar{x} : \varphi] \implies [\bar{x} : \psi]$$

and $[\bar{x} : (\forall x)\varphi]$ is the universal quantification of $[\bar{x}, y : \varphi]$ over the projection corresponding to y.

A formula φ is called *true* if its extension $[\bar{x} : \varphi]$ is all of $A_1 \times ... \times A_n$ when $\bar{x}$ lists exactly the variables free in φ. We say that a formula φ implies ψ if the extension of φ is contained in that of ψ. More generally, for any finite set of formulas Γ, we will write $[\bar{x} : \Gamma]$ for the intersection of the extensions over $\bar{x}$ of all formulas in Γ. In particular, $[\bar{x} : \ ] = A_1 \times ... \times A_n$ for the empty set of formulas. Then Γ *implies* φ iff $[\bar{x} : \Gamma] \subseteq [\bar{x} : \varphi]$ when $\bar{x}$ lists exactly the free variables in Γ and φ. A *sequent* is an expression $\Gamma : \varphi$, where Γ is a finite (possibly empty) set of formulas and φ is a formula. Think of $\Gamma : \varphi$ as a claim that the formula in Γ imply φ. The sequent is *true* iff Γ does imply φ. In particular, a sequent $: \varphi$ with empty left side is true iff $[\bar{x} : \varphi]$ is all of $A_1 \times ... \times A_n$ and thus iff φ is true. When we know a sequent $\Gamma : \varphi$ is true we write $\Gamma \vdash \varphi$. Topos logic, as it is known [4, p.129], can be formulated by way of the list of rules of inference for these sequents; that is, rules such that applying them to true sequents always yields true sequents. We would describe topos logic by means of the following rules:

$$\frac{*}{\varphi : \varphi} \qquad \frac{*}{: \top} \qquad \frac{*}{\bot : \varphi}$$

$$\frac{\Gamma : \varphi}{\Gamma, \psi : \varphi}(\text{Thinning} :)$$

$$\frac{\Gamma : \varphi}{\Gamma(x/s) : \varphi(x/s)}(\text{Substitution} :) \quad \begin{array}{l}\text{(for any term } s \text{ free for } x \text{ in} \\ \text{all the formulas)}\end{array}$$

$$\frac{\Gamma, \psi : \varphi \quad \Gamma : \psi}{\Gamma : \varphi}(\text{Cut} :) \quad \begin{array}{l}\text{(if every variable free in } \psi \text{ is} \\ \text{free in } \Gamma \text{ or in } \varphi)\end{array}$$

$$\frac{\Gamma, \varphi : \theta \quad \Gamma, \psi : \theta}{\Gamma, \varphi \vee \psi : \theta}(:\vee)$$

$$\frac{\Gamma : \varphi \quad \Gamma : \psi}{\Gamma : \varphi \wedge \psi}(:\wedge)$$

$$\frac{\Gamma, \varphi : \psi}{\Gamma : \varphi \to \psi}(:\to)$$

$$\frac{\Gamma, \varphi : \bot}{\Gamma : \neg \varphi}(:\neg)$$

$$\frac{\Gamma : \varphi}{\Gamma : (\forall x)\varphi}(:\forall) \quad \frac{\Gamma, \varphi : \psi}{\Gamma, (\exists x)\varphi : \psi}(\exists :) \quad \begin{array}{l}\text{(if the variable } x \text{ is} \\ \text{not free in } \Gamma \text{ or } \psi)\end{array}$$

An emergence of the complement topos introduced by Chris Mortensen (cf.[5]) raise the issue of the internal logic of it which version was proposed by L. Estrada-González in [1]. But if we take into account that complement-classifier is indistinguishable (via categorial methods) from a standard subobject classifier then topos, in fact, always can be considered as dual topos for which we have the Heyting algebra of subobjects of any object as well as co-Heyting (Brouwerian) algebra. Such "two-faced Janus" hypothetically will also have its own internal logic which reconstruction is the task of present paper.

In the second section as the first step on this way the typed internal language and logic of a complement topos is described in MacLarty's style. To that end some dual topos concepts are introduced and some troubles of such approach are analyzed.

Since Heyting logic and Brouwer logic always appear as Siamese twins and so do both the standard topos and the complement topos then in third section we discuss not the standard topos alone and not the complement topos alone but another type of category which, in a sense, contains them both. Such category - $H - B$ topos - is developed and for that topos the typed internal language and logic are described.

Finally, in the fourth section the proof of the correctness of $H - B$ topos logic is given. The question of completeness of this logic is disregarded in view of the fact that $H - B$ toposes are practically unexplored. So we have to postpone the analysis of this issue especially because even the standard topos logic itself is not complete for each individual topos,

2 Complement topos logic

Chris Mortensen in his book "Inconsistent Mathematics" [5] introduced the notion of complement topos which internal logic is dual to the usual logic of standard topos. A principal peculiarity of complemented topos lies in a presence of complement classifier in the latter. Its definiion is as follows.

Definition 1 *For a category $\mathcal{C}$ a complement classifier is a $\mathcal{C}$-arrow $false : 1 \to \Omega$ where for any monic $f : A \rightarrowtail B$ there is one and only one $\mathcal{C}$-arrow $B \to \Omega$, denoted $\bar{\chi}_f$, making the following a pullback in $\mathcal{C}$,*

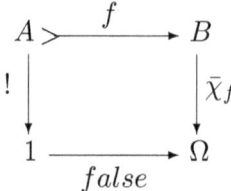

At the same time Mortensen shown that a complement classifier in a topos **Set** is indistinguishable (via categorical methods) from a standard subobject classifier, that they are isomorphic. Thus, in **Set** we always have paraconsistency because of the presence of both types of subobject classifiers. Moreover, the following proposition is obviously true:

Proposition 2 *Complement toposes support paraconsistency logic via Brouwerian algebra in a way exactly parallel to the way toposes support intuitionistic logic via Heyting algebras.*

Since toposes support intuitionistic logics due to reflecting the Heyting algebra structure by subobject classifier then in a complement topos complement classifier reflects the Brouwerian algebra structure respectively. Hence, to describe an internal logic of complement topos we have to proceed in a dual way.

Unfortunately, here immediately arises the problem of the "dual exponentiation". In standard topos the idea is that B^A represents arrows from A to B. But what is the "dual idea" in complement topos?

Let us turn usual definition inside out. Given objects A and B, a *co-exponential of B by A* will consist of an object I and an arrow $ev^o : B \to I + A$ with the following property. For any object C and arrow $g^o : B \to C + A$ there is a unique arrow $\overline{g^o} : I \to C$ that makes this triangle commute:

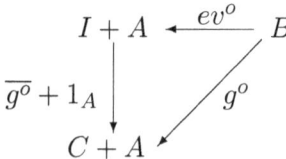

Co-exponentials are unique up to isomorphism. We say that the category is *co-Cartesian closed* if it has all finite co-products, and each two objects in it have a co-exponential. For any objects A and B we write $^A B$, and $ev^o : B \to {}^A B + A$ to indicate a co-exponential of B by A.

An arrow $B \to A$ is (up to the natural isomorphism $A \cong 0 + A$) an arrow $B \to 0 + A$, and that gives a unique arrow $^A B \to 0$. But we have to make this correspondence between arrows $B \to A$ and global elements of $^A B$ precise, and extend it to generalized elements of $^A B$.

For each $g^o : B \to C + A$, call the corresponding $\overline{g^o} : {}^A B \to C$ the *co-transpose* of g^o. Each $g^o : B \to C + A$ uniquely determines its co-transpose, by definition, but is also determined by its co-transpose since the definition says that $g^o = (\overline{g^o} + 1_A) \circ ev^o$. Furthermore, every $f : {}^A B \to C$ is the co-transpose of an arrow (and thus of a uniqe arrow) from B to $C + A$, namely $(f + 1_A) \circ ev^o$. We will also call $(f + 1_A) \circ ev^o$ the *co-transpose* of f, and write it as $f : B \to C + A$. So arrows from B to $C + A$ corresponds exactly to arrows from $^A B$ to C. We pass in either direction by forming the "co-transpose", and the co-transpose of co-transpose is the original arrow.

For any arrow $f : B \to A$ the *co-name* of f, $\llcorner f \lrcorner : {}^A B \to 0$, is defined to be the transpose of $B \xrightarrow{f} A \xrightarrow{\sim} 0 + A$ and it follows that every global co-element $x : {}^A B \to 0$ is the co-name of an arrow; specifically, $B \xrightarrow{\overline{x}} 0 + A \xrightarrow{\sim} A$.

Again, for any complement topos $\mathcal{C}$ the internal language is typed, with each $\mathcal{C}$-object as one type. Co-terms and their types are defined inductively:

(LT5) Each $\mathcal{C}$-object A has a list of variables over A, $x_1, x_2, x_3, ...$ Every variable over A is a term of type A.

LT6 For any arrow $f : A \to B$ and a term s of type B an expression $(fs)^o = f^o s$ is a term of type A. Each arrow $c : A \to 0$ with 0 as its domain is itself a term of type A (we call it a co-constant of type A). Let ? means co-constant of identity arrow for 0.

(LT7) For every terms s_1 of type A and s_2 of type B there is a term $[s_1, s_2]$ of type $A + B$.

(LT8) For every term s of type B and the variable y of type A there is a term $s(\lambda y)$ of type $^A B$.

A variable x is *free* unless it is *bound* by a "co-lambda operator" (λx). We will write $(\lambda x.A)$ to indicate the type of the variable. A co-term with no free variables is *closed*.

For a term s of type A and a list $\bar{x}$ including all the free variables of s we write $||s||_{\bar{x}} : A \to A_1 + ... + A_k$ for the interpretation of s relative to the list $\bar{x}$. The definition of intepretation relative the lists contains the following points of "co-interpretation":

(I5) For any list $\bar{x}$ and variables x_i in the list, $||x_i||_{\bar{x}}$ is the ith injection
$$A_i \to A_1 + ... + A_n$$

(I6) For any arrow $f : A_1 \to A_2$, if s is a co-term of type A_2 then $||f^o s||_{\bar{x}}$ is
$$A_2 \xrightarrow{f^{-1}} A_1 \xrightarrow{||s||_{\bar{x}}} A_1 + ... + A_n$$
For any co-constant c, $||c||_{\bar{x}}$ is
$$A_2 \xrightarrow{c^O} \bot \to A_1 + ... + A_n$$

(I7) For any co-terms s_1 of type A_1 and s_2 of type A_2, $||[s_1, s_2]||_{\bar{x}}$ is the pair arrow from $A_1 + A_2$ induced by $||s_1||_{\bar{x}}$ and $||s_2||_{\bar{x}}$.

(I8) For any co-term s of type A_2, if the variable y over A_1 is not in the list $\bar{x}$ then $||s(\lambda y)||_{\bar{x}}$ is the *co-transpose* of $||s||_{\bar{x},y} : A_2 \to A_1 + ... + A_n + A_1$, i.e. an arrow $\overline{||s||_{\bar{x},y}} : {}^{A_1}A_2 \to A_1 + ... + A_n$.

If the bound variable y is in the list $\bar{x}$ that is an irrelevant co-incidence. Then we replace y in $s(\lambda y)$ by some variable over A' neither in s nor in the list $\bar{x}$.

According to (LT6) for any co-term g of type ${}^{A_1}A_2$ and s of type A_1 there is a co-term $ev^o([g, s])$ which we will abbreviate to $g^o s$. Also, we will use coset builder notation for lambda abstraction over Ω writing $\{x.A_1 : s\}^o$ instead of $s(\lambda x.A_1)$ when s has type Ω.

Formulas are the co-terms of type Ω. By (LT7)-(LT8) for any formulas φ and ψ there are formulas $\vee(\varphi, \psi)$ and $\leftarrow(\varphi, \psi)$ which we will write as $\varphi \vee \psi$ and $\varphi \leftarrow \psi$. There are also formulas $\top, \ulcorner \varphi, \varphi \wedge \psi$.

For any formula φ and variable y over any object A, we define the formula $(\forall^o y.A)\varphi$ as an abbreviation for $\forall_A \{y.A : \varphi\}^o$. Thus $(\forall^o y.A)\varphi$ says that $\{y.A : \varphi\}^o$ is all of A. Its interpretation over a list of variables $\bar{x}$ follows the definition. It is

$$\Omega \xrightarrow{\forall_A^o} {}^A\Omega \xrightarrow{\overline{||\varphi||_{\bar{x},y}}} A_1 + ... + A_n$$

if the variable y does not occur in the list $\bar{x}$. If y is in the list it is first replaced by some new variable. Note that y is bound in $(\forall^o y.A)\varphi$.

The co-existential quantification of formula φ at a variable x can be defined to be $(\forall^o w)((\forall^o x)(\varphi \leftarrow w) \leftarrow w)$, where w is a variable over Ω not free in φ.

For any sub-objects $q : Q \rightarrowtail A$ and $r : R \rightarrowtail A$, we define $Q \Leftarrow R$ to be the sub-object classified by $\bar{\chi}_q \leftarrow \bar{\chi}_r$ and call $Q \Leftarrow R$ the (material) *co-implicate* of R by Q (here $\bar{\chi}_q \leftarrow \bar{\chi}_r$ abbreviates $\leftarrow \circ \langle \bar{\chi}_q, \bar{\chi}_r \rangle$ and $\leftarrow: \Omega \times \Omega \to \Omega$ is the *material co-conditional* arrow).

The *co-extension* of φ over a list of variables $\bar{x}$ is the subobject $A_1 + ... + A_n$ classified by $|\varphi|_{\bar{x}}$. We will write $[\bar{x} : \varphi]^o$ for this co-extension intending 'all $\bar{x}$ dually such that φ'. For example, we have:

$$[\bar{x} : \top]^o = A_1 + ... + A_n$$
$$[\bar{x} : \varphi \wedge \psi]^o = [\bar{x} : \varphi]^o \cup [\bar{x} : \psi]^o$$
$$[\bar{x} : \varphi \vee \psi]^o = [\bar{x} : \varphi]^o \cap [\bar{x} : \psi]^o$$
$$[\bar{x} : \varphi \leftarrow \psi]^o = [\bar{x} : \varphi]^o \Leftarrow [\bar{x} : \psi]^o$$

and $[\bar{x} : (\forall x)\varphi]^o$ is the co-universal quantification of $[\bar{x}, y : \varphi]^o$ over the injection corresponding to y.

A formula φ is called *false* if its co-extension $[\bar{x} : \varphi]^o$ is all of $A_1 + ... + A_n$ when $\bar{x}$ lists exactly the variables free in φ.

We say that a formula φ *co-implies* ψ if the co-extension of ψ is contained in that of φ. More generally, for any finite set of formulas Γ, we will write $[\bar{x} : \Gamma]^o$ for the join of the co-extensions over $\bar{x}$ of all formulas in Γ. In particular, $[\bar{x} : \ ]^o = A_1 + ... + A_n$ for the empty set of formulas. Then φ *co-implies* Γ iff $[\bar{x} : \varphi]^o \subseteq [\bar{x} : \Gamma]^o$ when $\bar{x}$ lists exactly the free variables in Γ and φ.

A *sequent* is an expression $\varphi : \Gamma$, where Γ is a finite (possibly empty) set of formulas and φ is a formula. Think of $\varphi : \Gamma$ as a claim that the formula in φ co-imply Γ. The sequent is *false* iff φ does co-imply Γ. In particular, a sequent $\varphi :$ with empty right side is false iff $[\bar{x} : \varphi]^o$ is all of $A_1 + ... + A_n$ and thus iff φ is false. When we know a sequent $\varphi : \Gamma$ is false we write $\varphi \vdash \Gamma$. Following C. Rauszer [7, p.64], we say that φ is *formally rejected* from a set Γ of formulas. In Shütte's style $\varphi \vdash \Gamma$ means the proof of the alternatives Γ from hypothesis φ.

L. Estrada-González in [1] present a sequent calculus for the zero-order complement topos logic. A sequent calculus for the first-order complement topos logic is obtained by adding rules for substitution and quantification:

$$\frac{*}{\varphi:\varphi} \qquad \frac{*}{\varphi:\top} \qquad \frac{*}{\bot:}$$

$$\frac{\varphi:\Gamma}{\varphi:\Gamma,\psi}(:\text{Thinning})$$

$$\frac{\varphi:\Gamma}{\varphi(x/s):\Gamma(x/s)}(:\text{Substitution}) \quad \text{(for any term } s \text{ free for } x \text{ in all the formulas)}$$

$$\frac{\varphi:\Gamma,\psi \quad \psi:\Gamma}{\varphi:\Gamma}(\text{Cut})$$

$$\frac{\varphi:\Gamma \quad \psi:\Gamma}{\varphi\vee\psi:\Gamma}(\vee:)$$

$$\frac{\theta:\Gamma,\varphi \quad \theta:\Gamma,\psi}{\theta:\Gamma,\varphi\wedge\psi}(:\wedge)$$

$$\frac{\psi:\varphi,\Gamma}{\psi\leftarrow\varphi:\Gamma}(\leftarrow:)$$

$$\frac{\top:\Gamma,\varphi}{\ulcorner\varphi:\Gamma}(\ulcorner:)$$

$$\frac{\varphi:\Gamma}{(\forall^\circ x)\varphi:\Gamma}(\forall^\circ :) \quad \frac{\varphi:\Gamma,\psi}{\varphi:\Gamma,(\exists^\circ x)\psi}(:\exists^\circ) \quad \text{(if the variable } x \text{ is not free in } \Gamma \text{ or } \psi)$$

Ttroubles with the complement topos (and thus with its internal logic) are caused by the lack of the "pure" example being independent of the standard topos structure. According to the legend, Saunders Mac Lane, in response to Mortensen and Laver's paper on complement topos which they sent him, said that complement toposes are just standard toposes, that they are indistinguishable because they have the same categorical structure. But it is not generally accepted point of view. L.Estrada-Gonzáles writes thereupon: "There is a 'bare' or 'abstract' categorial structure of toposes that can filled in at least two ways (the standard way and the way suggested by Mortensen and Lavers). Said otherwise, there are underlying universal properties in topos logic dissembled by certain intuitive conceptualizations of the categorial structure of toposes, yet not necessitated by this" [2].

3 $H - B$ topos logic

But it seems that this 'abstract' categorical structure of toposes is principally twofold by its nature. Actually, Heyting logic and Brouwer logic always appear as Siamese twins - if one is given then the second might be reconstructed. So do the standard topos and the complement topos too. They are not dual in the traditional categorial sense. So, maybe in this case we, in fact, discuss not the standard topos alone and not the complement topos alone but another type of category which, in a sense, contains them both?

One of such would-be categories is known as *bi-Heyting topos*. Its definition is the next [9, p.36]:

Definition 3 *A bi-Heyting topos is a topos for which the Heyting algebra of subobjects of any object is a co-Heyting algebra (and hence a bi-Heyting algebra).*

Unfortunately, the internal logic of a bi-Heyting topos is never dual to an intuitionistic logic. It is just a co-Heyting algebra wthin a Boolean algebra. It is known that Boolean algebra is a self-dual algebra and hence we arrive at a Boolean topos and not at the topos we search for.

Probably, required category should be a topos which algebra of subobjects of any object is an algebra somehow combining both Heyting and Brouwerian algebras. One of the pretendnt is a *semi-Boolean algebra* whose definition is the following: an abstract algebra $\langle A, \vee, \wedge, \Longrightarrow, \Longleftarrow, \neg, _\ulcorner \rangle$ will be called a semi-Boolean algebra provided that $\langle A, \vee, \wedge, \Longrightarrow, \neg \rangle$ is a Heyting algebra and $\langle A, \vee, \wedge, \Longleftarrow, _\ulcorner \rangle$ is a Brouwerian algebra [8, p.8]. It seems to be that semi-Boolean algebra might be regarded as the join of Heyting and Brouwerian algebras. Anyway, respective an $B - H$ topos is defined as follows:

Definition 4 *An $H - B$ topos is a topos for which an algebra of subobjects of any object is a semi-Boolean algebra.*

Since a semi-Boolean algebra is regarded as an extension of Heyting algebra then an $H - B$ topos also to be, in a sense, an extension of standard topos. Such $H - B$ topos will be cartesian and co-cartesian closed, to wit, having both exponentals and co-exponentials.

For an $H - B$ topos one can also define a language which would be employed as a convenient mean for yielding statements on objects and arrows of such topos in question. And obviously an internal language of $H - B$ topos will contains elements both of kinds of toposes above.

So, for any $H - B$ topos $\mathcal{C}$ the internal language is typed, with each $\mathcal{C}$-object as one type. Terms and their types are defined inductively:

(**LT9**) Each $\mathcal{C}$-object A has a list of variables over A, $x_1, x_2, x_3, ...$ Every variable over A is a term of type A.

(**LT10**) For any arrow $f : A \to B$ and term s type A, fs is a term of type B, and term r type B, $(fr)^o (= f^o r)$ is a term of type A. Every arrow $c : 1 \to A$ with domain type 1 is itself a term of type A (we will call it a constant of type A). ! means a constant for identity arrow on 1.

(**LT11**) For every term s_1 of type A and s_2 of type B there is a term $\langle s_1, s_2 \rangle$ of type $A \times B$ and a term $[s_1, s_2]$ of type $A + B$.

(**LT12**) For every term s of type B and variable y of type A there is a term $(\lambda y)s$ of type B^A and a term $s(\lambda y)$ of type $^A B$.

Now let us a term s has type B and all its free variables are in the list $y_1, ..., y_k$, where the y's are variables over $A_1, ..., A_k$ respectively. Then s refers to an arrow $|s| : A_1 \times ... \times A_k \to B$, which we call the interpretation of s. The arrow $|s|$ actually depends on the list of variables involved, therefore we should show the list in the notation.

Let us use $\bar{x}$ to abbreviate a list $x_1, ..., x_n$. Then $A_1, ..., A_n$ is the list of types of the variables in the same order. A variable can only appear in a list, but an object A will appear as many times as there are variables over A in the list. For a term s of type B and a list $\bar{x}$ including all the free variables of s we write $|s|_{\bar{x}} : A_1 \times ... \times A_k \to B$ for the interpretation of s relative to the list $\bar{x}$ and $|s|_{\bar{x}}^o : B \xrightarrow{|s|_{\bar{x}}^o} A_1 + ... + A_n$ for co-interpretation of s relative to the list $\bar{x}$. We always assume that lists of variables include all those that are free in the terms we apply them to. If s has no free variables then $\bar{x}$ can be an empty list of variables, and of course the product of an empty list of types is $\top$.

Now we define inductively the interpretation relative to the lists:

(**I9**) For any list $\bar{x}$ and variables x_i in the list, $|x_i|_{\bar{x}}$ is the ith projection

$$A_1 \times ... \times A_n \to A_i$$

(I10) For any arrow $f : A \to B$, if s is a term of type A then $|fs|_{\bar{x}}$ is

$$A_1 \times ... \times A_n \xrightarrow{|s|_{\bar{x}}} A \xrightarrow{f} B$$

and $|f^o s|_{\bar{x}}$ is

$$B \xrightarrow{f^{-1}} A \xrightarrow{|s|_{\bar{x}}^o} A_1 + ... + A_n$$

For any constant c, $|c|_{\bar{x}}$ is

$$A_1 \times ... \times A_n \to 1 \xrightarrow{c} B$$

(I11) For any terms s_1 of type A and s_2 of type B, $|\langle s_1, s_2 \rangle|_{\bar{x}}$ is the pair arrow to $A \times B$ induced by $|s_1|_{\bar{x}}$ and $|s_2|_{\bar{x}}$ while $[s_1, s_2]_{\bar{x}}$ is the pair arrow from $A + B$ induced by $|s_1|_{\bar{x}}^o$ and $|s_2|_{\bar{x}}^o$.

(I12) For any term s of type B, if the variable y over A is not in the list $\bar{x}$ then $|(\lambda y)s|_{\bar{x}}$ is the *transpose* of $|s|_{\bar{x},y} : A_1 \times ... \times A_n \times A \to B$, i.e. an arrow $\overline{|s|_{\bar{x},y}} : A_1 \times ... \times A_n \to B^A$, and $|s(\lambda y)|_{\bar{x}}$ is the *co-transpose* of $|s|_{\bar{x},y}^o : B \to A_1 + ... + A_n + A$, i.e. an arrow $\overline{|s|_{\bar{x},y}^o} : {}^A B \to A_1 + ... + A_n$.

If the bound variable y is in the list $\bar{x}$ that is an irrelevant coincidence. Then we replace y in $(\lambda y)s$ and $s(\lambda y)$ by some variable over A neither in s nor in the list $\bar{x}$.

According to (LT12) for any term g of type B^A and s of type A there is a term $ev(\langle g, s \rangle)$ which will be abbreviated to $g(s)$. Also for any term g of type ${}^{A_1}A_2$ and s of type A_1 there is a term $ev^o([g, s])$ which will be abbreviated to $g^o s$. We will use set builder notation for lambda abstraction over Ω writing $\{x.A : s\}$ instead of $(\lambda x.A)s$ and $\{x.A : s\}^o$ instead of $s(\lambda x.A)$ when s has type Ω.

Formulas are the terms of type Ω. By (LT11)-(LT12) for any formulas φ and ψ there are formulas $\wedge(\varphi, \psi), \vee(\varphi, \psi), \to (\varphi, \psi), \leftarrow (\varphi, \psi)$ which we will write as $\varphi \wedge \psi, \varphi \vee \psi, \varphi \to \psi$ and $\varphi \leftarrow \psi$. There are also formulas $\top, \bot, \neg \varphi, \lceil \varphi$.

For any formula φ and variable y over any object A, we define the formula $(\forall y.A)\varphi$ as an abbreviation for $\forall_A \{y.A : \varphi\}$ and $(\forall^o y.A)\varphi$ as $\forall_A \{y.A : \varphi\}^o$. Thus $(\forall y.A)\varphi$ says that $\{y.A : \varphi\}$ is all of A and $(\forall^o y.A)\varphi$ says that $\{y.A : \varphi\}^o$ is dually all of A. Its interpretation over a list of variables $\bar{x}$ follows the definitions. It is

$$A_1 \times ... \times A_n \xrightarrow{\overline{|\varphi|_{\bar{x},y}}} \Omega^A \xrightarrow{\forall_A} \Omega$$

$$\Omega \xrightarrow{\forall_A^o} {}^A\Omega \xrightarrow{\overline{\overline{||\varphi||_{\bar{x},y}}}} A_1 + ... + A_n$$

if the variable y does not occur in the list $\bar{x}$. If y is in the list it is first replaced by some new variable. Note that y is bound in $(\forall y.A)\varphi$ and in $(\forall^o y.A)\varphi$ too.

The existential quantification of formula φ at a variable x can be defined to be $(\forall w)((\forall x)(\varphi \to w) \to w)$ while the co-existential quantification can be defined to be $(\forall w)((\forall x)(\varphi \leftarrow w) \leftarrow w)$, where w is a variable over Ω not free in φ.

The *extension* of φ over a list of variables $\bar{x}$ is the subobject $A_1 \times ... \times A_n$ classified by $|\varphi|_{\bar{x}}$. We will write $[\bar{x} : \varphi]$ for this extension intending 'all $\bar{x}$ such that φ'. The *co-extension* of φ over a list of variables $\bar{x}$ is the subobject $A_1 + ... + A_n$ classified by $|\varphi|_{\bar{x}}$. We will write $[\bar{x} : \varphi]^o$ for this co-extension intending 'all $\bar{x}$ dually such that φ'.

A formula φ is *true* if its extension $[\bar{x} : \varphi]$ is all of $A_1 \times ... \times A_n$ when $\bar{x}$ lists exactly the variables free in φ. A formula φ is called *false* if its co-extension $[\bar{x} : \varphi]^o$ is all of $A_1 + ... + A_n$ when $\bar{x}$ lists exactly the variables free in φ.

A formula φ *implies* ψ if the extension of φ is contained in that of ψ. A formula φ *co-implies* ψ if the co-extension of φ is contained in that of ψ. More generally, for any finite set of formulas Γ, we will write $[\bar{x} : \Gamma]$ for the intersection of the extensions over $\bar{x}$ of all formulas in Γ and $[\bar{x} : \Gamma]^o$ for the join of the co-extensions over $\bar{x}$ of all formulas in Γ. In particular, $[\bar{x} : \ ] = A_1 \times ... \times A_n$ for the empty set of formulas and $[\bar{x} : \ ]^o = A_1 + ... + A_n$ for the empty set of formulas. Then Γ *implies* φ iff $[\bar{x} : \Gamma] \subseteq [\bar{x} : \varphi]$ and φ, while φ *co-implies* Γ iff $[\bar{x} : \Gamma]^o \subseteq [\bar{x} : \varphi]^o$, when $\bar{x}$ lists exactly the free variables in Γ and φ.

A *sequent* is an expression $\Gamma : \Delta$ where Γ, Δ are a finite (possibly empty) sets of formulas. In the sequents $\Gamma : \Delta$ the antecedent Γ and the succedent Δ cannot be sequences more than one-element simultaneously, i.e., if the sequence Γ is a sequence more than one-element then the sequence Δ can consists of at most one formula only and the other way round. Think of $\Gamma : \Delta$ as a claim that either Γ imply the one-element Δ or the one-element Γ co-imply Δ. At the same time both if in Γ and Δ formulas as the main connectives contain $\leftarrow$ and $\vdash$ ($\to$ and $\neg$) then, firstly, we should apply rules of inference to the formulas of Γ (Δ).

Finally, we will describe $H-B$ topos logic by means of the following rules:

$$\frac{*}{\varphi : \varphi} \qquad \frac{*}{: \top} \qquad \frac{*}{\bot : \varphi}$$

$$\frac{\Gamma : \Delta}{\Gamma : \Delta, \varphi}(: \text{Thinning}) \qquad \frac{\Gamma : \varphi}{\Gamma, \psi : \varphi}(\text{Thinning} :)$$

$$\frac{\Gamma : \Delta}{\Gamma(x/s) : \Delta(x/s)} \text{(Substitution :)} \quad \begin{array}{l}\text{(for any term } s \text{ free for } x \text{ in} \\ \text{all the formulas)}\end{array}$$

$$\frac{\Gamma : \Theta, \varphi \quad \varphi, \Delta : \Sigma}{\Gamma, \Delta : \Theta, \Sigma}(\text{Cut}) \quad \begin{array}{l}\text{(if every variable free in } \varphi \text{ is} \\ \text{free in } \Gamma \text{ or in } \varphi)\end{array}$$

$$\frac{\Gamma, \varphi : \Delta \quad \Gamma, \psi : \Delta}{\Gamma, \varphi \vee \psi : \Delta}(\vee :)$$

$$\frac{\Gamma : \Delta, \varphi \quad \Gamma : \Delta, \psi}{\Gamma : \Delta, \varphi \wedge \psi}(: \wedge)$$

$$\frac{\Gamma, \varphi : \psi}{\Gamma : \varphi \to \psi}(: \to)$$

$$\frac{\varphi : \Gamma, \psi}{\varphi \leftarrow \psi : \Gamma}(\leftarrow :)$$

$$\frac{\Gamma, \varphi : \bot}{\Gamma : \neg \varphi}(:\neg)$$

$$\frac{\top : \Gamma, \varphi}{\ulcorner \varphi : \Gamma}(\ulcorner :)$$

$$\frac{\Gamma : \varphi}{\Gamma : (\forall x)\varphi}(: \forall) \qquad \frac{\Gamma, \varphi : \psi}{\Gamma, (\exists x)\varphi : \psi}(\exists :) \quad \begin{array}{l}\text{(if the variable } x \text{ is} \\ \text{not free in } \Gamma \text{ or } \psi)\end{array}$$

$$\frac{\varphi : \Gamma}{(\forall^o x)\varphi : \Gamma}(\forall^o :) \qquad \frac{\varphi : \Gamma, \psi}{\varphi : \Gamma, (\exists^o x)\psi}(: \exists^o) \quad \begin{array}{l}\text{(if the variable } x \text{ is} \\ \text{not free in } \Gamma \text{ or } \psi)\end{array}$$

4 Correctness of $H-B$ topos logic

Variables that are not free in a term have no effect on its interpretation exept in determining the domain, as it follows from lemma below.

Lemma 5 (the superfluous variable lemma) *Let s be ay term of type A. Let $\bar{x}$ be a list of variables over $A_1 \times ... \times A_n$ and $A_1 + ... + A_n$, including those free in s. Let $\bar{y}$ be a list of variables over $B_1 \times ... \times B_k$*

and $B_1 + ... + B_k$, including all the variables of $\bar{x}$. Then $|s|_{\bar{y}}$ is the arrow

$$B_1 \times ... \times B_k \overset{proj}{\to} A_1 \times ... \times A_n \overset{|s|_{\bar{y}}}{\to} A$$

and $||s||_{\bar{y}}$ is the arrow

$$A \overset{||s||_{\bar{y}}}{\to} A_1 + ... + A_n \overset{inj}{\to} B_1 + ... + B_k$$

where $proj$ is the projection of $B_1 \times ... \times B_k$ on to those factors that are in $A_1 \times ... \times A_n$, while inj is the injection from $A_1 + ... + A_n$ in to those factors that are in $B_1 + ... + B_k$.

Proof. This is immediate if s is a variable and easy for terms given by (LT2),(LT3) and (LT6),(LT7). The result holds for terms $(\lambda y)s$ since the obvious projection of $B_1 \times ... \times B_k \times A$ to $A_1 \times ... \times A_n$ is just $(proj) \times 1_A$ and the transpose of any $h \circ (k \times 1_A)$ is $\bar{h} \circ k$. In a dual case the result holds for terms $(\lambda y)s$ since the injection $A_1 + ... + A_n$ to $B_1 + ... + B_k + A$ is $(inj) + 1_A$ and the co-transpose of any $(k + 1_A) \circ h^o$ is $k \circ \overline{h^o}$. ∎

Corollary 6 *For any formula φ and list $\bar{x}$ including all its free variables, and list $\bar{y}$ including all variables in $\bar{x}$, the extension $[\bar{y} : \varphi]$ is the pullback along $proj$ of $[\bar{x} : \varphi]$ while the co-extension $[\bar{y} : \varphi]^o$ is the pushout along inj of $[\bar{x} : \varphi]^o$. Thus, if $\Gamma : \varphi$ is true then $[\bar{y} : \Gamma] \subseteq [\bar{y} : \varphi]$ for every list $\bar{y}$ including all variables free in the sequent. Respectively, if $\varphi : \Gamma$ is false then $[\bar{y} : \varphi]^o \subseteq [\bar{y} : \Gamma]^o$ for every list $\bar{y}$ including all variables free in the sequent.*

Lemma 7 (the substitution lemma) *Consider a term s of type B and suppose that $\bar{x}$ contains every variable free in s. For each variable x_i in $\bar{x}$ let c_i be a term free for x_i in s. Let $s(\bar{x}/\bar{c})$ denote the result of substituting each c_i for x_i in s. Suppose that a list $\bar{y}$ of variables over $B_1 \times ... \times B_k$ and $B_1 + ... + B_k$ includes all variables free in $s(\bar{x}/\bar{c})$. Then $|s(\bar{x}/\bar{c})|_{\bar{y}}$ is the arrow*

$$B_1 \times ... \times B_k \overset{\langle c_1,...,c_n \rangle_{\bar{y}}}{\to} A_1 \times ... \times A_n \overset{|s|_{\bar{x}}}{\to} B$$

where $\langle c_1, ..., c_n \rangle_{\bar{y}}$ is the n-tuple of the arrows $|c_i|_{\bar{y}}$, and $||s(\bar{x}/\bar{c})||_{\bar{y}}$ is the arrow

$$B \overset{||s||_{\bar{x}}}{\to} A_1 + ... + A_n \overset{[c_1,...,c_n]_{\bar{y}}}{\to} B_1 + ... + B_k$$

where $[c_1, ..., c_n]_{\bar{y}}$ is the n-tuple of the arrows $||c_i||_{\bar{y}}$.

Proof. If s is a variable then the theorem merely repeats the definitions of $\langle c_1, ..., c_n \rangle_{\bar{y}}, [c_1, ..., c_n]_{\bar{y}}$. From there, the proof is virtually the same as that of the superfluous variable lemma. ∎

Corollary 8 *The extension $[\bar{y} : \varphi(\bar{x}/\bar{c})]$ is the pullback of $[\bar{x} : \varphi]$ along $\langle c_1, ..., c_n \rangle_{\bar{y}}$, the extension $[\bar{y} : \varphi(\bar{x}/\bar{c})]^o$ is the pushout of $[\bar{x} : \varphi]^o$ along $[c_1, ..., c_n]_{\bar{y}}$, The same holds for a finite set of formulas, so if $[\bar{x} : \Gamma] \subseteq [\bar{x} : \varphi]$ then*

$$[\bar{y} : \Gamma(\bar{x}/\bar{c})] \subseteq [\bar{y} : \varphi(\bar{x}/\bar{c})]$$

and if $[\bar{x} : \varphi]^o \subseteq [\bar{x} : \Gamma]^o$ then

$$[\bar{y} : \varphi(\bar{x}/\bar{c})]^o \subseteq [\bar{y} : \Gamma(\bar{x}/\bar{c})]^o$$

The structural rules are correct by simple properties of inclusion and intersection. We have $[\bar{x} : \varphi] \subseteq [\bar{x} : \varphi], [\bar{x} : \varphi]^o \subseteq [\bar{x} : \varphi]^o$ as well as $[\bar{x} : \;] \subseteq [\bar{x} : \top], [\bar{x} : \;]^o \subseteq [\bar{x} : \bot]^o$ respectively and if $[\bar{x} : \Gamma] \subseteq [\bar{x} : \varphi], [\bar{x} : \varphi]^o \subseteq [\bar{x} : \Gamma]^o$ then $[\bar{x} : \Gamma, \psi] \subseteq [\bar{x} : \varphi], [\bar{x} : \varphi]^o \subseteq [\bar{x} : \Gamma, \psi]^o$ respectively. The preceding corollary shows that substitution is correct.

For the cut rules the free variables restriction guarantees that the list $\bar{x}$ of variables free in the sequent $\Gamma : \varphi, \varphi : \Gamma$ is also the list of variables free in $\Gamma, \psi : \varphi$ and $\varphi : \Gamma, \psi$ respectively and includes all variables free in $\Gamma : \psi, \psi : \Gamma$ respectively. Given $[\bar{x} : \Gamma] \subseteq [\bar{x} : \psi], [\bar{x} : \psi]^o \subseteq [\bar{x} : \Gamma]$ it is easy to see that $[\bar{x} : \Gamma, \psi] \equiv [\bar{x} : \Gamma], [\bar{x} : \Gamma, \psi]^o \equiv [\bar{x} : \Gamma]^o$. And in that case $[\bar{x} : \Gamma, \psi] \subseteq [\bar{x} : \varphi], [\bar{x} : \varphi]^o \subseteq [\bar{x} : \Gamma, \psi]^o$ itself says that $[\bar{x} : \Gamma] \subseteq [\bar{x} : \varphi], [\bar{x} : \varphi]^o \subseteq [\bar{x} : \Gamma]^o$.

For the connective rule for conjunction we have $[\bar{x} : \Gamma] \subseteq [\bar{x} : \varphi], [\bar{x} : \varphi]^o \subseteq [\bar{x} : \Gamma]^o$ and $[\bar{x} : \Gamma] \subseteq [\bar{x} : \psi], [\bar{x} : \psi]^o \subseteq [\bar{x} : \Gamma]^o$ iff we have $[\bar{x} : \Gamma] \subseteq [\bar{x} : \varphi] \cap [\bar{x} : \psi], [\bar{x} : \varphi]^o \cup [\bar{x} : \psi]^o \subseteq [\bar{x} : \Gamma]^o$. For the connective rule for the disjunction we have $[\bar{x} : \Gamma, \psi] \subseteq [\bar{x} : \varphi], [\bar{x} : \varphi]^o \subseteq [\bar{x} : \Gamma, \psi]^o$ and $[\bar{x} : \Gamma, \chi] \subseteq [\bar{x} : \varphi], [\bar{x} : \varphi]^o \subseteq [\bar{x} : \Gamma]^o$ iff we have $[\bar{x} : \Gamma, \psi] \cup [\bar{x} : \Gamma, \chi] \subseteq [\bar{x} : \varphi], [\bar{x} : \varphi]^o \cup [\bar{x} : \psi]^o \subseteq [\bar{x} : \Gamma]^o$.

For the correctness of $\rightarrow$- and $\leftarrow$- rules we need the following theorem:

Theorem 9 *For any sub-objects $Q, R,$ and $s : S \rightarrowtail A$, we have $(S \cap Q) \subseteq R$ iff $S \subseteq (Q \implies R)$ and $(S \cup Q) \subseteq R$ iff $S \subseteq (Q \impliedby R)$.*

Proof. Since for any sub-objects $q : Q \rightarrowtail A$ and $r : R \rightarrowtail A$, we define $Q \implies R$ to be the sub-object classified by $\chi_q \rightarrow \chi_r$, then $S \subseteq (Q \implies R)$ is equivalent to $\rightarrow \circ \langle \chi_q, \chi_r \rangle \circ s = \top_s$. This is equivalent to $(\chi_q \circ s) \leq_1 (\chi_r \circ s)$ and so to $(S \cap Q) \subseteq (S \cap R)$, which is easily equivalent to $(S \cap Q) \subseteq R$.

$S \subseteq (Q \Leftarrow R)$ is equivalent to $\leftarrow \circ \langle \bar{\chi}_q, \bar{\chi}_r \rangle \circ s = \perp_s$ and this is equivalent to $(\bar{\chi}_q \circ s) \geq (\bar{\chi}_r \circ s)$ ($\geq$ is the dual of $\leq_1$) and so to $(S \cup Q) \subseteq (S \cup R)$, which is easily equivalent to $(S \cup Q) \subseteq R$. ∎

The correctness of $(: \forall)$- and $(\forall^o :)$-rules is by the following theorem:

Theorem 10 *For any $s : S \rightarrowtail B$, $S \subseteq (\forall a)R$ iff $S \times A \subseteq R$ and $(\forall^o a)R \subseteq S$ iff $R \subseteq S + A$.*

Proof. For any E, the name $'T'_E : 1 \to \Omega^E$ represents the maximal sub-object $1_E : E \rightarrowtail E$. Because every arrow from 1 is monic, $'T'_E$ itself has a classifying arrow $\forall_E : \Omega^E \to \Omega$; that is $\forall_E$ takes a sub-object of E to *true* iff it is the maximal sub-object.

Given any relation $r : R \to B \times A$ the universal quantification of r over A gives us the largest sub-object of B, $(\forall a)r : (\forall a)R \rightarrowtail B$ that is r-related to all of A and $(\forall a)R$ is the sub-object classified by $\forall_A \circ \hat{\chi}_r$ (where $\hat{\chi}_r$ is an arrow $\hat{\chi}_r : B \to \Omega^A$). We have $S \subseteq (\forall a)R$ iff $\hat{\chi}_r \circ s = 'T'_A \circ !_S$. Transposing both sides gives $\chi_r \circ (s \times 1_A) = \top_{S \times A}$. Since for any sub-object $u : U \rightarrowtail A$ with classifying arrow χ_u, and $v : V \rightarrowtail A$, we have $V \subseteq U$ iff $\chi_u \circ v = \top_V$, then we obtain that $S \times A \subseteq R$.

In a dual case for any E, the co-name $'\perp_E' :^E \Omega \to 1$ represents the minimal super-object $1_E : E \rightarrowtail E$. Because every arrow from 1 is monic, $'\perp_E'$ itself has a classifying arrow $\forall^o_E : \Omega \to {}^E\Omega$; that is $\forall^o_E$ takes a super-object of E to *false* iff it is the minimal super-object.

Given any co-relation $r^o : R \to B + A$ the dual universal quantification of r^o over A gives us the smallest super-object of B, $(\forall a)^o r^0 : B \rightarrowtail (\forall^o a)R$ that is r^o-related to all of A and $(\forall^o a)R$ is the super-object classified by $\hat{\chi}^o_r \circ \forall^o_A$ (where $\hat{\chi}^o_r$ is an arrow $\hat{\chi}^o_r : B \to {}^A\Omega$). We have $(\forall a)R \subseteq S$ iff $s^{-1} \circ \hat{\chi}^o_r = !^o_S \circ {}'\perp_A{}'$ (where $!^o_S : S \to 0$). Co-transposing both sides gives $(s^{-1} + 1_A) \circ \bar{\chi}_{r^o} = \perp_{S+A}$. Since dually for any super-object $u : A \rightarrowtail U$ with classifying arrow $\bar{\chi}_u$, and $v : A \rightarrowtail V$, we have $U \subseteq V$ iff $v \circ \bar{\chi}_u = \perp_V$, then we obtain that $R \subseteq S + A$. ∎

Since the rules for $\neg$, $\lceil$ and $(\exists x)$, $(\exists^o x)$ can be derived from rules that we have proved correct, they are correct too.

Tte final question is concerned with the completeness of internal logic of $H - B$-topos. The completeness of the structural rules and connective rules means that they are complete in this sense: if in every $H - B$ topos every sequent with the logical form of $\Gamma : \varphi$ or $\varphi : \Gamma$ is deducible from the rules and axioms of $H-B$ topos logic. But taking into account that $H-B$ toposes are practically unexplored then we have to postpone the analysis of this issue especially because even the standard topos logic itself is not complete for each individual topos (cf.[4, p.139]).

References

[1] *L. Estrada-González.* Complement-topoi and Dual Intuitionistic Logic. Australasian J. Log. 9, 26–44, 2010.

[2] *L. Estrada-González.* From (Paraconsistent) Topos Logic to Universal (Topos) Logic, A.Koslow, A. Buchsbaum (eds.), The Road to Universal Logic, Volume II, Springer International Publishing Switzerland, 263-296, 2015.

[3] P.T. Johnstone. Topos Theory. Academic Press, London, New York, San Fransisco, 1977.

[4] *McLarty.* Elementary Categories, Elementary Toposes, Clarendon Press, Oxford. 1992.

[5] *C. Mortensen.* InconsistentMathematics. Kluwer Mathematics and Its Applications Series, Kluwer, Dordrecht, 1995.

[6] *C. Rauszer.* A Formalization of the Propositional Calculus of H-B-logic, Studia Logica, 33, 1 (1973), 23-34.

[7] *C. Rauszer.* Applications of Kripke Models to Heyting-Brouwer Logic, Studia Logica, 36, 1-2 (1977), 61-71.

[8] *C. Rauszer.* An algebraic and Kripke-style approach to a certain extension of intuionistic logic, Dissertationes Mathematicae, CLXVII. PWN, Warszawa, 1-61,1980.

[9] *G.Reyes, H. Zolfaghari.* Bi-Heyting algebras, toposes and modalities. J. Philos. Log. 25(1), 25–43, 1996.

On categorial combination of logics

Marcelo E. Coniglio

Institute of Philosophy and the Humanities (IFCH), and Centre for Logic,
Epistemology and the History of Science (CLE), University of Campinas
(UNICAMP), Brazil

Dedicated to Edelcio G. de Souza on the occasion of his 60th birthday

Abstract

We propose in this paper a generalization of fibring of propositional logic systems within the framework of category theory. Specifically, we generalize the categorial construction of fibring by using arbitrary colimits and limits. We prove that limits and colimits preserve completeness (under reasonable conditions) in the category of propositional Hilbert calculi endowed with general algebraic semantics, generalizing a well-known result in the literature of combining logics.

1 Introduction

Combining logics is still a young subject in contemporary logic.[1] It offers a natural philosophical interest, given the possibility of defining mixed logic systems in which the logical operators satisfy laws of different nature. Besides the interest coming from Philosophy, there are also pragmatical and methodological reasons which justify considering combined logics. For instance, in Computer Science it can be required the integration of several logic systems into a homogeneous environment, in the context of knowledge representation.

Several interesting questions in the philosophy of logic naturally arise concerning this topic: by asuming a pluralist position, are there logics which are incompatible? Is it possible to combine different logics by producing new coherent logic systems? If it is possible to compose logics, would it also be possible to decompose them? By decomposing a given logic into several fragments, would it be possible to recover the original

---

[1]For general references on combining logics see [13] and [14].

logic by combining such fragments? What kind of metaproperties of given logics can be transferred to their combinations?

As observed in [12], an illustrative and early example of combining logics can be found in the well-known "ought-implies-can" thesis attributed to I. Kant, according to which, if an agent ought to do an act, then it has to be logically possible to do it. This problem could be analyzed from the perspective of combining logics, specifically connected to accepting properties of combining deontic and alethic logics. In formal terms, the "ought-implies-can" thesis concerns sentences of the form $Op \to \Diamond p$, where O represents the deontic "obligatory" operator, the diamond $\Diamond$ denotes the alethic "possibly" operator, and p is a proposition (representing an action). Thus, this principle means that if an action is obligatory then it must be possible. According to another interpretations, what Kant allegedly believed is that we cannot be obliged to do something if we are not capable of acting in that way. This would be formalized by the contrapositive of the previous formula, namely $\neg \Diamond p \to \neg Op$, meaning that "cannot-implies-has no duty to". Formulas involving modalities of different kind (or, more generally, connectives from different logics) naturally arise when combining logics, and are called *bridge principles* in [12].[2]

The first methods for combining logic systems were *products of logics*, independently introduced by K. Segerberg in [27] and by V. Šehtman in [33]; *fusion*, introduced by R. Thomason in [32]; and *fibring*, introduced by D. Gabbay in [24], Observe that all of these methods were defined exclusively to combine modal logics. It should be mentioned that M. Fitting in 1969 already gave early examples of fusion of modal logics, anticipating the notion of fusion (see [23]).

Other combination mechanisms where afterwards introduced: in the context of formal software specification, M. Finger and D. Gabbay introduced *temporalization* ([22]), which was generalized in [9] towards the method called *parameterization*.

All of these methods are designed for creating new logic systems from given ones, with the aim of integrating different aspects of them. This situation can appear in Computer Science, for instance in software engineering and security. Specifically, in formal specification and verification of algorithms and protocols it is useful and convenient to work

---

[2]This name has been introduced in the literature to denote a statement that binds factualities to norms, which appears in the context of David Hume's "is-ought problem".

with several logics. This direction (or approach) to combining logic is what is called in [11] a process of *splicing logics*. In the terminology introduced in [14], it would be a *synthesis process*, in which a logic is synthesized from given ones.

However, it would be reasonable to expect that a method for combining logics would work in the opposite directions: hence, a logic that one wants to investigate could be decomposed into factors of lesser complexity. To give an example, a bimodal alethic-deontic logic could be decomposed into its alethic and deontic fragments. It would be interesting to see whether the given logic is the least extension of its factors, or if additional bridge principles would have to be added in order to recover the original system, a problem which has been analyzed in [16]. This approach to combining logics, in which a given logic is decomposed into (possibly) simpler factors, is what is called in [11] a process of *splitting logics*. According to the terminology introduced in [14], it would correspond to an *analysis process*, in which a given logic is analyzed into simpler components. An important method for splitting logics is *possible-translations semantics*, introduced by W. Carnielli in [10].

Many of the early splicing methods for combining logics mentioned above have been generalized by the *categorial* (a.k.a. *algebraic*) notion of fibring introduced by A. Sernadas, C. Sernadas and C. Caleiro in [28]. Indeed, this framework dramatically improved the scope of these techniques by means of (universal) categorial constructions. From this, it is possible to combine wider classes of logics besides modal logics, see for instance [28, 34, 30, 31, 7, 18, 21, 8, 16, 19, 29, 17].

The interplay between the general framework of category theory and abstract logic has shown to be extremely useful. For instance, E. de Souza, A. Costa-Leite and D. Dias consider in [20] a functor between categories of suitable logics in order to provide, in an uniform way, a natural paraconsistent expansion of any given logic. In the realm of combining logics, the work of A. Sernadas and his collaborators in formalizing fibring for several classes of logic systems by using category theory clarifies the fact that fibring can be seen as a particular kind of colimit in the category in which the fibred logics are represented.

In fact, in the 'classical' approach to categorial fibring, there are two possibilities for fibring logics: to perform a free (or *unconstrained*) fibring, without sharing of connectives, or to perform a *constrained* fibring, by sharing some connectives. In terms of category theory, the former is characterized as a coproduct in the underlying category of logic systems.

The latter is constructed by a cocartesian lifting from the category of signatures. In most cases it is possible to substitute the cocartesian lifting by a pushout in the underlying category of logic systems (cf. [34]). Therefore, fibring two logics, from the point of view of category theory, consists basically of coproducts or pushouts in the category in which the logic systems are represented.

The main goal of this paper is to extend the standard notion of categorial fibring to arbitrary colimits and limits in appropriate categories of logic systems. In particular, when the category of logic systems is composed by systems with both semantic and syntactic consequence relations, it is desirable to preserve completeness through the combination process. This important question will be addressed here.

As a first approach to the question of extending fibring to other categorial constructions, in this paper we concentrate our attention in three categories of logic systems: **Hil**, **Int**, and **Lsp**, which were introduced in [34]. The first consists of (propositional) Hilbert calculi in which the notions of *local* (derivations) and *global* (proofs) inferences are distinguished. The second consists of the semantic (algebraic) counterpart of Hilbert calculi: the category of interpretation systems, which generalizes Kripke frames. The third category consists of logic system presentations (l.s.p.'s) where the objects are simultaneously a Hilbert calculus as well as an interpretation system. Clearly, a l.s.p. is interesting when both semantic and syntactic entailments coincide.

The framework for representing logics proposed in [34], both at the proof-theoretical level (by means of Hilbert calculi) and at the semantical level (by means of interpretation systems) is cleary oriented to modal logics. Indeed, as mentioned above, two notions of semantical entailment are considered: the global ones, which deal with global truths (formulas valid in every state or world), and the local ones, whose trueness is preserved pointwise (from world/state to world/state). Accordingly, the Hilbert calculi have two kind of inference rules: the proof rules, which represent theoremhood, and the derivation rules, apt to deal with inferences from premises. This difference is clear in the context of modal logics, which are usually presented in terms of global semantics and theoremhood instead of local semantics and derivations from premises. The kind of semantical structures adopted in [34] is formed by algebras defined over subsets of the powerset $\wp(U)$ of a given universe U. This framework is slightly more general than the one proposed in [28], in which the algebras are defined exclusively over the powerset $\wp(U)$ of

U. It is worth noting that, in [14, Chapter 3], general algebras were considered instead of algebras defined over powersets.

In [34], general conditions which guarantee the preservation of completeness of l.s.p.'s by fibring were found. The main result of this paper states that the same conditions ensure the preservation of completeness by arbitrary colimits. On the other hand, it is shown that completeness is preserved by arbitrary limits. These stimulating results suggest that it is possible to obtain complex systems of logics through specifications by diagrams, taking limits or colimits, and completeness will be preserved under reasonable assumptions.

2 The category of signatures

Propositional signatures constitute the formal basis to describe terms in abstract algebras, as well as propositional languages for logic systems. We begin by briefly analyze the category **Sig** of propositional-based signatures, on which all the logic systems considered in this paper are based. We assume that the reader is familiar with the (very) basic notions from category theory, such as limits, colimits and functors.[3] In what follows, **Set** (**Cls**, respectively) denotes the category of sets (classes, resp.) and functions between them.

Definition 2.1 (Propositional signatures). A *propositional signature* is a denumerable family of sets $C = \{C_k\}_{k \in \mathbb{N}}$ such that $C_k \cap C_i = \emptyset$ for every $i \neq k$. A *signature morphism* $h : C \longrightarrow C'$ between signatures is a family $h = \{h_k\}_{k \in \mathbb{N}}$ of functions $h_k : C_k \longrightarrow C'_k$. The composition $h' \circ h : C \longrightarrow C''$ of two signature morphisms $h : C \longrightarrow C'$ and $h' : C' \longrightarrow C''$ is given by the family $\{h'_k \circ h_k\}_{k \in \mathbb{N}}$. For any signature C the identity arrow $id_C : C \longrightarrow C$ is given by the family $\{id_{C_k}\}_{k \in \mathbb{N}}$, where $id_{C_k} : C_k \to C_k$ is the indentity function.

Observe that propositional signatures and their morphisms, with composition and identity morphisms as described in Definition 2.1, constitute the comma category **Set**/$\mathbb{N}$, which is (small) complete and cocomplete. That is, it contains the limits and colimits of every (small) diagram.

In order to describe propositional logic systems and their combinations, it will be convenient to consider *schema formulas*. This will

---
[3]Good general references to category theory are [25] and [26].

be done by following the approach introduced by A. Sernadas et al. in [28], by using a set of variables (called *schema variables*) which acts as metavariables for denoting arbitary (concrete) propositional formulas. Observe that schema variables *are not* the same as propositional variables: while the latter are concrete formulas (usually called *atomic fomulas*), the former are not formulas, but *schema formulas*. The intended meaning of a schema formula is that it represents an arbitary formula of a given language. The distinction between schema formulas and formulas is useful in order to describe process for combining logics, but this is also interesting for describing propositional logic systems: it avoids, for instance, the necessity of using the uniform substitution inference rule in the context of Hilbert calculi (see Definition 3.4 below).[4]

Definition 2.2 (The category **Sig** of propositional signatures over Ξ). Let us fix from now on a denumerable set $\Xi = \{\xi_k : k \in \mathbb{N}\}$ of symbols called *schema variables*. The category **Sig** of *signatures over* Ξ is the full subcategory of the category of signatures described in Definition 2.1, by considering as objects signatures C such that $C_k \cap \Xi = \emptyset$ for every $k \in \mathbb{N}$.

It is easy to see that **Sig** is (small) complete and cocomplete. This is a fundamental feature of **Sig** given that the limits and colimits in all the categories of logics to be considered along this paper are based on limits and colimits, respectively, of the underlying signatures over Ξ. When there is no risk of confusion, given a signature morphism $h = \{h_k\}_{k \in \mathbb{N}}$, the subscript k in h_k will be omitted. Clearly a morphism $h : C \longrightarrow C'$ is monic iff each h_k is an injective map, and it is epic iff each h_k is a surjective map.

Definition 2.3 (Schema formulas). Let C be a signature in **Sig**. Let C_Ξ be the propositional signature obtained from C by adding Ξ to C_0.[5] The set of *schema formulas over* C is the C_Ξ-algebra freely generated by $C_0 \cup \Xi$, which will be denoted by $L(C, \Xi)$. That is, $L(C, \Xi)$ is the least C_Ξ-algebra satisfying:

- $C_0 \cup \Xi \subseteq L(C, \Xi)$;

---

[4]In [16] the notion of schema variables for schema formulas was generalized to schema variables for contexts in formal sequent calculi, in order to consider their combination by fibring. This technique was extended in [17] to fibring of formal hypersequent calculi.

[5]Observe that C_Ξ is not an object in **Sig**, given that $(C_\Xi)_0 \cap \Xi = \Xi \neq \emptyset$.

- if $c \in C_k$ (for $k > 0$) and $\gamma_1, \ldots, \gamma_k \in L(C, \Xi)$ then $c(\gamma_1, \ldots, \gamma_k) \in L(C, \Xi)$.

The set of *formulas over C* (which is the C-algebra freely generated by C_0) will be denoted by $L(C)$.

As usual, a morphism $h : C \longrightarrow C'$ induce a (unique) function
$$\widehat{h} : L(C, \Xi) \longrightarrow L(C', \Xi)$$
defined inductively as follows:

- $\widehat{h}(\xi) = \xi$ if $\xi \in \Xi$;
- $\widehat{h}(c) = h_0(c)$ if $c \in C_0$;
- $\widehat{h}(c(\gamma_1, \ldots, \gamma_k)) = h_k(c)(\widehat{h}(\gamma_1), \ldots, \widehat{h}(\gamma_k))$, for $c \in C_k$ and $k > 0$.

2.1 Limits in Sig

Recall that a (small) *diagram* in a category $\mathbf{C}$ is a pair $\mathcal{D} = \langle \mathcal{O}, \mathcal{M} \rangle$, where $\mathcal{O} = \{O_i\}_{i \in I}$ is a small (possibly empty) family of objects of $\mathbf{C}$, and $\mathcal{M}$ is a (possibly empty) set of morphisms in $\mathbf{C}$ contained in $\bigcup_{i,j \in I} \mathbf{C}(O_i, O_j)$.

Let $\mathcal{D} = \langle \{C^i\}_{i \in I}, \mathcal{M} \rangle$ be a diagram in **Sig**. The limit of $\mathcal{D}$ in **Sig** is $\langle C, \{h^i\}_{i \in I} \rangle$ where

- $C_k = \{(c_i)_{i \in I} \in \prod_{i \in I} C_k^i \ : \ \text{if } C^i \xrightarrow{h} C^j \text{ is in } \mathcal{M} \text{ then } c_j = h_k(c_i)\}$ for all $k \in \mathbb{N}$.
- $h^i : C \longrightarrow C^i$, $h_k^i((c_i)_{i \in I}) = c_i$.

In particular, the terminal object in **Sig** is $\mathbf{1}$ given by $\mathbf{1}_k = \{*_k\}$ for every $k \in \mathbb{N}$. For any signature C, the unique morphism $!_C : C \longrightarrow \mathbf{1}$ is given by $!_C(c) = *_k$ if $c \in C_k$. The product C of a family $\{C^i\}_{i \in I}$ is given by $C_k = \prod_{i \in I} C_k^i$.

2.2 Colimits in Sig

Let $\mathcal{D} = \langle \{C^i\}_{i \in I}, \mathcal{M} \rangle$ be a diagram in **Sig**. For each $k \in \mathbb{N}$ consider the set $\overline{C}_k = \bigcup_{i \in I} C_k^i \times \{i\}$ and the equivalence relation: $(c, i) \sim_k (c', j)$ iff there exist $f^1, \ldots, f^n, g^1, \ldots, g^m$ in $\mathcal{M}$ (possibly $n = 0$ or $m = 0$) with $Dom(f^n) = Dom(g^m)$ and there exists $\overline{c} \in Dom(f_k^n) = Dom(g_k^m)$ such that $c = (f_k^1 \circ \cdots \circ f_k^n)(\bar{c})$ and $c' = (g_k^1 \circ \cdots \circ g_k^m)(\bar{c})$. For (c, i) in $\overline{C}_k$ let $(c, i)/{\sim_k}$ be its equivalence class under $\sim_k$ and let $C_k = \overline{C}_k/{\sim_k}$ be the quotient of $\overline{C}_k$ under $\sim_k$. The colimit of $\mathcal{D}$ in **Sig** is $\langle C, \{h^i\}_{i \in I}\rangle$ where

- $C = \{C_k\}_{k \in \mathbb{N}}$;
- $h^i : C^i \longrightarrow C$, $h_k^i(c) = (c, i)/{\sim_k}$ for all $i \in I$, $k \in \mathbb{N}$ and $c \in C_k^i$.

Therefore $h^i((f^1 \circ \cdots \circ f^n)(\bar{c}))$ and $h^j((g^1 \circ \cdots \circ g^m)(\bar{c}))$ always coincide in C_k, whenever $f^1, \ldots, f^n, g^1, \ldots, g^m$ belongs to $\mathcal{M}$ and $n, m \geq 0$.

3 The category of Hilbert calculi

In this section we analyze the category **Hil** of propositional-based Hilbert calculi as defined in [34]. In this context there exist two kinds of inferences: the local entailment and the global entailment. The latter uses *proof* rules (those in the set P below) while the former uses *derivation* rules (those in the set D below) plus proof rules applied to theorems. These two kinds of inferences appears frequently in complex inference systems such as modal logic and first-order logic, in which the *necessitation* rule or the *generalization* rule only can be applied to theorems (see for instance [28, 34]). On the other hand, in some cases (propositional classical logic, for instance) there is no distinction between proofs and derivations. A detailed discussion on local and global inferences can be found in [14, Section 3.1].

From now on, $\wp_{\text{fin}}(X)$ denotes the set of finite subsets of a given set X.

Definition 3.1. A *Hilbert calculus* is a triple $\langle C, P, D\rangle$ such that C is a signature, $P \subseteq \wp_{\text{fin}}(L(C, \Xi)) \times L(C, \Xi)$ and $D \subseteq P \cap ((\wp_{\text{fin}}(L(C, \Xi)) \setminus \emptyset) \times L(C, \Xi))$.

As observed at the beginning of Section 2, elements in Ξ play the rôle of "arbitrary" formulas, which can be replaced through *substitution maps* $\sigma : \Xi \longrightarrow L(C, \Xi)$ when an inference rule is applied (see Definition 3.2 below). Observe that any substitution $\sigma : \Xi \longrightarrow L(C, \Xi)$ can be extended to a unique endomorphism $\bar{\sigma} : L(C, \Xi) \longrightarrow L(C, \Xi)$ defined inductively as follows:

- $\bar{\sigma}(\xi) = \sigma(\xi)$ if $\xi \in \Xi$;
- $\bar{\sigma}(c) = c$ if $c \in C_0$;

- $\bar{\sigma}(c(\gamma_1,\dots,\gamma_k)) = c(\bar{\sigma}(\gamma_1),\dots,\bar{\sigma}(\gamma_k))$, for $c \in C_k$ and $k > 0$.

Definition 3.2. Let $\Gamma \cup \{\delta\} \subseteq L(C,\Xi)$ be a set of schema formulas.

(1) We say that δ is *provable* from Γ in the Hilbert calculus $\langle C, P, D \rangle$, denoted by $\Gamma \vdash^p_{\langle C,P,D \rangle} \delta$, if there exists a finite sequence $\gamma_1, \dots, \gamma_n$ in $(L(C,\Xi)$ such that γ_n is δ and, for every $1 \le i \le n$, either

- $\gamma_i \in \Gamma$, or

- there exists a rule $\langle \Delta, \psi \rangle$ in P and a substitution map $\sigma: \Xi \longrightarrow L(C,\Xi)$ such that $\bar{\sigma}(\Delta) \subseteq \{\gamma_1, \dots, \gamma_{i-1}\}$ and $\bar{\sigma}(\psi) = \gamma_i$.

We say that δ is *provable* in $\langle C, P, D \rangle$ if $\emptyset \vdash^p_{\langle C,P,D \rangle} \delta$.

(2) We say that δ is *derivable* from Γ in the Hilbert calculus $\langle C, P, D \rangle$, denoted by $\Gamma \vdash^d_{\langle C,P,D \rangle} \delta$, if there exists a finite sequence $\gamma_1, \dots, \gamma_n$ in $(L(C,\Xi)$ such that γ_n is δ and, for every $1 \le i \le n$, either

- $\gamma_i \in \Gamma$, or

- δ_i is provable in $\langle C, P, D \rangle$, or

- there exists a rule $\langle \Delta, \psi \rangle$ in D and a substitution map $\sigma: \Xi \longrightarrow L(C,\Xi)$ such that $\bar{\sigma}(\Delta) \subseteq \{\gamma_1, \dots, \gamma_{i-1}\}$ and $\bar{\sigma}(\psi) = \gamma_i$.

Example 3.3 (Modal Hilbert calculi). Let C be a *modal signature* such that $C_0 = VAR = \{p_n : n \in \mathbb{N}\}$ (a denumerable set of *propositional variables*); $C_1 = \{\neg, \Box\}$; $C_2 = \{\Rightarrow\}$; $C_n = \emptyset$ for $n > 2$.

(1) A Hilbert calculus for the alethic modal system K is $H_K = \langle C, P_K, D_K \rangle$ such that P_K consists of the following rules:[6]

- $\langle \emptyset, \xi_1 \Rightarrow (\xi_2 \Rightarrow \xi_1) \rangle$

- $\langle \emptyset, (\xi_1 \Rightarrow (\xi_2 \Rightarrow \xi_3)) \Rightarrow ((\xi_1 \Rightarrow \xi_2) \Rightarrow (\xi_1 \Rightarrow \xi_3)) \rangle$

- $\langle \emptyset, (\neg \xi_1 \Rightarrow \xi_2) \Rightarrow ((\neg \xi_1 \Rightarrow \neg \xi_2) \Rightarrow \xi_1) \rangle$

- $\langle \emptyset, \Box(\xi_1 \Rightarrow \xi_2) \Rightarrow (\Box \xi_1 \Rightarrow \Box \xi_2) \rangle$

- $\langle \{\xi_1\}, \{\Box \xi_1\} \rangle$

- $\langle \{\xi_1, (\xi_1 \Rightarrow \xi_2)\}, \xi_2 \rangle$

---

[6] As usual, we will adopt infix notation, writing $(\psi \Rightarrow \varphi)$, or even $\psi \Rightarrow \varphi$, instead of $\Rightarrow (\psi, \varphi)$. Moreover, we will write $\neg \psi$ and $\Box \psi$ instead of $\neg(\psi)$ and $\Box(\psi)$.

and $D_K = \{\langle\{\xi_1, (\xi_1 \Rightarrow \xi_2)\}, \xi_2\rangle\}$.

(2) A Hilbert calculus for the standard deontic system KD is $H_{KD} = \langle C, P_{KD}, D_{KD}\rangle$ such that $P_{KD} = P_K \cup \{\langle \emptyset, \Box\xi_1 \Rightarrow \neg\Box\neg\xi_1\rangle\}$ and $D_{KD} = D_K$.

Definition 3.4. The category **Hil** of *Hilbert calculi* is defined as follows:

- Objects: Hilbert calculi $\langle C, P, D\rangle$.
- Morphisms: A morphism $h : \langle C, P, D\rangle \to \langle C', P', D'\rangle$ in **Hil** is given by a morphism $h : C \to C'$ in **Sig** such that:
  - $\langle \Gamma, \delta\rangle \in P$ implies that $\widehat{h}(\Gamma) \vdash^p_{\langle C', P', D'\rangle} \widehat{h}(\delta)$;
  - $\langle \Gamma, \delta\rangle \in D$ implies that $\widehat{h}(\Gamma) \vdash^d_{\langle C', P', D'\rangle} \widehat{h}(\delta)$.
- Composition and identity arrows: Inherited from **Sig**.

Definition 3.5. The forgetful functor $\mathsf{N} : \mathbf{Hil} \to \mathbf{Sig}$ is given by $\mathsf{N}(\langle C, P, D\rangle) = C$ and $\mathsf{N}(h) = h$.

Proposition 3.6. *The forgetful functor N has a left adjoint $\overline{\mathsf{N}} : \mathbf{Sig} \to \mathbf{Hil}$.*

Proof. For every signature C consider the Hilbert calculus $\overline{\mathsf{N}}(C) = \langle C, \emptyset, \emptyset\rangle$. Then $\langle id_C, \overline{\mathsf{N}}(C)\rangle$ is N-universal for C. For if H' is a Hilbert calculus and $h : C \to \mathsf{N}(H')$ is a morphism in **Sig** then $h^* = h : \overline{\mathsf{N}}(C) \to H'$ is the unique morphism in **Hil** such that

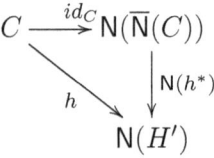

commutes in **Sig**. □

This means that, if $H_i = \langle C^i, P^i, D^i\rangle$ is a Hilbert calculus and $f_i : C \to C^i$ is a monic in **Sig** (for $i = 1, 2$) then the pushout in **Hil** of the diagram

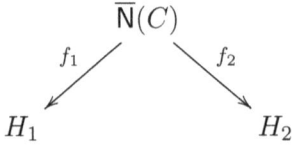

is the fibring of H_1 and H_2 constrained by $f_i : C {\longrightarrow} C^i$ ($i = 1, 2$) (cf. [6] and [34]). Thus, unconstrained fibrings (when $C = \emptyset$) are coproducts, and constrained fibrings are pushouts (cf. [28] and [34]). We will prove now the existence in **Hil** of arbitrary (small) limits and colimits, extending the notion of fibring in **Hil**.

3.1 Limits in Hil

As discussed in Section 1, there is a way of decomposing a logic system by a splitting or an analysis process. This kind of process could be represented by a limit in the corresponding category of logics. In particular, this could be done in the category **Hil** of Hilbert calculi. With this aim in mind, in this section it will be proven that **Hil** is (small) complete, by showing how to construct the limit of any given diagram.

Let $\mathcal{D} = \langle \{H_i\}_{i \in I}, \mathcal{M} \rangle$ be a diagram in **Hil**, where $H_i = \langle C^i, P^i, D^i \rangle$ for each $i \in I$. The limit of $\mathcal{D}$ in **Hil** is $\langle H, \{h^i\}_{i \in I}\rangle$ where $H = \langle C, P, D \rangle$ is defined as follows:

- $\langle C, \{h^i\}_{i \in I}\rangle$ is the limit in **Sig** of $\langle\{C^i\}_{i \in I}, \mathcal{M}\rangle$;

- $P = \{\langle \Gamma, \delta \rangle \in \wp_{\text{fin}}(L(C, \Xi)) \times L(C, \Xi) \ : \ \widehat{h}^i(\Gamma) \vdash^p_{H_i} \widehat{h}^i(\delta)$ for all $i \in I\}$;

- $D = \{\langle \Gamma, \delta \rangle \in (\wp_{\text{fin}}(L(C, \Xi)) \setminus \emptyset) \times L(C, \Xi) \ : \ \widehat{h}^i(\Gamma) \vdash^d_{H_i} \widehat{h}^i(\delta)$ for all $i \in I\}$.

Observe that $D \subseteq P$, hence H is indeed a Hilbert calculus.

Proposition 3.7. *The pair $\langle H, \{h^i\}_{i \in I}\rangle$ defined above is the limit in **Hil** of $\mathcal{D}$.*

Proof. Note that each $h^i : H {\longrightarrow} H_i$ is in fact a morphism in **Hil** such that $h \circ h^i = h^j$ for all $h : H_i {\longrightarrow} H_j$ in $\mathcal{M}$. Let $H' = \langle C', P', D' \rangle$ be a Hilbert calculus and let $g^i : H' {\longrightarrow} H_i$ for every $i \in I$ such that $h \circ g^i = g^j$ for all $h : H_i {\longrightarrow} H_j$ in $\mathcal{M}$. Since C is the limit in **Sig** of $\langle\{C^i\}_{i \in I}, \mathcal{M}\rangle$ there exists an unique morphism $h' : C' {\longrightarrow} C$ in **Sig** such that $h^i \circ h' = g^i$ for all $i \in I$, therefore

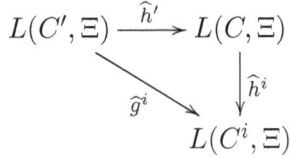

commutes in **Set** for all $i \in I$. Let $\langle \Gamma, \delta \rangle \in P'$. Since $g^i : H' \longrightarrow H_i$ in **Hil** then $\widehat{g}^i(\Gamma) \vdash^p_{H_i} \widehat{g}^i(\delta)$ for each $i \in I$, that is, $\widehat{h}^i(\widehat{h}'(\Gamma)) \vdash^p_{H_i} \widehat{h}^i(\widehat{h}'(\delta))$ for each $i \in I$. By definition of P we infer that $\widehat{h}'(\Gamma) \vdash^p_H \widehat{h}'(\delta)$. Analogously, if $\langle \Gamma, \delta \rangle \in D'$ then $\widehat{h}'(\Gamma) \vdash^d_H \widehat{h}'(\delta)$. Hence $h' : H' \longrightarrow H$ is a morphism in **Hil** such that

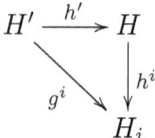

commutes in **Hil** for all $i \in I$. Clearly h' is unique, by the universal property of C, thus H is the limit in **Hil** of $\mathcal{D}$. □

Corollary 3.8. Let $\mathcal{D}$ be a diagram in **Hil** and let $H = \langle C, P, D \rangle$ be its limit with morphisms h^i, for $i \in I$. Let $\Gamma \cup \{\delta\} \subseteq L(C, \Xi)$ be a set of schema formulas. Then $\Gamma \vdash^p_H \delta$ iff $\widehat{h}^i(\Gamma) \vdash^p_{H_i} \widehat{h}^i(\delta)$ for all $i \in I$, and $\Gamma \vdash^d_H \delta$ iff $\widehat{h}^i(\Gamma) \vdash^d_{H_i} \widehat{h}^i(\delta)$ for all $i \in I$.

Limits of Hilbert calculi can be interesting, as the following examples show.

Example 3.9. As a particular case of the construction of limits given in the proof of Proposition 3.7, the product of a family $\{H_i\}_{i \in I}$ in **Hil** is $\langle H, \{h^i\}_{i \in I} \rangle$ where $H = \langle C, P, D \rangle$ is defined as follows:

- $\langle C, \{h^i\}_{i \in I} \rangle$ is the product in **Sig** of $\{C^i\}_{i \in I}$;

- $P = \{\langle \Gamma, \delta \rangle \in \wp_{\text{fin}}(L(C, \Xi)) \times L(C, \Xi) \; : \; \widehat{h}^i(\Gamma) \vdash^{H_i}_p \widehat{h}^i(\delta) \text{ for all } i \in I\}$;

- $D = \{\langle \Gamma, \delta \rangle \in (\wp_{\text{fin}}(L(C, \Xi)) \setminus \emptyset) \times L(C, \Xi) \; : \; \widehat{h}^i(\Gamma) \vdash^{H_i}_d \widehat{h}^i(\delta) \text{ for all } i \in I\}$.

In particular, the terminal object in **Hil** is $H_1 = \langle \mathbf{1}, P_1, D_1 \rangle$ where $\mathbf{1}$ is the terminal signature in **Sig** (see the end of Subsection 2.1) and

- $P_1 = \wp_{\text{fin}}(L(\mathbf{1}, \Xi)) \times L(\mathbf{1}, \Xi)$;

- $D_1 = (\wp_{\text{fin}}(L(\mathbf{1}, \Xi)) \setminus \emptyset) \times L(\mathbf{1}, \Xi)$.

For any $H = \langle C, P, D \rangle$, the unique morphism $!_H : H \longrightarrow H_1$ is given by $!_C : C \longrightarrow \mathbf{1}$.

Remark 3.10. From the perspective of combining logics, a product of Hibert calculi could be seen as a splitting of the product into its factors, the canonical projections being *translations* between them. The relationship between products of logics and splitting logics by means of traslations between them was already investigated in [4] (see also [5]). The latter establishes a link between category theory and the splitting method of possible-translations semantics mentioned in Section 1.

Example 3.11. Consider the well-known hierarchy of da Costa's paraconsistent calculi $\mathcal{C} = \{\mathcal{C}_n\}_{n\in\mathbb{N}}$, where $\mathcal{C}_0$ is propositional classical logic (see, for instance, [15]). By the very definition, $Th(\mathcal{C}_0) \supset Th(\mathcal{C}_1) \supset \cdots$, where $Th(\mathcal{C}_n)$ denotes the set of theorems of $\mathcal{C}_n$. An interesting question is how to axiomatize the so-called *limit* of the hierarchy $\mathcal{C}$, a calculus $\mathcal{C}_{lim}$ such that $Th(\mathcal{C}_{lim}) = \bigcap_{n\in\mathbb{N}} Th(\mathcal{C}_n)$ (see [15]). However, it is interesting to note that in fact $\mathcal{C}_{lim}$ appears in **Hil** as the limit of $\mathcal{C}$ (together with the respective embeddings). Consider for each $n \in \mathbb{N}$ the signature $C_0^n = \{p_k^n : k \in \mathbb{N}\}$, $C_1^n = \{\neg^n\}$, $C_2^n = \{\Rightarrow^n, \wedge^n, \vee^n\}$, and $C_k^n = \emptyset$ if $k > 2$, as well as the morphism $g^n : C^{n+1} \longrightarrow C^n$, $g_0^n(p_k^{n+1}) = p_k^n$, $g_1^n(\neg^{n+1}) = \neg^n$ and $g_2^n(c^{n+1}) = c^n$ for all $c^{n+1} \in C_2^{n+1}$. Define a Hilbert calculus $H_n = \langle C^n, P^n, D^n \rangle$ corresponding to $\mathcal{C}_n$ (identifying local and global inferences, cf. [7]). Thus $g^n : H_{n+1} \longrightarrow H_n$ is a morphism in **Hil** and $\mathcal{D} = \langle \{H_n\}_{n\in\mathbb{N}}, \{g^n\}_{n\in\mathbb{N}} \rangle$ is a diagram with limit $H = \langle C, P, D \rangle$ such that $C_0 = \{p_k : k \in \mathbb{N}\}$, $C_1 = \{\neg\}$, $C_2 = \{\Rightarrow, \wedge, \vee\}$, $C_k = \emptyset$ if $k > 2$, with the obvious morphisms $h^n : H \longrightarrow H_n$, $c \mapsto c^n$. Clearly the limit H represents $\mathcal{C}_{lim}$ (compare with Corollary 7.3 in [11]).

3.2 Colimits in Hil

By duality, and taking into account the discussion in Section 1, it makes sense to consider now the other way of combining logics, by a splicing or a synthesis process. Such a process could be represented by a colimit in the corresponding category of logics, generalizing so the notion of categorial fibring. In particular, this process could be done in the category **Hil**. Then, as a natural counterpart of what was done in Subsection 3.1, in this section it will be proven that **Hil** is (small) cocomplete. The proof will be constructive, that is, by showing how to construct the colimit of any given diagram.

Let $\mathcal{D} = \langle \{H_i\}_{i\in I}, \mathcal{M} \rangle$ be a diagram in **Hil**, where $H_i = \langle C^i, P^i, D^i \rangle$ for each $i \in I$. The colimit of $\mathcal{D}$ in **Hil** is $\langle H, \{h^i\}_{i\in I} \rangle$ where $H = \langle C, P, D \rangle$ is defined as follows:

- $\langle C, \{h^i\}_{i\in I}\rangle$ is the colimit in **Sig** of $\langle\{C^i\}_{i\in I}, \mathcal{M}\rangle$;
- $P = \{\langle\widehat{h^i}(\Gamma), \widehat{h^i}(\delta)\rangle \in \wp_{\text{fin}}(L(C,\Xi)) \times L(C,\Xi) \;:\; i \in I \text{ and } \langle\Gamma,\delta\rangle \in P^i\}$;
- $D = \{\langle\widehat{h^i}(\Gamma), \widehat{h^i}(\delta)\rangle \in (\wp_{\text{fin}}(L(C,\Xi)) \setminus \emptyset) \times L(C,\Xi) \;:\; i \in I \text{ and } \langle\Gamma,\delta\rangle \in D^i\}$.

Clearly $D \subseteq P$, and so H is indeed a Hilbert calculus.

Proposition 3.12. $\langle H, \{h^i\}_{i\in I}\rangle$ defined above is the colimit in **Hil** of $\mathcal{D}$.

Proof. Note that each $h^i : H_i \longrightarrow H$ is in fact a morphism in **Hil**, by definition of H. Moreover, if $h : H_i \longrightarrow H_j$ is in $\mathcal{M}$ then $h^j \circ h = h^i$ in **Sig** and then, by definition of **Hil**, the diagram

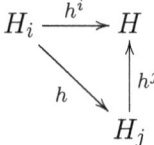

commutes in **Hil**. Let $H' = \langle C', P', D'\rangle$ be a Hilbert calculus and let $g^i : H_i \longrightarrow H'$ for every $i \in I$ such that $g^j \circ h = g^i$ for all $h : H_i \longrightarrow H_j$ in $\mathcal{M}$. Since C is the colimit in **Sig** of $\langle\{C^i\}_{i\in I}, \mathcal{M}\rangle$ there exists an unique morphism $h' : C \longrightarrow C'$ in **Sig** such that $h' \circ h^i = g^i$ for all $i \in I$, therefore $\widehat{h'} \circ \widehat{h^i} = \widehat{g^i}$ for all $i \in I$. Let $\langle\widehat{h^i}(\Gamma), \widehat{h^i}(\delta)\rangle \in P$ such that $\langle\Gamma,\delta\rangle \in P^i$. Since $g^i : H_i \longrightarrow H'$ in **Hil** then $\widehat{g^i}(\Gamma) \vdash^p_{H'} \widehat{g^i}(\delta)$, that is, $\widehat{h'}(\widehat{h^i}(\Gamma)) \vdash^p_{H'} \widehat{h'}(\widehat{h^i}(\delta))$. Analogously, if $\langle\widehat{h^i}(\Gamma), \widehat{h^i}(\delta)\rangle \in D$ such that $\langle\Gamma,\delta\rangle \in D^i$ then $\widehat{h'}(\widehat{h^i}(\Gamma)) \vdash^{H'}_d \widehat{h'}(\widehat{h^i}(\delta))$, thus $h' : H \longrightarrow H'$ is a morphism in **Hil** such that

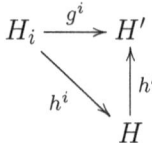

commutes in **Hil** for all $i \in I$. Clearly h' is unique, by the universal property of C, thus H is the colimit in **Hil** of $\mathcal{D}$. □

Corollary 3.13. *The category* **Hil** *is (small) complete and cocomplete.*

4 The category of interpretation systems

In this section we analyze the category **Int** of Interpretation systems as defined in [34]. This is the semantic counterpart of **Hil**.

Definition 4.1. Let C be a signature. A *C-structure* is a triple $S = \langle U, \mathcal{B}, \nu \rangle$ such that U is a non-empty set, $\mathcal{B}$ is a non-empty subset of $\wp(U)$ and $\nu = \{\nu_k\}_{k \in \mathbb{N}}$ is a family of maps $\nu_k : C_k \longrightarrow \mathcal{B}^{(\mathcal{B}^k)}$. We denote by $Str(C)$ the class of all C-structures. If $\langle U, \mathcal{B}, \nu \rangle, \langle U', \mathcal{B}', \nu' \rangle \in Str(C)$ we say that they are *isomorphic* if there exists a bijection $f : U \longrightarrow U'$ such that $f(\mathcal{B}) = \mathcal{B}'$ and $f(\nu_k(c)(\vec{b})) = \nu'_k(c)(f(\vec{b}))$ for all $k \in \mathbb{N}$, $c \in C_k$ and $\vec{b} \in \mathcal{B}^k$ (here, $f(\vec{b})$ denotes $(f(b_1), \ldots, f(b_k))$ whenever $\vec{b} = (b_1, \ldots, b_k)$). In this case we write $\langle U, \mathcal{B}, \nu \rangle \cong \langle U', \mathcal{B}', \nu' \rangle$.

Example 4.2 (Kripke C-structures). Recall the modal Hilbert calculi K and KD defined over the modal signature C in Example 3.3. Let $m = \langle W, R, V \rangle$ be a Kripke model, that is: W is a non-empty set (of *worlds*) and $R \subseteq W \times W$ is a relation over W (the *accessibility relation*). Then, m induces a C-structure $S_m = \langle W, \wp(W), \nu \rangle$ called *Kripke C-structure*, which is defined as follows:

- $\nu_0(p) = V(p)$ for every propositional variable $p \in VAR$;
- $\nu_1(\neg)(b) = W \setminus b$ for every $b \subseteq W$;
- $\nu_1(\Box)(b) = \{w \in W : wRw'$ implies that $w' \in b$, for every $w' \in W\}$ for every $b \subseteq W$;
- $\nu_2(\Rightarrow)(b, b') = (W \setminus b) \cup b'$ for every $b, b' \subseteq W$.

If m is a Kripke model for KD (that is, R is *serial*, meaning that for every $w \in W$ there exists some $w' \in W$ such that wRw') then the induced C-structure S_m is called a *Kripke C-structure for KD*.

Definition 4.3. The category **Int** of *Interpretation systems* is given by:

- Objects: Triples $\langle C, M, A \rangle$ where C is a signature, M is a class and $A : M \longrightarrow Str(C)$ is an injective map. We also assume that $\langle C, M, A \rangle$ is closed under isomorphic images, disjoint unions and subalgebras, that is:

    - if $A(m) \cong S$ for $S \in Str(C)$ and $m \in M$ then there exists $m' \in M$ such that $S = A(m')$;

- assume the notation $A(m) = \langle U_m, \mathcal{B}_m, \nu^m \rangle$ for each $m \in M$, and let N be a subset of M such that $U_n \cap U_{n'} = \emptyset$ for every $n \neq n'$ in N. Then there exists $m \in M$ such that $U_m = \bigcup_{n \in N} U_n$, $\mathcal{B}_m = \{b \in \wp(U_m) : b \cap U_n \in \mathcal{B}_n$ for all $n \in N\}$ and, for every $k \in \mathbb{N}$, $c \in C_k$ and $\vec{b} \in \mathcal{B}_m^k$, $\nu_k^m(c)(\vec{b}) = \bigcup_{n \in N} \nu_k^n(c)(b_1 \cap U_n, \ldots, b_k \cap U_n)$;
- let $m \in M$ and suppose that $\mathcal{B} \subseteq \mathcal{B}_m$ is a ν^m-subalgebra of $\mathcal{B}_m$, that is: $\mathcal{B}$ is closed under every operation ν_k^m for all $k \in \mathbb{N}$. Then there exists $m' \in M$ such that $U_{m'} = U_m$, $\mathcal{B}_{m'} = \mathcal{B}$ and $\nu_k^{m'}(c) = \nu_k^m(c)|_{\mathcal{B}^k}$.

- Morphisms: A morphism $\langle h, F \rangle : \langle C, M, A \rangle \longrightarrow \langle C', M', A' \rangle$ in **Int** is given by a morphism $h : C \longrightarrow C'$ in **Sig** and a function $F : M' \longrightarrow M$ such that:

  - $U_{F(m')} = U_{m'}$ and $\mathcal{B}_{F(m')} = \mathcal{B}_{m'}$ for every $m' \in M'$;
  - $\nu_k^{F(m')}(c) = \nu_k^{m'}(h_k(c))$ for every $k \in \mathbb{N}$, $c \in C_k$ and $m' \in M'$.

- Composition: $\langle h, F \rangle \circ \langle h', F' \rangle = \langle h \circ h', F' \circ F \rangle$.

- Identity arrows: $id_{\langle C, M, A \rangle} = \langle id_C, id_M \rangle$.

Remark 4.4 (Modal interpretation systems). Recall the modal signature C introduced in Example 3.3, as well as the notion of Kripke C-structure given in Example 4.2. In [34, Proposition 4.8] it was shown that every C-structure in the interpretation system $\langle C, M, A \rangle$ given by the closure by isomorphic images, disjoint unions and subalgebras of a given class $\mathbb{M}$ of Kripke C-structures is isomorphic to a Kripke C-structure. Such interpretation system will be called the *modal interpretation system generated by* $\mathbb{M}$. Thus, for every $m \in M$ there exists a Kripke model m' such that $A(m)$ is the Kripke C-structure $S_{m'}$ (up to isomorphisms).

Definition 4.5. The forgetful functor $\mathsf{O} : \mathbf{Int} \longrightarrow \mathbf{Sig}$ is given by $\mathsf{O}(\langle C, M, A \rangle) = C$ and $\mathsf{O}(\langle h, F \rangle) = h$.

Proposition 4.6. *The forgetful functor O has a left adjoint $\overline{\mathsf{O}} : Sig \longrightarrow \mathbf{Int}$.*

Proof. Given C define $\overline{\mathsf{O}}(C) = \langle C, Str(C), id_{Str(C)} \rangle$. Then $\langle id_C, \overline{\mathsf{O}}(C) \rangle$ is O-universal for C. For if $\langle C', M', A' \rangle$ is an interpretation system and

$h : C \longrightarrow C'$ is a morphism in **Sig** then there exists an unique morphism $h^* : \overline{O}(C) \longrightarrow \langle C', M', A' \rangle$ in **Int** such that the diagram below commutes in **Sig**.

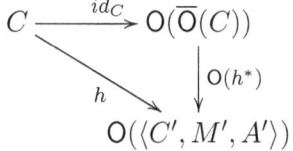

In fact $h^* = \langle h, F \rangle$ such that $F : M' \longrightarrow Str(C)$ is given as follows: let $m' \in M'$ and $A'(m') = \langle U_{m'}, \mathcal{B}_{m'}, \nu^{m'} \rangle$. Then $F(m') = \langle U_{m'}, \mathcal{B}_{m'}, \overline{\nu}^{m'} \rangle$ such that $\overline{\nu}_k^{m'}(c) = \nu_k^{m'}(h(c))$ for every $k \in \mathbb{N}$ and $c \in C_k$. □

This means that, if $\langle C^i, M^i, A^i \rangle$ is an interpretation system and $f_i : C \longrightarrow C^i$ is a monic in **Sig** (for $i = 1, 2$) then the pushout in **Int** of the diagram

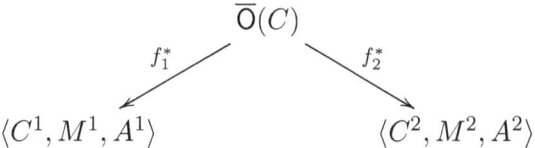

is the fibring of $\langle C^1, M^1, A^1 \rangle$ and $\langle C^2, M^2, A^2 \rangle$ constrained by $f_i : C \longrightarrow C^i$, for $i = 1, 2$ (cf. [6] and [34]). Thus, unconstrained fibrings (i.e., when $C = \emptyset$) are coproducts, and constrained fibrings are pushouts (cf. [34]). We will prove in the next subsection the existence in **Int** of arbitrary (small) limits and colimits, extending so the notion of fibring in **Int**.

4.1 Limits in Int

As it was done in Subsection 3.1 for Hilbert calculi, in this section it will be proven that the category **Int** of interpretation systems is (small) complete. According to the discussions above, limits constitute the categorial realization of a splitting or an analysis process of combining logics, in this case involving logics presented semantically as interpretation systems.

Let $\mathcal{D} = \langle \{\langle C^i, M^i, A^i \rangle\}_{i \in I}, \mathcal{M} \rangle$ be a diagram in **Int** and let $\mathcal{M}_1 = \{h : \langle h, F \rangle \in \mathcal{M} \text{ for some } F\}$, $\mathcal{M}_2 = \{F : \langle h, F \rangle \in \mathcal{M} \text{ for some } h\}$. Consider the pair $\langle \langle C, M, A \rangle, \{\langle h^i, F^i \rangle\}_{i \in I} \rangle$ defined as follows:

- $\langle C, \{h^i\}_{i \in I} \rangle$ is the limit in **Sig** of $\langle \{C^i\}_{i \in I}, \mathcal{M}_1 \rangle$;

- $\langle M, \{F^i\}_{i\in I}\rangle$ is the colimit in **Cls** of $\langle\{M^i\}_{i\in I}, \mathcal{M}_2\rangle$;

- $A : M \longrightarrow Str(C)$ is a map such that, if $i \in I$, $m \in M^i$ and $A^i(m) = \langle U_m, \mathcal{B}_m, \nu^m\rangle$ then $A(F^i(m)) = \langle U_m, \mathcal{B}_m, \nu^{F^i(m)}\rangle$ is given by

$$\nu_k^{F^i(m)}((c_i)_{i\in I}) = \nu_k^m(c_i) \text{ for all } k \in \mathbb{N} \text{ and } (c_i)_{i\in I} \in C_k.$$

Proposition 4.7. Let $\mathcal{D}$ be a diagram and consider $\langle\langle C, M, A\rangle, \{\langle h^i, F^i\rangle\}_{i\in I}\rangle$ as defined above. The limit of $\mathcal{D}$ in **Int** is $\langle\langle C, M/\sim, \overline{A}\rangle, \{\langle h^i, \overline{F^i}\rangle\}_{i\in I}\rangle$ where

- $M/\sim$ is the quotient of M under the equivalence relation $F^i(m) \sim F^j(m')$ iff $A(F^i(m)) = A(F^j(m'))$;

- $\overline{A}(F^i(m)/\sim) = A(F^i(m))$; and

- $\overline{F^i}(m) = F^i(m)/\sim$.

Proof. First we must prove that A is well-defined. Suppose that $F^i((F_1 \circ \cdots \circ F_r)(\overline{m})) = F^j((G_1 \circ \cdots \circ G_s)(\overline{m}))$ in M for some $\langle h_1, F_1\rangle, \ldots, \langle h_r, F_r\rangle$, $\langle k_1, G_1\rangle, \ldots, \langle k_s, G_s\rangle \in \mathcal{M}$ and $\overline{m} \in Dom(F_r) = Dom(G_s) = M^{i_1}$. Let $m = (F_1 \circ \cdots \circ F_r)(\overline{m}) \in M^i$, $m' = (G_1 \circ \cdots \circ G_s)(\overline{m}) \in M^j$ and $(c_i)_{i\in I} \in C_k$. By definition of **Int** we have that, letting $c = (h_r \circ \cdots \circ h_1)(c_i)$ and $c' = (k_s \circ \cdots \circ k_1)(c_j)$ then $\nu_k^m(c_i) = \nu_k^{\overline{m}}(c)$ and $\nu_k^{m'}(c_j) = \nu_k^{\overline{m}}(c')$. Since $(c_i)_{i\in I} \in C_k$ then $c = c_{i_1} = c'$, therefore $\nu_k^m(c_i) = \nu_k^{m'}(c_j)$ and thus A is well-defined. Clearly $\overline{A}$ is well-defined, and it is injective. Moreover, $\langle C, M/\sim, \overline{A}\rangle$ is closed under isomorphic images, disjoint unions and subalgebras, therefore it is an interpretation system. If $(c_i)_{i\in I} \in C_k$ and $m \in M^i$ then $\nu_k^{F^i(m)}((c_i)_{i\in I}) = \nu_k^m(c_i) = \nu_k^m(h^i((c_i)_{i\in I}))$, therefore $\langle h^i, \overline{F^i}\rangle : \langle C, M/\sim, \overline{A}\rangle \longrightarrow \langle C^i, M^i, A^i\rangle$ is a morphism in **Int** such that

$$\langle C, M/\sim, \overline{A}\rangle \xrightarrow{\langle h^i, \overline{F^i}\rangle} \langle C^i, M^i, A^i\rangle$$
$$\searrow_{\langle h^j, \overline{F^j}\rangle} \qquad \downarrow^{\langle h, F\rangle}$$
$$\langle C^j, M^j, A^j\rangle$$

commutes in **Int** for each $\langle h, F\rangle : \langle C^i, M^i, A^i\rangle \longrightarrow \langle C^j, M^j, A^j\rangle$ in $\mathcal{M}$. Now consider a family of **Int**-morphisms $\langle g^i, G^i\rangle : \langle C', M', A'\rangle \longrightarrow \langle C^i, M^i, A^i\rangle$ for $i \in I$ such that $\langle h, F\rangle \circ \langle g^i, G^i\rangle = \langle g^j, G^j\rangle$ for each $\langle h, F\rangle :$

$\langle C^i, M^i, A^i\rangle \longrightarrow \langle C^j, M^j, A^j\rangle$ in $\mathcal{M}$. Since $\langle C, \{h^i\}_{i\in I}\rangle$ is the limit in **Sig** of $\langle\{C^i\}_{i\in I}, \mathcal{M}_1\rangle$ there exists in **Sig** an unique morphism $\overline{h} : C' \longrightarrow C$ given by $\overline{h}_k(c) = (g^i_k(c))_{i\in I}$, such that $h^i \circ \overline{h} = g^i$ for all $i \in I$. Since $\langle M, \{F^i\}_{i\in I}\rangle$ is the colimit in **Cls** of $\langle\{M^i\}_{i\in I}, \mathcal{M}_2\rangle$ there exists in **Cls** an unique map $G : M \longrightarrow M'$ given by $G(F^i(m)) = G^i(m)$, such that $G \circ F^i = G^i$ for all $i \in I$. Suppose now that $A(F^i(m)) = A(F^j(m')) = \langle U, \mathcal{B}, \nu\rangle$. Then

$$(*) \qquad \nu_k^m(c_i) = \nu_k((c_i)_{i\in I}) = \nu_k^{m'}(c_j) \qquad \text{for all } k \in \mathbb{N} \text{ and } (c_i)_{i\in I} \in C_k.$$

By definition of morphism in **Int** we have that $A'(G^i(m)) = \langle U, \mathcal{B}, \nu^1\rangle$ and $A'(G^j(m')) = \langle U, \mathcal{B}, \nu^2\rangle$ such that $\nu^1_k(c) = \nu^m_k(g^i_k(c))$ and $\nu^2_k(c) = \nu^{m'}_k(g^j_k(c))$. But $\overline{h}_k(c) = (g^i_k(c))_{i\in I} \in C_k$ thus, using $(*)$ above, $\nu^1_k(c) = \nu^2_k(c)$ for all $c \in C'_k$, that is, $A'(G^i(m)) = A'(G^j(m'))$. Since A' is injective we infer that $G^i(m) = G^j(m')$, therefore the function $\overline{G} : M/_\sim \longrightarrow M'$ given by $G(F^i(m)/_\sim) = G^i(m)$ is well-defined, and then $\langle \overline{h}, \overline{G}\rangle : \langle C', M', A'\rangle \longrightarrow \langle C, M/_\sim, \overline{A}\rangle$ is a morphism in **Int** which commutes in **Int** all the diagrams below.

$$\begin{array}{ccc} \langle C', M', A'\rangle & \xrightarrow{\langle \overline{h}, \overline{G}\rangle} & \langle C, M/_\sim, \overline{A}\rangle \\ & \searrow{\langle g^i, G^i\rangle} & \downarrow{\langle h^i, \overline{F^i}\rangle} \\ & & \langle C^i, M^i, A^i\rangle \end{array}$$

Clearly $\langle \overline{h}, \overline{G}\rangle$ is the unique morphism with this property, therefore the pair $\langle\langle C, M/_\sim, \overline{A}\rangle, \{\langle h^i, \overline{F^i}\rangle\}_{i\in I}\rangle$ constitutes the limit of $\mathcal{D}$. □

4.2 Colimits in Int

Now, the results obtaned in the previous subsection will be dualized. In [34, Prop/Definition 3.9] it was defined (and proved the existence of) the constrained and unconstrained fibring of two given interpretation systems $\langle C^i, M^i, A^i\rangle$ (for $i = 1, 2$). In Proposition 3.13 of that paper it was provided a categorical characterization of unconstrained fibring, namely, when no connective is shared by the two given interpretation systems: it corresponds to the coproduct of $\langle C^1, M^1, A^1\rangle$ and $\langle C^2, M^2, A^2\rangle$ in the category **Int**. If some connectives are to be shared, then the unconstrained fibring, presented in categorical terms, is obtained through the coequalizer in **Int** of a suitable pair of parallel arrows (see [34, Proposition 3.14]). Taking into consideration that coproducts and coequalizers are special cases of colimits, it is a natural question whether these constructions can be generalized to arbitrary colimits, as we already done for the category **Hil**. This should provide new ways of combining interpretation systems by a splicing or synthesis process, as observed in Subsection 3.2 for Hilbert calculi. We will prove in this section that **Int** is (small) cocomplete, showing how to construct the colimits.

Let $\mathcal{D} = \langle \{\langle C^i, M^i, A^i \rangle\}_{i \in I}, \mathcal{M} \rangle$ be a diagram in **Int** and let $\mathcal{M}_1$ and $\mathcal{M}_2$ given by

$$\mathcal{M}_1 = \{h : \langle h, F \rangle \in \mathcal{M} \text{ for some function } F\},$$

$$\mathcal{M}_2 = \{F : \langle h, F \rangle \in \mathcal{M} \text{ for some } \mathbf{Sig}\text{-morphism } h\}.$$

Consider the pair $\langle \langle C, M, A \rangle, \{\langle h^i, F^i \rangle\}_{i \in I} \rangle$ defined as follows:

- $\langle C, \{h^i\}_{i \in I} \rangle$ is the colimit in **Sig** of the diagram $\langle \{C^i\}_{i \in I}, \mathcal{M}_1 \rangle$;

- let $\langle \overline{M}, \{\overline{F^i}\}_{i \in I} \rangle$ be the limit in **Cls** of the diagram $\langle \{M^i\}_{i \in I}, \mathcal{M}_2 \rangle$; then $M = \{(m_i)_{i \in I} \in \overline{M} : U_{m_i} = U_{m_j} \text{ and } B_{m_i} = B_{m_j} \text{ for all } i, j \in I\}$, and $F^i : M \longrightarrow M^i$ is the function given by $F^i((m_i)_{i \in I}) = \overline{F^i}((m_i)_{i \in I}) = m_i$ for all $i \in I$;

- $A : M \longrightarrow Str(C)$ is the map defined as follows: let $(m_i)_{i \in I} \in M$ and $A^i(m_i) = \langle U_{m_i}, B_{m_i}, \nu^{m_i} \rangle$ for each $i \in I$; then $A((m_i)_{i \in I}) = \langle U_{m_i}, B_{m_i}, \nu^{(m_i)_{i \in I}} \rangle$ where

$$\nu_k^{(m_i)_{i \in I}}(h_k^i(c)) = \nu_k^{m_i}(c) \text{ for all } k \in \mathbb{N}, i \in I \text{ and } c \in C_k^i.$$

Proposition 4.8. Let $\mathcal{D}$ be a diagram. Then $\langle \langle C, M, A \rangle, \{\langle h^i, F^i \rangle\}_{i \in I} \rangle$ as defined above is the colimit of $\mathcal{D}$ in **Int**.

Proof. First we need to prove that A is well-defined and it is injective. Suppose that $h^i((f^1 \circ \cdots \circ f^r)(\overline{c})) = h^j((g^1 \circ \cdots \circ g^s)(\overline{c}))$ in C_k where $\overline{c} \in C_k^{i_1}$ and $Dom(f^r) = C^{i_1} = Dom(g^s)$. Let $c = (f^1 \circ \cdots \circ f^r)(\overline{c})$, $c' = (g^1 \circ \cdots \circ g^s)(\overline{c})$, $m = (F^r \circ \cdots \circ F^1)(m_i)$ and $m' = (G^s \circ \cdots \circ G^1)(m_j)$. By definition of **Int** we have that $\nu_k^{m_i}(c) = \nu_k^m(\overline{c})$, and $\nu_k^{m_j}(c') = \nu_k^{m'}(\overline{c})$. But $(m_i)_{i \in I} \in M$, thus $m = m_{i_1} = m'$ and then $\nu_k^{m_i}(c) = \nu_k^{m_j}(c')$, therefore A is well-defined. If $A((m_i)_{i \in I}) = A((m_i')_{i \in I})$ then, by definition of A,

$$A^i(m_i) = A((m_i)_{i \in I})|_{h^i} = A((m_i')_{i \in I})|_{h^i} = A^i(m_i')$$

for each $i \in I$, therefore $m_i = m'_i$ for all $i \in I$, that is, $(m_i)_{i \in I} = (m'_i)_{i \in I}$ and then A is injective. Now we will check that $\langle C, M, A \rangle$ is closed under isomorphic images, disjoint unions and subalgebras. Suppose that $A((m_i)_{i \in I}) = \langle U_{m_i}, \mathcal{B}_{m_i}, \nu^{(m_i)_{i \in I}} \rangle \cong \langle U, \mathcal{B}, \nu \rangle$, and let $f : U_{m_i} \to U$ a bijection such that $f(\mathcal{B}_{m_i}) = \mathcal{B}$ and $\nu_k^{m_i}(c)(b) = \nu_k^{(m_i)_{i \in I}}(h_k^i(c))(b) = \nu_k(h_k^i(c))(f(b))$. This means that $\langle U, \mathcal{B}, \nu \rangle|_{h^i} \cong A((m_i)_{i \in I})|_{h^i} = A^i(m_i)$ for all $i \in I$. Since $\langle C^i, M^i, A^i \rangle$ is closed under isomorphic images, there exists $m'_i \in M^i$ such that $\langle U, \mathcal{B}, \nu \rangle|_{h^i} = A^i(m'_i)$. It suffices to prove that $(m'_i)_{i \in I} \in M$. In order to do this, consider a morphism $\langle h, F \rangle : \langle C^i, M^i, A^i \rangle \to \langle C^j, M^j, A^j \rangle$ in $\mathcal{M}$. Then $\nu^{F(m'_j)}(c) = \nu^{m'_j}(h(c)) = \nu(h^j(h(c))) = \nu(h^i(c)) = \nu^{m'_i}(c)$, thus $A^i(m'_i) = A^i(F(m'_j))$, therefore $m'_i = F(m'_j)$, because A^i is injective. This shows that $(m'_i)_{i \in I} \in M$ such that $A((m'_i)_{i \in I}) = \langle U, \mathcal{B}, \nu \rangle$, then $\langle C, M, A \rangle$ is closed under isomorphic images. Consider now a subset N of M such that $U_n \cap U_{n'} = \emptyset$ for every $n \neq n'$ in N, and let $N_i = \{n \in M^i : n = n_i \text{ for some } (n_j)_{j \in I} \in N\}$ for each $i \in I$. Since $U_n \cap U_{n'} = \emptyset$ for every $n \neq n'$ in N_i there exists $m_i \in M^i$ such that $U_{m_i} = \bigcup_{n \in N_i} U_n$, $\mathcal{B}_{m_i} = \{b \in \wp(U_{m_i}) : b \cap U_n \in \mathcal{B}_n \text{ for all } n \in N_i\}$ and, for every $k \in \mathbb{N}$, $c \in C_k^i$ and $b \in \mathcal{B}_{m_i}^k$, $\nu_k^{m_i}(c)(b) = \bigcup_{n \in N_i} \nu_k^n(c)(b \cap U_n^k)$. In order to prove that $(m_i)_{i \in I} \in M$, consider a morphism $\langle h, F \rangle : \langle C^i, M^i, A^i \rangle \to \langle C^j, M^j, A^j \rangle$ in $\mathcal{M}$. If $(n_i)_{i \in I} \in N$ then, by definition of colimit and morphism in **Int**, $F(n_j) = n_i$ and, moreover, $\nu^{F(n_j)}(c) = \nu^{n_j}(h(c))$ and $U_{n_j} = U_{n_i}$. By definition of N_i and N_j it follows that $F(N_j) = N_i$. Thus, $\nu^{F(m_j)}(c)(b) = \nu^{m_j}(h(c))(b) = \bigcup_{n' \in N_j} \nu^{n'}(h(c))(b \cap U_{n'}^k) = \bigcup_{n' \in N_j} \nu^{F(n')}(c)(b \cap U_{n'}^k) = \bigcup_{n \in N_i} \nu^n(c)(b \cap U_n^k) = \nu^{m_i}(c)(b)$. Therefore $A^i(F(m_j)) = A^i(m_i)$, then $F(m_j) = m_i$ because A^i is injective. This shows that $m = (m_i)_{i \in I} \in M$ such that $U_m = \bigcup_{n \in N} U_n$, $\mathcal{B}_m = \{b \in \wp(U_m) : b \cap U_n \in \mathcal{B}_n \text{ for all } n \in N\}$ and, for every $k \in \mathbb{N}$, $h^i(c) \in C_k$ and $b \in \mathcal{B}_m^k$, $\nu_k^m(h^i(c))(b) = \bigcup_{n \in N} \nu_k^n(h^i(c))(b \cap U_n^k)$. That is, $\langle C, M, A \rangle$ is closed under disjoint unions. Finally we show that $\langle C, M, A \rangle$ is closed under subalgebras. Let $m = (m_i)_{i \in I} \in M$ and $\mathcal{B}' \subseteq \mathcal{B}_m$ a ν^m-subalgebra. Then $\mathcal{B}' \subseteq \mathcal{B}_{m_i}$ is a ν^{m_i}-subalgebra, thus there exists $m'_i \in M^i$ such that $A^i(m'_i) = \langle U_{m_i}, \mathcal{B}', \nu^{m'_i} \rangle$, where $\nu^{m'_i}(c) = \nu^{m_i}(c)|_{\mathcal{B}'}$ for all $i \in I$. We prove now that $m' = (m'_i)_{i \in I}$ is in M. Consider a morphism $\langle h, F \rangle : \langle C^i, M^i, A^i \rangle \to \langle C^j, M^j, A^j \rangle$ in $\mathcal{M}$. Then $\nu^{F(m'_j)}(c) = \nu^{m'_j}(h(c)) = \nu^{m_j}(h(c))|_{\mathcal{B}'} = \nu^m(h^j(h(c)))|_{\mathcal{B}'} = \nu^m(h^i(c))|_{\mathcal{B}'} = \nu^{m_i}(c)|_{\mathcal{B}'} = \nu^{m'_i}(c)$. From this we infer that $A^i(F(m'_j)) = A^i(m'_i)$ and so $F(m'_j) = m'_i$, since A^i is injective. Thus $m' \in M$ such that $A(m') = \langle U_m, \mathcal{B}', \nu^{m'} \rangle$ with

$\nu^{m'}(h^i(c)) = \nu^m(h^i(c))|_{\mathcal{B}'}$, that is, $\langle C, M, A\rangle$ is closed under subalgebras. This shows that $\langle C, M, A\rangle$ is an interpretation system. Clearly every $\langle h^i, F^i\rangle$ is a morphism in **Int** such that $\langle h^j, F^j\rangle \circ \langle h, F\rangle = \langle h^i, F^i\rangle$ for each $\langle h, F\rangle : \langle C^i, M^i, A^i\rangle \longrightarrow \langle C^j, M^j, A^j\rangle$ in $\mathcal{M}$. Now consider a family of **Int**-morphisms $\langle g^i, G^i\rangle : \langle C^i, M^i, A^i\rangle \longrightarrow \langle C', M', A'\rangle$ for $i \in I$ such that $\langle g^j, G^j\rangle \circ \langle h, F\rangle = \langle g^i, G^i\rangle$ for each $\langle h, F\rangle : \langle C^i, M^i, A^i\rangle \longrightarrow \langle C^j, M^j, A^j\rangle$ in $\mathcal{M}$. Since $\langle C, \{h^i\}_{i\in I}\rangle$ is the colimit in **Sig** of $\langle \{C^i\}_{i\in I}, \mathcal{M}_1\rangle$ there exists in **Sig** an unique morphism $\bar{h} : C \longrightarrow C'$ given by $\bar{h}_k(h_k^i(c)) = g_k^i(c)$, such that $\bar{h} \circ h^i = g^i$ for all $i \in I$. Since $\langle M, \{F^i\}_{i\in I}\rangle$ is the limit in **Cls** of $\langle \{M^i\}_{i\in I}, \mathcal{M}_2\rangle$ there exists in **Cls** an unique map $\overline{G} : M' \longrightarrow \overline{M}$ given by $\overline{G}(m') = (G^i(m'))_{i\in I}$, such that $\overline{F^i} \circ \overline{G} = G^i$ for all $i \in I$. Note that $U_{G^i(m')} = U_{m'}$ and $B_{G^i(m')} = B_{m'}$ for all $i \in I$ and $m' \in M'$, by definition of **Int**. Therefore, there exists a map $G : M' \longrightarrow M$ defined by $G(m') = (G^i(m'))_{i\in I}$ such that $F^i \circ G = G^i$ for all $i \in I$. It is clear from the definitions that $\langle \bar{h}, G\rangle$ is the unique morphism in **Int** such that

$$\langle C^i, M^i, A^i\rangle \xrightarrow{\langle g^i, G^i\rangle} \langle C', M', A'\rangle$$
$$\searrow_{\langle h^i, F^i\rangle} \quad \uparrow_{\langle \bar{h}, G\rangle}$$
$$\langle C, M, A\rangle$$

commutes in **Int** for each $i \in I$. This concludes the proof. $\square$

Corollary 4.9. *The category **Int** is (small) complete and cocomplete.*

4.3 Entailment in Int

The notion of (semantic) inference can be naturally defined in **Int**. As in the case of **Hil**, there are two kind of entailments in **Int**: a local entailment (corresponding to derivations) and a global entailment (corresponding to proofs). Recall from [34] the following notions. A *variable assignment* over a C-structure $S = \langle U, \mathcal{B}, \nu\rangle$ is a map $\alpha : \Xi \longrightarrow \mathcal{B}$. Given S and α, the *interpretation map* of $L(C, \Xi)$ in S with assignment α is the function $[\![\cdot]\!]_\alpha^S : L(C, \Xi) \longrightarrow \mathcal{B}$ defined inductively by

- $[\![\xi]\!]_\alpha^S = \alpha(\xi)$, if $\xi \in \Xi$;
- $[\![c]\!]_\alpha^S = \nu_0(c)$, if $c \in C_0$;
- $[\![c(\delta_1, \ldots, \delta_k)]\!]_\alpha^S = \nu_k(c)([\![\delta_1]\!]_\alpha^S, \ldots, [\![\delta_k]\!]_\alpha^S)$.

The global and local semantic entailment are respectively defined as follows:

$$\Gamma \models^p_S = \{\delta : (\forall \alpha)((\bigcap_{\gamma \in \Gamma} [\![\gamma]\!]^S_\alpha = U) \Rightarrow ([\![\delta]\!]^S_\alpha = U))\};$$

$$\Gamma \models^d_S = \{\delta : (\forall \alpha)(\bigcap_{\gamma \in \Gamma} [\![\gamma]\!]^S_\alpha \subseteq [\![\delta]\!]^S_\alpha)\}.$$

Finally, given an interpretation system $\langle C, M, A \rangle$ it is defined

$$\Gamma \models^p_{\langle C,M,A \rangle} = \bigcap_{m \in M} \Gamma \models^p_{A(m)} ;$$

$$\Gamma \models^d_{\langle C,M,A \rangle} = \bigcap_{m \in M} \Gamma \models^d_{A(m)} .$$

Observe that the notions of local and global entailment are similar to the ones usually considered for Kripke semantics in modal logics (see, for instance, [2, Definitions 1.35 and 1.37]). By its turn, these notions correspond, respectively, to the idea of degree-preserving and truth-preserving reasoning considered by [3] in the context of algebraic semantics.

5 The category of logic system presentations

We study here the combined category of **Hil** and **Int**, called **Lsp**. Its objects are endowed with both a syntactic component (via a Hilbert calculi) and a semantic component (via an interpretation system). Of course, it is a desirable property that both entailments (syntactic and semantic) coincide. From an intuitive point of view, it would be reasonable to require that logic systems should be at least sound, that is, syntactic derivations should be valid semantically. As we shall see along this section, we can restrict to sound logic systems since this feature is preserved by arbitrary limits and colimits. On the other hand, in Section 6 it will be shown that, in order to preserve completeness by colimits in **Lsp**, the involved logic systems must be necessarily sound (see Remark 6.2). Then, for the purposes of the present study, the relevant logic systems are the sound ones.

Definition 5.1. The category **Lsp** of *logic system presentations* (in short, l.s.p.'s) is defined as follows:

- Objects: Tuples $\langle C, M, A, P, D\rangle$ where $\langle C, P, D\rangle$ is a Hibert calculus and $\langle C, M, A\rangle$ is an interpretation system.
- Morphisms: A morphism $\lambda : \langle C, M, A, P, D\rangle \longrightarrow \langle C', M', A', P', D'\rangle$ in **Lsp** is a pair $\lambda = \langle h, F\rangle$ such that $h : \langle C, P, D\rangle \longrightarrow \langle C', P', D'\rangle$ is a morphism in **Hil** and $\langle h, F\rangle : \langle C, M, A\rangle \longrightarrow \langle C', M', A'\rangle$ is a morphism in **Int**.
- Composition and identity arrows: Inherited from **Int**.

Definition 5.2. The forgetful functor $\mathsf{F} : \mathbf{Lsp} \longrightarrow \mathbf{Sig}$ is given by

- $\mathsf{F}(\langle C, M, A, P, D\rangle) = C$;
- $\mathsf{F}(\langle h, F\rangle) = h$.

Proposition 5.3. The forgetful functor F has a left adjoint $\overline{\mathsf{F}} : \mathbf{Lsp} \longrightarrow \mathbf{Sig}$.

Proof. For each signature C consider the l.s.p.

$$\overline{\mathsf{F}}(C) = \langle C, Str(C), id_{Str(C)}, \emptyset, \emptyset\rangle.$$

Using Proposition 4.6 it is easy to prove that $\langle id_C, \overline{\mathsf{F}}(C)\rangle$ is F-universal for C. In fact, if $\langle C', M', A', P', D'\rangle$ is an logic system presentation and $h : C \longrightarrow C'$ is a morphism in **Sig** then, by Proposition 4.6, there exists an unique morphism $h^* : \overline{\mathsf{O}}(C) \longrightarrow \langle C', M', A'\rangle$ in **Int** such that the diagram below commutes in **Sig**.

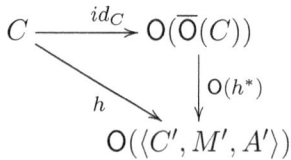

By definition of **Hil** we have that $h^* : \overline{\mathsf{F}}(C) \longrightarrow \langle C', M', A', P', D'\rangle$ is a morphism in **Lsp** which commutes the following diagram in **Lsp**.

$$\begin{array}{ccc} C & \xrightarrow{id_C} & \mathsf{F}(\overline{\mathsf{F}}(C)) \\ & \searrow^{h} & \downarrow^{\mathsf{F}(h^*)} \\ & & \mathsf{F}(\langle C', M', A', P', D'\rangle) \end{array}$$

□

This means that, if $\langle C^i, M^i, A^i, P^i, D^i \rangle$ is a logic system presentation and $f_i : C \longrightarrow C^i$ is a monic in **Sig** (for $i = 1, 2$) then the pushout in **Lsp** of the diagram

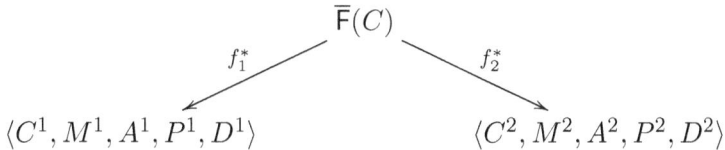

is the fibring of $\langle C^1, M^1, A^1, P^1, D^1 \rangle$ and $\langle C^2, M^2, A^2, P^2, D^2 \rangle$ constrained by $f_i : C \longrightarrow C^i$, for $i = 1, 2$ (cf. [6] and [34]). Thus, unconstrained fibrings are coproducts, and constrained fibrings are pushouts. Now we will extend the notion of fibring in **Lsp** to arbitrary (small) limits and colimits.

Proposition 5.4. The category **Lsp** is (small) complete and cocomplete.

Proof. By definition of **Lsp**, limits in **Lsp** are limits in both **Hil** and **Int**, and colimits in **Lsp** are colimits in both **Hil** and **Int**. The result follows from Corollaries 3.13 and 4.9. □

Let $\mathfrak{L} = \langle C, M, A, P, D \rangle$ be a l.s.p.. In accordance with [34], and using the notation from Subsection 4.3, we define $\Gamma \vDash^p_{\mathfrak{L}} = \Gamma \vDash^p_{\langle C, M, A \rangle}$ and $\Gamma \vDash^d_{\mathfrak{L}} = \Gamma \vDash^d_{\langle C, M, A \rangle}$, as well as the following.

Definition 5.5. Let $\mathfrak{L} = \langle C, M, A, P, D \rangle$ be a l.s.p.. We say that $\mathfrak{L}$ is

- *p-sound* if, for all $\Gamma \subseteq L(C, \Xi)$, $\Gamma \vdash^p_{\mathfrak{L}} \subseteq \Gamma \vDash^p_{\mathfrak{L}}$.
- *p-complete* if, for all $\Gamma \subseteq L(C, \Xi)$, $\Gamma \vDash^p_{\mathfrak{L}} \subseteq \Gamma \vdash^p_{\mathfrak{L}}$.
- *d-sound* if, for all $\Gamma \subseteq L(C, \Xi)$, $\Gamma \vdash^d_{\mathfrak{L}} \subseteq \Gamma \vDash^d_{\mathfrak{L}}$.
- *d-complete* if, for all $\Gamma \subseteq L(C, \Xi)$, $\Gamma \vDash^d_{\mathfrak{L}} \subseteq \Gamma \vdash^d_{\mathfrak{L}}$,

where $\Gamma \vdash^q_{\mathfrak{L}} = \Gamma \vdash^q_{\langle C, P, D \rangle} = \{\delta : \Gamma \vdash^q_{\langle C, P, D \rangle} \delta\}$ for $q \in \{p, d\}$.

Definition 5.6 (Modal logic system presentations). Recall the modal Hilbert calculi K and KD defined over the modal signature C in Example 3.3, as well as the notion of modal interpretation systems introduced in Remark 4.4. A l.s.p. $\mathfrak{L} = \langle C, M, A, P, D \rangle$ over the modal signature C

is said to be a *modal logic system presentation* if $\langle C, M, A \rangle$ is a modal interpretation system and $\langle C, P, D \rangle$ is a Hilbert calculus such that $P_K \subseteq P$ and $D = D_K$.

Example 5.7 (Modal logics K and KD as modal l.s.p.'s). From the previous definitions, it is easy to consider modal l.s.p.'s $\mathcal{L}_K$ and $\mathcal{L}_{KD}$ for modal logics K and KD, respectively. Indeed, consider $\mathcal{L}_K = \langle C, M_K, A_K, P_K, D_K \rangle$ such that $\langle C, M_K, A_K \rangle$ is the modal interpretation system generated by the class of Kripke C-structures, as described in Remark 4.4, and $\langle C, P_K, D_K \rangle$ is the modal Hilbert calculus H_K for K introduced in Example 3.3. On the other hand, let $\mathcal{L}_{KD} = \langle C, M_{KD}, A_{KD}, P_{KD}, D_{KD} \rangle$ such that $\langle C, M_{KD}, A_{KD} \rangle$ is the modal interpretation system generated by the class of Kripke C-structures for KD (that is, where the accessibility relation is serial, recall Example 4.2), and $\langle C, P_{KD}, D_{KD} \rangle$ is the modal Hilbert calculus H_{KD} for KD introduced in Example 3.3. It is easy to prove, by adapting well-known results of modal logic, that $\mathcal{L}_K$ and $\mathcal{L}_{KD}$ are both q-sound and q-complete for $q \in \{p, d\}$.

The preservation of soundness by limits and colimits in **Lsp** is immediate.

Proposition 5.8. *Soundness is preserved by limits in **Lsp**.*

Proof. With same notation as in Propositions 3.7 and 4.7, let $\mathcal{D}$ be a diagram in **Lsp** and let $\mathcal{L}$ be its limit with morphisms $\langle h^i, \overline{F^i} \rangle$. Since $A(F^i(m)) = A^i(m)|_{h^i}$ then $[\![\widehat{h^i}(\delta)]\!]_\alpha^{A^i(m)} = [\![\delta]\!]_\alpha^{A(F^i(m))}$ for all $\delta \in L(C, \Xi)$ and $\alpha : \Xi \longrightarrow \mathcal{B}_m$, and the result follows. $\square$

Proposition 5.9. *Soundness is preserved by colimits in **Lsp**.*

Proof. With same notation as in Propositions 3.12 and 4.8, let $\mathcal{D}$ be a diagram in **Lsp** and let $\mathcal{L}$ be its colimit with morphisms $\langle h^i, F^i \rangle$. Since $A^i(m_i) = A((m_i)_{i \in I})|_{h^i}$ then $[\![\delta]\!]_\alpha^{A^i(m_i)} = [\![\widehat{h^i}(\delta)]\!]_\alpha^{A((m_i)_{i \in I})}$ for every $\delta \in L(C_i, \Xi)$ and $\alpha : \Xi \longrightarrow \mathcal{B}_{m_i}$, and the result follows. $\square$

The latter results suggest considering the full subcategory **SLsp** of **Lsp** whose objects are the l.s.p.'s which are both p-sound and d-sound. Conceptually, sound l.s.p.'s are more interesting structures than arbitrary l.s.p.'s. Moreover, the category **SLsp** is well-behaved with respect to categorial combinations, as the following result (an immediate consequence of the previous ones) shows:

Corollary 5.10. The category **SLsp** is (small) complete and cocomplete.

As we shall see in the next section, sound l.s.p.'s will be also relevant in order to preserve completeness by colimits in **Lsp** (see Remark 6.2).

Remark 5.11. The modal l.s.p.'s $\mathfrak{L}_K$ and $\mathfrak{L}_{KD}$ for K and KD introduced in Example 5.7 are both p-sound and d-sound. Hence, they belong to the category **SLsp**.

Example 5.12 (The bimodal alethic-deontic logic $K \oplus KD$). Consider the modal signature C', which is obtained from the modal signature C (recall Example 3.3) by replacing $\Box$ by the modal operator O (for "obligatory"). Let $H'_{KD} = \langle C', P'_{KD}, D'_{KD} \rangle$ be the version of the Hilbert calculus H_{KD} for KD, now presented in the signature C' (observe that $D'_{KD} = D_{KD}$). Analogously, given a Kripke model $m = \langle W, R, V \rangle$ for KD, let $S'_m = \langle W, \wp(W), \nu' \rangle$ be the Kripke C'-structure for KD induced by m as in Example 4.2. Finally, let $\mathfrak{L}'_{KD}$ be the version of the l.s.p. $\mathfrak{L}_{KD}$ over C'. That is, $\mathfrak{L}'_{KD} = \langle C', M'_{KD}, A'_{KD}, P'_{KD}, D'_{KD} \rangle$ such that $\langle C', M'_{KD}, A'_{KD} \rangle$ is the modal interpretation system generated by the class of Kripke C'-structures for KD. Let C'' be the classical signature underlying C and C', i.e., $C'''_0 = VAR$, $C'''_1 = \{\neg\}$, $C'''_2 = \{\Rightarrow\}$ and $C'''_n = \emptyset$ for $n > 2$. By the results presented in [34] (see Corollary 4.1 and the Examples below it, on page 428), the constrained fibring of $\mathfrak{L}_K$ and $\mathfrak{L}'_{KD}$ by sharing C'' is the l.s.p. $\mathfrak{L} = \mathfrak{L}_K \overset{C''}{\oplus} \mathfrak{L}'_{KD} = \langle \bar{C}, M, A, P, D \rangle$ such that:

- $\bar{C}$ is the bimodal signature given by $\bar{C}_0 = VAR$, $\bar{C}_1 = \{\neg, \Box, O\}$, $\bar{C}_2 = \{\Rightarrow\}$ and $\bar{C}_n = \emptyset$ for $n > 2$;
- $P = P_K \cup P'_{KD}$ and $D = D_K = D_{KD} = D'_{KD}$;
- $m \in M$ iff m is a Kripke model for the bimodal alethic-deontic logic $K \oplus KD$; that is, $m = \langle W, R, R', V \rangle$ such that $m' = \langle W, R, V \rangle$ and $m'' = \langle W, R', V \rangle$ are Kripke models for K and KD, respectively (equivalently, $m = \langle m', m'' \rangle$ where m' and m'' are as above);
- $A : M \longrightarrow Str(\bar{C})$ is the map defined as follows: $A(m) = S_m = \langle W, \wp(W), \nu \rangle$ is the Kripke $\bar{C}$-structure induced by m as in Example 4.2; in particular,
  $\nu_1(\Box)(b) = \{w \in W : wRw' \text{ implies that } w' \in b, \text{ for every } w' \in W\}$

and $\nu_1(O)(b) = \{w \in W : wR'w'$ implies that $w' \in b$, for every $w' \in W\}$, for every $b \subseteq W$ (equivalently, $A(m) = S_m$ such that the respective reducts of S_m to C and C' are $S_{m'}$ and $S_{m''}$).

By using standard arguments of modal logic, it is easy to see that $\mathfrak{L}$ is sound and complete. Observe that the bimodal alethic-deontic logic $K \oplus KD$, formally represented by the l.s.p. $\mathfrak{L}$, is the least modal logic capable to express the "ought-implies-can" thesis $Op \to \Diamond p$ as well as the "cannot-implies-has no duty to" thesis $\neg \Diamond p \to \neg Op$ briefly discussed in Section 1. Here, $\Diamond p$ denotes, as usual, the formula $\neg \Box \neg p$. It can be proven that none of these two bridge principles (which are equivalent in $K \oplus KD$, since the underlying propositional logic is classical) are valid in $\mathfrak{L}$. This is coherent with the fact that the fibring $\mathfrak{L}$ is the least l.s.p. which extends both components $\mathfrak{L}_K$ and $\mathfrak{L}'_{KD}$ while sharing the classical connectives.

6 Completeness preservation by colimits in Lsp

Finally, in this section the results on completeness preservation by fibrings in the category **Lsp** obtained in [34] will be generalized to colimits. Thus, it will be proven that completeness is preserved by arbitrary colimits in **Lsp**, under the same assumptions for the given logic systems stated in [34] in order to prove completeness preservation by fibring. From the results obtained in [34], we know that the following conditions are sufficient to preserve completeness by fibring: (1) the l.s.p.'s are full, with implication $\Rightarrow$ and with equivalence $\Leftrightarrow$; and (2) both connectives are shared. We recall in Definitions 6.1, 6.3 and 6.4 the basic concepts of [34] concerning this question.

Definition 6.1. A l.s.p. $\mathfrak{L} = \langle C, M, A, P, D \rangle$ is *full* if $\{A(m) : m \in M\}$ is the class of all the C-structures S satisfying: $\delta \in \Gamma \vDash^p_S$ for all $\langle \Gamma, \delta \rangle \in P$, and $\delta \in \Gamma \vDash^d_S$ for all $\langle \Gamma, \delta \rangle \in D$.

Remark 6.2. By [34, Proposition 4.3], if $\mathfrak{L}$ is a full l.s.p. then it is both p-sound and d-sound.

Definition 6.3. A Hilbert calculus $H = \langle C, P, D \rangle$ is *with implication* if C_2 contains a connective $\Rightarrow$ satisfying:

$$\Gamma \vdash^d_H (\delta_1 \Rightarrow \delta_2) \text{ iff } \Gamma, \delta_1 \vdash^d_H \delta_2$$

for every $\Gamma \cup \{\delta_1, \delta_2\} \subseteq L(C, \Xi)$ such that $\Gamma = \Gamma^{\vdash_H^p}$. A l.s.p. $\mathfrak{L} = \langle C, M, A, P, D \rangle$ is *with implication* if $\langle C, P, D \rangle$ is with implication.

Definition 6.4. A Hilbert calculus $H = \langle C, P, D \rangle$ with implication $\Rightarrow$ is *with equivalence* if C_2 contains a connective $\Leftrightarrow$ satisfying

- $\Gamma \vdash_H^d (\delta_1 \Rightarrow \delta_2)$ and $\Gamma \vdash_H^d (\delta_2 \Rightarrow \delta_1)$ iff $\Gamma \vdash_H^d (\delta_1 \Leftrightarrow \delta_2)$;
- $\Gamma \vdash_H^d (\delta_1 \Leftrightarrow \delta_2)$ iff $\Gamma \vdash_H^d (\gamma \Leftrightarrow \gamma')$, where γ' is obtained from γ by replacing some ocurrences of δ_1 by δ_2

for every $\Gamma \cup \{\delta_1, \delta_2\} \subseteq L(C, \Xi)$ such that $\Gamma = \Gamma^{\vdash_H^p}$. A l.s.p. $\mathfrak{L} = \langle C, M, A, P, D \rangle$ is *with equivalence* if $\langle C, P, D \rangle$ is with equivalence.

Using the left adjoint to F (cf. Proposition 5.3), it is easy to share a signature C through a diagram, that is, without imposing any properties to the connectives in C (as, for example, being an implication or being an equivalence). We will prove now that the completeness preservation theorem stated in [34] can be generalized to any colimit, by extending any diagram by the free sharing of both implication and equivalence. Indeed, by Theorems 6.6 and 5.7 in [34] we know that every full l.s.p. with equivalence is q-complete ($q = p, d$). The first step for obtaining a general theorem of completeness preservation is to introduce the notion of sharing of connectives through a diagram in **Lsp**.

Definition 6.5. Let $\mathcal{D} = \langle \{\mathfrak{L}_i\}_{i \in I}, \mathcal{M} \rangle$ be a diagram in **Lsp**. A *free sharing* for $\mathcal{D}$ is a family $\mathcal{F} = \{f_i : C \longrightarrow \mathsf{F}(\mathfrak{L}_i)\}_{i \in I}$ of monics in **Sig**. The diagram $\mathcal{D}$ restricted to $\mathcal{F}$ is $\mathcal{D}_\mathcal{F} = \langle \{\mathfrak{L}_i\}_{i \in I}, \mathcal{M} \cup \{f_i^* : \overline{\mathsf{F}}(C) \longrightarrow \mathfrak{L}_i\}_{i \in I} \rangle$.

Note that we cannot assume any logical properties for the connectives in C because $\overline{\mathsf{F}}(C)$ is free. For example, $\overline{\mathsf{F}}(C)$ is without implication and equivalence. On the other hand, since $\overline{\mathsf{F}}(C)$ has no rules, the colimit of any non-empty diagram of l.s.p.'s with implication and equivalence restricted to the sharing of these connectives, is with implication and equivalence, as the following proposition states:

Proposition 6.6. *Let $\mathcal{D} = \langle \{\mathfrak{L}_i\}_{i \in I}, \mathcal{M} \rangle$ be a non-empty diagram in **Lsp** such that every $\mathfrak{L}_i$ is with implication $\Rightarrow^i$ and with equivalence $\Leftrightarrow^i$. Consider the signature C' defined as follows: $C_2' = \{\Rightarrow', \Leftrightarrow'\}$, and $C_k' = \emptyset$ if $k \neq 2$. For every $i \in I$ let $f_i : C' \longrightarrow \mathsf{F}(\mathfrak{L}_i)$ be the monomorphism in **Sig** given by $f_i(\Rightarrow') = \Rightarrow^i$ and $f_i(\Leftrightarrow') = \Leftrightarrow^i$. Let $\mathcal{F} = \{f_i : C' \longrightarrow \mathsf{F}(\mathfrak{L}_i)\}_{i \in I}$ and $\mathcal{D}_\mathcal{F}$ as in Definition 6.5. Then, the colimit $\mathfrak{L}$ of $\mathcal{D}_\mathcal{F}$ is a l.s.p. with implication and equivalence.*

Proof. Let $\mathfrak{L} = \langle C, M, A, P, D \rangle$ be the colimit of $\mathcal{D}_\mathcal{F}$, with morphisms $h^i : \mathfrak{L}_i \longrightarrow \mathfrak{L}$ (for $i \in I$) and $h' : \overline{\mathsf{F}}(C') \longrightarrow \mathfrak{L}$. Let $\Rightarrow = h^i(\Rightarrow^i) = h'(\Rightarrow')$ and $\Leftrightarrow = h^i(\Leftrightarrow^i) = h'(\Leftrightarrow')$ (for $i \in I$). By Proposition 6.2 in [34] we have that $\Rightarrow$ is an implication in $\mathfrak{L}$ iff it holds:

1. $\{(\xi_1 \Rightarrow \xi_2)\}^{\vdash_p^\mathfrak{L}}, \xi_1 \vdash_d^\mathfrak{L} \xi_2$;

2. $\vdash_d^\mathfrak{L} (\xi_1 \Rightarrow \xi_1)$;

3. $\{\xi_1\}^{\vdash_p^\mathfrak{L}} \vdash_d^\mathfrak{L} (\xi_2 \Rightarrow \xi_1)$;

4. if $r = \langle \{\gamma_1, \ldots, \gamma_k\}, \gamma \rangle$ is in D and $\xi \in \Xi$ does not occur in r then

$$\{(\xi \Rightarrow \gamma_1), \ldots, (\xi \Rightarrow \gamma_k)\}^{\vdash_p^\mathfrak{L}} \vdash_d^\mathfrak{L} (\xi \Rightarrow \gamma).$$

Let $i \in I$. Since $\mathfrak{L}_i$ is with implication then $\{(\xi_1 \Rightarrow^i \xi_2)\}^{\vdash_p^{\mathfrak{L}_i}}, \xi_1 \vdash_d^{\mathfrak{L}_i} \xi_2$, therefore $\widehat{h}^i(\{(\xi_1 \Rightarrow^i \xi_2)\}^{\vdash_p^{\mathfrak{L}_i}}), \xi_1 \vdash_d^\mathfrak{L} \xi_2$. But clearly $\widehat{h}^i(\{(\xi_1 \Rightarrow^i \xi_2)\}^{\vdash_p^{\mathfrak{L}_i}}) \subseteq \{(\xi_1 \Rightarrow \xi_2)\}^{\vdash_p^\mathfrak{L}}$, then $\mathfrak{L}$ satisfies item 1 above. Items 2 and 3 are proved analogously. Finally, observe that any $r \in D$ is of the form $\langle \{\widehat{h}^i(\gamma_1), \ldots, \widehat{h}^i(\gamma_k)\}, \widehat{h}^i(\gamma) \rangle$ for some $i \in I$ and some $r_i = \langle \{\gamma_1, \ldots, \gamma_k\}, \gamma \rangle \in D_i$. In fact $\overline{\mathsf{F}}(C')$ has no inference rules, therefore the derivation (and proof) rules of $\mathfrak{L}$ are just those inherited from the l.s.p.'s $\mathfrak{L}_i$. Let $r = \langle \{\widehat{h}^i(\gamma_1), \ldots, \widehat{h}^i(\gamma_k)\}, \widehat{h}^i(\gamma) \rangle \in D$ and $\xi \in \Xi$ not occurring in r. Since $\mathfrak{L}_i$ is with implication then $\{(\xi \Rightarrow^i \gamma_1), \ldots, (\xi \Rightarrow^i \gamma_k)\}^{\vdash_p^{\mathfrak{L}_i}} \vdash_d^{\mathfrak{L}_i} (\xi \Rightarrow^i \gamma)$, thus $\widehat{h}^i(\{(\xi \Rightarrow^i \gamma_1), \ldots, (\xi \Rightarrow^i \gamma_k)\}^{\vdash_p^{\mathfrak{L}_i}}) \vdash_d^\mathfrak{L} (\xi \Rightarrow \widehat{h}^i(\gamma))$. But $\widehat{h}^i(\{(\xi \Rightarrow^i \gamma_1), \ldots, (\xi \Rightarrow^i \gamma_k)\}^{\vdash_p^{\mathfrak{L}_i}}) \subseteq \{(\xi \Rightarrow \widehat{h}^i(\gamma_1)), \ldots, (\xi \Rightarrow \widehat{h}^i(\gamma_k))\}^{\vdash_p^\mathfrak{L}}$, therefore

$$\{(\xi \Rightarrow \widehat{h}^i(\gamma_1)), \ldots, (\xi \Rightarrow \widehat{h}^i(\gamma_k))\}^{\vdash_p^\mathfrak{L}} \vdash_d^\mathfrak{L} (\xi \Rightarrow \widehat{h}^i(\gamma))$$

and then $\mathfrak{L}$ satisfies item 4. With respect to equivalence, by Proposition 6.5 in [34] we have that $\Leftrightarrow$ is an equivalence in $\mathfrak{L}$ iff it holds:

1. $\{(\xi_1 \Rightarrow \xi_2), (\xi_2 \Rightarrow \xi_1)\}^{\vdash_p^\mathfrak{L}} \vdash_d^\mathfrak{L} (\xi_1 \Leftrightarrow \xi_2)$;

2. $\{(\xi_1 \Leftrightarrow \xi_2)\}^{\vdash_p^\mathfrak{L}} \vdash_d^\mathfrak{L} (\xi_1 \Rightarrow \xi_2)$;

3. $\{(\xi_1 \Leftrightarrow \xi_2)\}^{\vdash_p^\mathfrak{L}} \vdash_d^\mathfrak{L} (\xi_2 \Rightarrow \xi_1)$;

4. $\{(\xi_1 \Leftrightarrow \xi_1'), \ldots, (\xi_k \Leftrightarrow \xi_k')\}^{\vdash_p^\mathfrak{L}} \vdash_d^\mathfrak{L} (c(\xi_1, \ldots, \xi_k) \Leftrightarrow c(\xi_1', \ldots, \xi_k'))$ for any $c \in C_k$ (where $k > 0$).

Since any $c \in C_k$ is of the form $h^i(c_i)$ for some $i \in I$ and $c_i \in C_k^i$, and since every $\mathfrak{L}_i$ is with equivalence, item 4 is proved as above. The proof of the other items is also as above, therefore $\mathfrak{L}$ is with equivalence, concluding the proof. □

Finally, we state the following results, omitting the proofs (which are easy).

Proposition 6.7. The l.s.p. $\overline{\mathsf{F}}(C) = \langle C, Str(C), id_{Str(C)}, \emptyset, \emptyset \rangle$ is full, for any signature C.

Proposition 6.8. Fullness is preserved by colimits in **Lsp**.

As an immediate consequence of the results above we obtain a generalization of the Theorem 6.7 of completeness preservation of fibring stated in [34].

Theorem 6.9. Let $\mathcal{D} = \langle \{\mathfrak{L}_i\}_{i \in I}, \mathcal{M} \rangle$ be a non-empty diagram in **Lsp** such that every $\mathfrak{L}_i$ is with implication $\Rightarrow^i$ and with equivalence $\Leftrightarrow^i$, and let $\mathcal{D}_{\overline{\mathsf{F}}}$ be as in Proposition 6.6. Assume additionally that every $\mathfrak{L}_i$ is full. Then the colimit of $\mathcal{D}_{\overline{\mathsf{F}}}$ is p-complete and d-complete.

Proof. Let $\mathfrak{L}$ be the colimit of $\mathcal{D}_{\overline{\mathsf{F}}}$. By Proposition 6.6 we obtain that $\mathfrak{L}$ is with equivalence. By Propositions 6.7 and 6.8 we get that $\mathfrak{L}$ is full. Therefore, by Theorems 6.6 and 5.7 in [34] we obtain that $\mathfrak{L}$ is p-complete and d-complete. □

In particular, if both implication and equivalence are already shared in $\mathcal{D}$, is no longer necessary to extend the diagram $\mathcal{D}$ to $\mathcal{D}_{\overline{\mathsf{F}}}$, and we obtain immediately the following completeness preservation result.

Theorem 6.10. Let $\mathcal{D} = \langle \{\mathfrak{L}_i\}_{i \in I}, \mathcal{M} \rangle$ be a diagram in **Lsp** such that every $\mathfrak{L}_i$ is with implication $\Rightarrow^i$ and with equivalence $\Leftrightarrow^i$ which are shared through $\mathcal{D}$, that is, in the colimit of $\mathcal{D}$ it holds: $h^i(\Rightarrow^i) = h^j(\Rightarrow^j)$ and $h^i(\Leftrightarrow^i) = h^j(\Leftrightarrow^j)$ for all $i, j \in I$. Assume additionally that every $\mathfrak{L}_i$ is full. Then the colimit of $\mathcal{D}$ is p-complete and d-complete.

Proof. Clearly, in this case the colimit $\mathfrak{L}$ of $\mathcal{D}$ is the colimit of $\mathcal{D}_{\overline{\mathsf{F}}}$. The result follows from Theorem 6.9. □

The preservation of completeness by limits in **Lsp** is easier to state and it holds in general, without any assumptions on the l.s.p.'s involved.

Theorem 6.11. Let $\mathcal{D} = \langle \{\mathfrak{L}_i\}_{i \in I}, \mathcal{M} \rangle$ be a diagram in **Lsp** such that every $\mathfrak{L}_i$ is p-complete (d-complete, resp.). Then the limit of $\mathcal{D}$ is p-complete (d-complete, resp.).

Proof. Let $\mathfrak{L} = \langle C, M/\!\sim, \overline{A}, P, D \rangle$ be the limit of $\mathcal{D}$ with morphisms $\langle h^i, \overline{F^i} \rangle$, obtained from F^i and A as in Proposition 4.7. Since $[\![\widehat{h}^i(\delta)]\!]_\alpha^{A^i(m)} = [\![\delta]\!]_\alpha^{A(F^i(m))}$ for all $\delta \in L(C, \Xi)$ and $\alpha : \Xi \longrightarrow \mathcal{B}_m$, we obtain the following:

$\Gamma \models^p_{A(F^i(m))} \delta$ iff $(\forall \alpha)[(\bigcap_{\gamma \in \Gamma} [\![\gamma]\!]_\alpha^{A(F^i(m))} = U_m) \Rightarrow ([\![\delta]\!]_\alpha^{A(F^i(m))} = U_m)]$

iff $(\forall \alpha)[(\bigcap_{\gamma \in \Gamma} [\![\widehat{h}^i(\gamma)]\!]_\alpha^{A^i(m)} = U_m) \Rightarrow ([\![\widehat{h}^i(\delta)]\!]_\alpha^{A^i(m)} = U_m)]$

iff $\widehat{h}^i(\Gamma) \models^p_{A^i(m)} \widehat{h}^i(\delta)$

for all $\Gamma \cup \{\delta\} \subseteq L(C, \Xi)$, $i \in I$ and $m \in M^i$. In other words, $\Gamma^{\models^p_{A(F^i(m))}} = (\widehat{h}^i)^{-1}(\widehat{h}^i(\Gamma)^{\models^p_{A^i(m)}})$ for all $\Gamma \subseteq L(C, \Xi)$, $i \in I$ and $m \in M^i$. Let $\Gamma \subseteq L(C, \Xi)$, and suppose that every $\mathfrak{L}_i$ is p-complete. Then

$$\Gamma^{\models^p_\mathfrak{L}} = \bigcap_{i \in I} \bigcap_{m \in M^i} \Gamma^{\models^p_{A(F^i(m))}} = \bigcap_{i \in I} \bigcap_{m \in M^i} (\widehat{h}^i)^{-1}(\widehat{h}^i(\Gamma)^{\models^p_{A^i(m)}})$$

$$= \bigcap_{i \in I} (\widehat{h}^i)^{-1}(\bigcap_{m \in M^i} \widehat{h}^i(\Gamma)^{\models^p_{A^i(m)}}) = \bigcap_{i \in I} (\widehat{h}^i)^{-1}(\widehat{h}^i(\Gamma)^{\models^p_{\mathfrak{L}_i}})$$

$$\subseteq \bigcap_{i \in I} (\widehat{h}^i)^{-1}(\widehat{h}^i(\Gamma)^{\vdash^p_{\mathfrak{L}_i}}) = \Gamma^{\vdash^p_\mathfrak{L}}$$

by Corollary 3.8. Therefore, $\mathfrak{L}$ is p-complete. The proof for d-completeness is similar. □

7 Concluding remarks

We show, through the analysis of three categories, that it is possible to generalize the basic construction of fibrings of logics — which are colimits of a simple kind — to arbitrary colimits and limits. The main result of this paper states that, under reasonable conditions, arbitrary colimits preserve completeness, thus generalizing the result for fibring presented in [34]. On the other hand, we prove that no requirements are necessary in order to preserve completeness by arbitrary limits. This suggests that it is possible to obtain complex systems of logics through specifications given by diagrams, by taking limits or colimits of such diagrams, and completeness will be preserved under reasonable assumptions.

In this paper, following the approach in [34], we have studied categorial combinations of propositional logic in which the syntactic entailment is expressed by means of Hilbert calculus, and the semantics is truth-functional, characterized by classes of algebras defined over powersets. The study of limits and colimits in other categories of logic systems such as the ones introduced in [7] and [29], with the aim of defining sophisticated forms of composing and decomposing logics of many kinds, is an issue that could be addressed in the future research. In particular, it could be interesting to analyze the generation of relevant bridge principles by means of colimits in the categories of sequent calculi introduced in [16], given that they are suitable to this end, as discussed in Section 5 of that paper (see also [1]). This issue is relevant to questions such as the "ought-implies-can" thesis mentioned in Section 1 which, as discussed in Example 5.12, has a negative answer in the context of (standard) fibring in the categories studied here.

Besides being interesting and useful from the technical and pragmatical points of view, the generalization of fibring to arbitrary limits and colimits proposed here brings us several intriguing questions from the conceptual side. For instance, by looking at the dual constructions from the perspective of category theory, unconstrained fibrings are coproducts, and their duals correspond to products of logics, a way to split a logic into factors. On the other hand, constrained fibrings are pushouts of diagrams consisting of two morphisms with the same domain in the category in which the logic systems are represented. What would be the significance, from the point of view of abstract logics and combining logics, of pullbacks of diagrams consisting of two morphisms with a common codomain in such categories? Of course this question can be formulated for limits and colimits of diagrams in general.

We consider that category theory offers a extremely useful conceptual framework to deal with combining logics. We hope that the notions and results presented in this paper can contribute to that discussion.

Acknowledgment

The author was financially supported by an individual research grant from CNPq (Brazil) number 306530/2019-8.

References

[1] J.-Y. Béziau and M.E. Coniglio. To distribute or not to distribute? *Logic Journal of the IGPL*, 19(4):566–583, 2011.

[2] P. Blackburn, M. de Rijke, and Y. Venema. *Modal Logic*. Cambridge University Press, 2002.

[3] F. Bou, F. Esteva, J.M. Font, A. Gil, L. Godo, A. Torrens, and V. Verdú. Logics preserving degrees of truth from varieties of residuated lattices. *Journal of Logic and Computation*, 19(6):1031–1069, 2009.

[4] J. Bueno-Soler, M.E. Coniglio, and W.A. Carnielli. Finite algebraizability via possible-translations semantics. In W.A. Carnielli, F.M. Dionísio, and P. Mateus, editors, *Proceedings of CombLog'04 - Workshop on Combination of Logics: Theory and Applications*, pages 79–86. Departamento de Matemática, Instituto Superior Técnico, Lisbon, 2004.

[5] J. Bueno-Soler, M.E. Coniglio, and W.A. Carnielli. Possible-translations algebraizability. In J.-Y. Béziau, W.A. Carnielli, and D.M. Gabbay, editors, *Handbook of Paraconsistency*, volume 9 of *Studies in Logic (Logic & Cognitive Systems)*, pages 321–340. College Publications, London, 2007.

[6] C. Caleiro. *Combining Logics*. PhD thesis, IST, Universidade Técnica de Lisboa, Portugal, 2000.

[7] C. Caleiro, W.A. Carnielli, M.E. Coniglio, A. Sernadas, and C. Sernadas. Fibring Non-Truth-Functional Logics: Completeness Preservation. *Journal of Logic, Language and Information*, 12(2):183–211, 2003.

[8] C. Caleiro and J. Ramos. From fibring to cryptofibring: A solution to the collapsing problem. *Logica Universalis*, 1(1):71–92, 2007.

[9] C. Caleiro, C. Sernadas, and A. Sernadas. Parameterisation of logics. In J. Fiadeiro, editor, *Recent Trends in Algebraic Development Techniques: 13th International Workshop, WADT'98, Lisbon, Portugal, April 2-4, 1998, Selected Papers*, volume 1589 of *Lecture Notes in Computer Science*, pages 48–62. Springer, Berlin, 1999.

[10] W.A. Carnielli. Many-valued logics and plausible reasoning. In G. Epstein, editor, *Proceedings of the Twentieth International Symposium on Multiple-Valued Logic, Charlotte, NC, USA*, pages 328–335. The IEEE Computer Society Press, 1990.

[11] W.A. Carnielli and M.E. Coniglio. A categorial approach to the combination of logics. *Manuscrito*, 22(2):69–94, 1999.

[12] W.A. Carnielli and M.E. Coniglio. Bridge principles and combined reasoning. In T. Müller and A. Newen, editors, *Logik, Begriffe, Prinzipien des Handelns (Logic, Concepts, Principles of Action)*, pages 32–48. Mentis Verlag, Paderborn, 2007.

[13] W.A. Carnielli and M.E. Coniglio. Combining logics. In E.N. Zalta, editor, *The Stanford Encyclopedia of Philosophy*. 2007. URL: http://plato.stanford.edu/entries/logic-combining/.

[14] W.A. Carnielli, M.E. Coniglio, D.M. Gabbay, P. Gouveia, and C. Sernadas. *Analysis and Synthesis of Logics: How to Cut and Paste Reasoning Systems*, volume 35 of *Applied Logic Series*. Springer-Verlag, 2008.

[15] W.A. Carnielli and J. Marcos. Limits for paraconsistent calculi. *Notre Dame Journal of Formal Logic*, 40(3):375–390, 1999.

[16] M.E. Coniglio. Recovering a logic from its fragments by meta-fibring. *Logica Universalis*, 1(2):1–39, 2007.

[17] M.E. Coniglio and M. Figallo. A formal framework for hypersequent calculi and their fibring. In A. Koslow and A. Buchsbaum, editors, *The Road to Universal Logic: Festschrift for 50th Birthday of Jean-Yves Béziau, Volume I*, Studies in Universal Logic Series, pages 73–93. Springer, Basel, 2015.

[18] M.E. Coniglio, A. Sernadas, and C. Sernadas. Fibring logics with topos semantics. *Journal of Logic and Computation*, 13(4):595–624, 2003.

[19] L. Cruz-Filipe, A. Sernadas, and C. Sernadas. Heterogeneous fibring of deductive systems via abstract proof systems. *Logic Journal of the IGPL*, 16(2):121–153, 2008.

[20] E.G. de Souza, A. Costa-Leite, and D.H.B. Dias. On a paraconsistentization functor in the category of consequence structures. *Journal of Applied Non-Classical Logics*, 26(3):240–250, 2016.

[21] V.L. Fernández and M.E. Coniglio. Fibring in the Leibniz hierarchy. *Logic Journal of the IGPL*, 15(5-6):475–501, 2007.

[22] M. Finger and D. Gabbay. Adding a temporal dimension to a logic system. *Journal of Logic, Language and Information*, 1(3):203–233, 1992.

[23] M. Fitting. Logics with several modal operators. *Theoria*, 35:259–266, 1969.

[24] D. Gabbay. Fibred semantics and the weaving of logics: part 1. *Journal of Symbolic Logic*, 61(4):1057–1120, 1996.

[25] S. Mac Lane. *Categories for the Working Mathematician*. Springer-Verlag, 1971.

[26] S. Mac Lane and I. Moerdjik. *Sheaves in Geometry and Logic*. Springer-Verlag, 1992.

[27] K. Segerberg. Two-dimensional modal logic. *Journal of Philosophical Logic*, 2(1):77–96, 1973.

[28] A. Sernadas, C. Sernadas, and C. Caleiro. Fibring of logics as a categorial construction. *Journal of Logic and Computation*, 9(2):149–179, 1999.

[29] A. Sernadas, C. Sernadas, J. Rasga, and M.E. Coniglio. On graph-theoretic fibring of logics. *Journal of Logic and Computation*, 19(6):1321–1357, 2009.

[30] A. Sernadas, C. Sernadas, and A. Zanardo. Fibring modal first-order logics: Completeness preservation. *Logic Journal of the IGPL*, 10(4):413–451, 2002.

[31] C. Sernadas, J. Rasga, and W.A. Carnielli. Modulated fibring and the collapsing problem. *The Journal of Symbolic Logic*, 67(4):1541–1569, 2002.

[32] R. Thomason. Combinations of tense and modality. In D.M. Gabbay and F. Guenthner, editors, *Handbook of Philosophical Logic*, volume 2, pages 135–165. D. Reidel, Dordrecht, 1984.

[33] V. Šehtman. Two-dimensional modal logics. *Akademiya Nauk SSSR. Matematicheskie Zametki*, 23(5):759–772, 1978.

[34] A. Zanardo, A. Sernadas, and C. Sernadas. Fibring: Completeness preservation. *Journal of Symbolic Logic*, 66(1):414–439, 2001.

GÖDEL'S INCOMPLETENESS THEOREMS FROM A PARACONSISTENT PERSPECTIVE

Walter Carnielli°
David Fuenmayor•

°Centre for Logic, Epistemology and the History of Science, Department of Philosophy, University of Campinas, Brazil
•Department of Mathematics and Computer Science, Freie Universität Berlin, Germany

Dedicated to Edelcio Gonçalves de Souza on the occasion of his 30th anniversary - in the vigesimal basis, just to reflect his youthful spirit.

Abstract

This paper explores the general question of the validity of Gödel's incompleteness theorems by examining the respective arguments from a paraconsistent perspective, while employing combinations of modal logics with Logics of Formal Inconsistency (**LFI**s). For this purpose, abstract versions of the incompleteness theorems, employing provability logic, need to be carefully crafted. This analysis considers distinct variants of the notion of consistency for formal systems, which, together with the lighter character of the negation operator of the **LFI**s, enable new formalization variants of the Gödelian arguments, eventually leading to some thought-provoking conclusions. We show that the standard formulation of Gödel's theorems is not valid under some weak **LFI**s: a valid reconstruction requires further premises corresponding to the consistency (in the sense of **LFI**s) of particular formulas. This readily leads us to a reformulation of Gödel's theorems as an existence claim. We counted with the assistance of the proof assistant *Isabelle/HOL* for verifying and falsifying certain hypotheses during the process of formal proof reconstruction.

1 How universal are Gödel's arguments?

In a rough and intuitive formulation, Gödel's first incompleteness theorem (G_1) says that for certain consistent formal systems there are (true) sentences that the systems cannot decide, i.e., neither prove nor disprove; in its turn, the second incompleteness theorem (G_2) says that such a system cannot prove its own consistency. A formidable amount of papers deal with explanations or interpretations of the Gödelian arguments, but few of them touch on their limits.

Gödel formulated his incompleteness theorems[1] employing notions such as axiomatic systems, primitive recursive arithmetic, arithmetization/numbering, representability/interpretability, consistency, completeness, diagonalization, etc; even a cursory literature survey reveals several different (and in some cases non-equivalent) 'formalizations', or more appropriately, *explications*, of these notions. In this paper we focus on the notions of *consistency* and *negation* from the point of view of paraconsistent logic. For this purpose, we will abstract away the complexities of Gödel's arithmetization procedure and assume the corresponding fixed-point (diagonalization) lemma as a premise. We will employ for this (a paraconsistent version of) provability logic [30, 3].

In view of the abundant literature and approaches towards Gödel's results, we are obliged to restrict ourselves to considering only a few sources. Succinct, but self-contained and fairly detailed, discussions emphasizing the most important points in Gödel's proofs can be found, for example, in the works of Smoryński [26, §1–§2], Epstein & Carnielli [18, Ch. 23–24], and in the Stanford Encyclopedia of Philosophy [23].

Section 2 raises the question of the range of Gödel's theorems, setting the stage for an analysis of the dependence of the Gödelian arguments on standard logical conventions. Section 3 introduces the paraconsistentist program, and foresees some difficulties as regards the validity of the Gödelian objection in more subtle logical scenarios. Section 4 discusses the Logics of Formal Inconsistency, and justifies the choice of the logic **RmbC** among eligible paraconsistent scenarios. Section 5 illustrates the mechanism by which we add a *provability operator* to **RmbC**, thus obtaining the logic **RmbC⊕K**. Section 6 reconstructs some proofs for paraconsistent variants of Gödel's theorems employing the logic **RmbC⊕K** (and extensions). Finally, Section 7 offers the main conclusions of this paper.

---

[1]There is some controversy in referring to Gödel's results as either 'a theorem' or 'theorems'. Saul Kripke (in private conversation with the first author) insists that we should refer jointly to both as 'Gödel's theorem', since G_2 is a corollary of G_1. We prefer to maintain the plural.

2 On the range of Gödel's theorems

Gödel's incompleteness theorems do not apply unrestrictedly to every mathematical system. G_1 does not apply, for instance, to Euclidean geometry. Tarski proved in 1948 [29] that the first-order theory of Euclidean geometry is complete and decidable,[2] in the sense that every statement in its language is either a theorem (i.e., provable) or its negation is a theorem, and that there is an algorithm to determine which is the case. Even non-Euclidean geometry falls outside the Gödelian barrier: if plane Euclidean geometry is a consistent theory, then so is plane hyperbolic geometry. Nor do Gödel's theorems apply to systems of arithmetic with the addition operation only, such as Presburger arithmetic. It is necessary to have a certain critical mass of mathematical strength to be attacked by the Gödelian objection, or, in other words: to qualify as what Gödel himself defined as 'a formal system'.

Our aim in this paper is to give some first steps towards the analysis, using non-classical logics, of the applicability of Gödel's theorems to formal systems. Very little investigation, if any, touches on the validity of Gödel's proofs in non-classical environments. For instance, [1] claims to be studying versions of Gödel's arguments under the umbrella of non-classical logics, but only concentrates on substructural logics: the authors show that the Gödelian reasoning presupposes a certain amount of contraction in the underlying logic, that is, the validity of the meta-rule 'from $\Gamma, \psi, \psi \vdash \varphi$ follows $\Gamma, \psi \vdash \varphi$'. They then exhibit a modal system without contraction that invalidates Gödel's argument. On the other hand, several authors including Kreisel, Feferman, Löb, Jeroslow, Bezboruah–Shepherdson, Pudlák, Wilkie–Paris, Adamowicz–Zdanowski, Willard, Friedman, and Visser, among others (see [1] for references) have studied abstract conditions that permit the incompleteness theorems to be derived in a way somewhat independent of logic, but without changing the underlying standard logic. The interested reader may further consult some book-length discussions on Gödel's results, e.g., [25] and [27]; for a more philosophically-oriented discussion, a good reference is [19].

The intuition behind Gödel's proof of G_1 is basically the following: assume that the formal system **F** is consistent (otherwise it proves every sentence by the Principle of Explosion of classical logic, and thus it is

---

[2]According to [22], the results were obtained in 1930 and published privately in their full development in 1948.

trivially complete). By Gödel's *diagonal* (or *fixed-point*) lemma, one can then construct a sentence G_F (hinging on **F**) that is neither provable nor refutable in **F**, and that can also be shown to be true. Thus **F** is incomplete, both in the sense that there is a sentence that it cannot 'decide', and in the sense that there is a true sentence that it cannot prove. G_2, stating that **F** cannot prove its own consistency, follows as a corollary of G_1. Alternatively, G_2 can be derived directly from Löb's theorem.

As is well recognized, the notion of consistency employed in G_1 is not the same in the different variants that have been presented; these variants range from the so-called ω-consistency originally introduced by Gödel, through the (weaker) 1-consistency commonly used in the literature, while also including a more classical notion of consistency as in the variant introduced by Rosser in 1936.

But what is the idea of consistency behind G_2? At first sight, one can say that consistency means simply the imperative not to derive a contradiction (absence of contradiction, or universal validity of non-contradiction), which is the same as non-triviality, in view of the *Principle of Explosion (PEx)* of traditional logic. However, the *PEx* is an unnecessary burden that classical logic carries uselessly: it is not used, except to mark the ban on contradictions. Contrary to what some unsuspecting people may think, the *PEx* is not even used in *reductio ad absurdum* proofs: a little thought will convince them that what is at stake in such proofs is the rule of *negation introduction*. Indeed, any time we have a bottom particle $\bot$, the rule $\alpha \to \bot \vdash \neg \alpha$ can be applied, as it involves the *introduction of negation*, and not any use of *PEx*. In classical logic the variant $\neg \alpha \to \bot \vdash \alpha$ also holds; notice, however, that this second variant does not hold in intuitionistic logic. This means that the beloved and useful *reductio ad absurdum* method of proof acquires its legitimacy independently of *PEx*.

Paraconsistent logics, particularly the Logics of Formal Inconsistency (**LFI**), liberate logical systems from this burden by weakening the *PEx*, and for a good reason. As is widely acknowledged (see, e.g., [17]), in many situations we have no other choice but to reason from contradictory premises, and this is a critical issue, since large knowledge bases or complex arguments almost inevitably include contradictions. A consequence of weakening the *PEx* is that consistency does not coincide with non-contradiction anymore, nor does it coincide with non-triviality. This is already a major philosophical difficulty for the classical stance on G_2, which also affects, to a lesser extent, G_1. As pointed out in the literature, e.g., in [1], we cannot easily pinpoint a class of formulas that expresses consistency. When we paraphrase G_2 as saying that 'a sufficiently strong consistent theory cannot prove its own consistency' we are forced to remain vague, and more so in the domain of paraconsistent logics.

A natural question thus emerges: is there a way to avoid Gödel's conclusions by changing the underlying logic? Admittedly, it is not very encouraging to know that G_1 and G_2 are both intuitionistically valid; this is so because the usual proof of G_1 is entirely constructive. Moreover, the usual proof of G_2 consists in coding the proof of G_1 using arithmetic, which is, again, constructive. This means that neither classical nor intuitionistic logic are free of the Gödelian challenge. We show that the situation is different for paraconsistent logics, while at the same time providing a means to recover Gödel's results by adding further premises that concern the consistency (in the paraconsistent sense to be explained below) of particular formulas.

3 The paraconsistentist program

The Logics of Formal Inconsistency (**LFI**s) are a broad family of paraconsistent logics, which constitute a wide generalization of da Costa's original hierarchy C_n by incorporating operators for consistency (∘) and inconsistency (•). **LFI**s turn out to be highly flexible logic systems (see, e.g., [12] for references and discussion).

The paraconsistent program is the investigation of logic systems endowed with a negation ¬, such that not every contradiction of the form α and $\neg\alpha$ entails everything; in other words, a paraconsistent logic does not suffer from *deductive trivialism*, in the sense that a contradiction does not necessarily trivialize the deductive machinery of the system by proving everything.[3]

Formalizing what has been said before, deductive trivialism stems from the fact that classical logic cannot stand contradictions, since it endorses the inference rule *ex contradictione sequitur quodlibet*, or *Principle of Explosion*:

$$(\text{PEx}) \quad \alpha, \neg\alpha \vdash \beta,$$

---

[3] *Deductive trivialism* should not be confused with *trivialism*, according to which everything is true.

which authorizes the derivation of any proposition β from a pair of contradictory propositions $\alpha, \neg\alpha$.[4] The challenge for paraconsistent logics is to shun such an 'explosive' negation, while still preserving resources for designing an expressive logic.

As mentioned above, the language of **LFI**s internalizes a notion of consistency at the formula-level, independent of (but related to) negation. *Consistency* thus becomes represented in the logic by a new unary connective ◦. In the same vein, some **LFI**s internalize a notion of *inconsistency* employing the connective •. In this setting, the notion of inconsistency (•) does not necessarily correspond to the negation of consistency (¬◦).

In the **LFI**s, consistent statements are those too rigid to admit contradictions, errors or vagueness, as exemplified by yes–no statements (e.g., whether or not you are pregnant, or have a certain disease). Those statements are rigid, or 'consistent', in the sense that they cannot stand any contradiction. On the other hand, more flexible, 'non-consistent' or 'inconsistent' statements (like whether it is hot today) have the ability to resist to contradictions (by not entailing everything).

This intuitive notion of consistency becomes expressed formally by means of the connective ◦, whose meaning is governed by axioms and rules. Analogous ideas can be found in the notion of *rigidity*, as employed in computational ontologies (cf. *OntoClean* and *UFO*), as well as in the notion of *rigid* or *stable* predicates found in quantified modal logics (cf. [20] for an application in the analysis of another Gödelian argument). This indicates that the idea of formally abstracting a notion of consistency is a natural desideratum, and has instances in other fields.

The basic intuition is that contradictions should not affect all sentences (or all judgments) in the same way, and this is why the sort of Principle of Explosion employed in the **LFI**s is restricted to a special set of *consistent* sentences. Hence a contradictory theory is not necessarily trivial, provided that the contradictions do not involve statements that have been tagged as 'consistent' by employing the connective ◦. This flexibility characterizing **LFI**s is expressed in the so-called *Principle of Gentle Explosion*, which is an essential part of the definition of **LFI**s,

---

[4] This is independent from the fact that classical logic endorses the validity of the *Principle of Non-Contradiction*: $\vdash \neg(\alpha \wedge \neg\alpha)$, see [11]. Also note that the principle *ex contradictione sequitur quodlibet* is often conflated with *ex falso sequitur quodlibet* ($\bot \vdash \beta$). As we will see, they are *not* equivalent from the point of view of the **LFI**s, where the former fails to hold while the latter continues to be valid.

which we present below.

Definition 3.1. *Let $\mathbf{L} = \langle \Theta, \vdash \rangle$ be a Tarskian, finitary and structural logic defined over a propositional signature Θ, which features a negation $\neg$ and a (primitive or defined) unary connective $\circ$. $\mathbf{L}$ is then said to be a Logic of Formal Inconsistency (**LFI**) with respect to $\neg$ and $\circ$, if the following holds:*

(1) $\varphi, \neg\varphi \nvdash \psi$ for some φ and ψ;

(2) $\circ\varphi, \varphi, \neg\varphi \vdash \psi$ for every φ and ψ;

(3) there are two formulas α and β such that

  (a) $\circ\alpha, \alpha \nvdash \beta$;
  (b) $\circ\alpha, \neg\alpha \nvdash \beta$.

Condition **(1)** signals the failure of the Principle of Explosion. Condition **(2)** represents the Principle of Gentle Explosion. Condition **(3)** is required in order to prevent condition **(2)** from being trivially satisfied.

As a consequence, and in contrast to classical logic, consistency in **LFI**s is not synonymous with freedom from contradiction, and here the role of negation is fundamental. The meaning of consistency in the **LFI**s is dictated by its axioms, as occurs with negation (and the other connectives). For conceptual clarifications the reader is referred to [10] and to [8]. The **LFI**-hierarchy starts from a logic called **mbC**, which extends positive (i.e., 'negation-less') classical logic $\mathbf{CPL}^+$ by adding a (paraconsistent) negation $\neg$ and a unary *consistency* operator $\circ$ satisfying some minimal requirements in order to define an **LFI** (as in Definition 3.1).

Gödel's theorems are, of course, crucially dependent on the properties of negation, and we will evaluate how this new perspective may affect their validity. It should from now on become clear that the statements $\circ\alpha$ (α is consistent) and $\neg(\alpha \wedge \neg\alpha)$ (α is non-contradictory) are not equivalent for a paraconsistent negation $\neg$, that is, for a negation subject to the Principle of Gentle Explosion (instead of the classical Principle of Explosion).

This separation between consistency and non-contradiction, contradiction and non-consistency, as well as inconsistency and non-consistency, together with the consequent distinction between contradiction and triviality, are the main tenets of **LFI**s. As we shall see in Sections 5 and

6, they will lead to distinct proposals for the formalization of Gödel's theorems, and thus affect the proofs accordingly.[5]

4 Choosing among paraconsistent scenarios

We start by introducing a 'negation-less' fragment of classical logic, or full classical positive logic:

Definition 4.1 (Classical Positive Logic). *The classical positive logic* **CPL**$^+$ *is defined over the language containing* $\{\wedge, \vee, \rightarrow\}$ *by the following axioms and inference rule:*

Axiom schemas:

$$\alpha \rightarrow (\beta \rightarrow \alpha) \tag{Ax1}$$

$$\big(\alpha \rightarrow (\beta \rightarrow \gamma)\big) \rightarrow \big((\alpha \rightarrow \beta) \rightarrow (\alpha \rightarrow \gamma)\big) \tag{Ax2}$$

$$\alpha \rightarrow \big(\beta \rightarrow (\alpha \wedge \beta)\big) \tag{Ax3}$$

$$(\alpha \wedge \beta) \rightarrow \alpha \tag{Ax4}$$

$$(\alpha \wedge \beta) \rightarrow \beta \tag{Ax5}$$

$$\alpha \rightarrow (\alpha \vee \beta) \tag{Ax6}$$

$$\beta \rightarrow (\alpha \vee \beta) \tag{Ax7}$$

$$\big(\alpha \rightarrow \gamma\big) \rightarrow \big((\beta \rightarrow \gamma) \rightarrow ((\alpha \vee \beta) \rightarrow \gamma)\big) \tag{Ax8}$$

$$(\alpha \rightarrow \beta) \vee \alpha \tag{Ax9}$$

Inference rule:

$$\frac{\alpha \quad \alpha \rightarrow \beta}{\beta} \tag{MP}$$

Starting from **CPL**$^+$ above as a base logic, we extend it with a (paraconsistent) negation $\neg$ and a (primitive or defined) consistency operator $\circ$ satisfying the conditions stated in Definition 3.1. We thus obtain the paraconsistent logic **mbC**, which is a basic **LFI** in the sense

---

[5]We will investigate these effects only at an abstract level and by employing provability logic. In particular, we will not consider Gödel's arithmetization procedure, nor the fixed-point (diagonalization) lemma, which we simply assume. One can thus describe our work as formally reconstructing 'the last mile' of Gödel's proofs.

that its negation and consistency operators enjoy the minimal properties in order to satisfy the definition of **LFI**s.[6]

Definition 4.2. *The logic* **mbC**, *defined over the language containing* $\{\wedge, \vee, \to, \neg, \circ\}$, *is an* **LFI** *obtained from* **CPL**$^+$ *by adding the connectives* $\neg, \circ$ *and the following axiom schemas:*

$$\alpha \vee \neg\alpha \qquad \text{(Ax10)}$$

$$\circ\alpha \to \big(\alpha \to (\neg\alpha \to \beta)\big) \qquad \text{(bc1)}$$

Moreover, we can use $\neg$ and $\circ$ to define bottom particles in the language of **mbC**, as well as *classical negation*, also called *strong negation*.

Definition 4.3. *For any sentences α and γ:*

- $\bot_\alpha := \circ\alpha \wedge \alpha \wedge \neg\alpha$ *act precisely as bottom particles, i.e., they satisfy* $\bot_\alpha \vdash_{\mathbf{mbC}} \beta$ *for every sentence β;*

- $\sim_\gamma \alpha := \alpha \to \bot_\gamma$ *act precisely as classical (strong) negations, i.e., they satisfy* $\vdash_{\mathbf{mbC}} \alpha \vee \sim_\gamma \alpha$, *and* $\alpha \wedge \sim_\gamma \alpha \vdash_{\mathbf{mbC}} \beta$ *for every sentence β.*

Since $\bot_\alpha$ are equivalent for all α, and $\sim_\gamma$ are equivalent for all γ (see [8, Ch. 2] for an elaborate discussion), we write simply $\bot$ and $\sim$. It is worth remarking that **mbC** (as introduced below) both extends and is extended by classical logic. Indeed, on the one hand **mbC** is obviously a subclassical logic by definition, and on the other hand the defined connectives $\bot$ and $\sim$, added to $\{\wedge, \vee, \to\}$, completely encode classical logic. **mbC** can be (equivalently) expounded as a direct extension of classical logic, by incorporating an additional negation $\neg$ to the language, thus defining the logic **mbC**$^\bot$. The equivalence between **mbC**$^\bot$ and **mbC** is shown in [8, Ch. 2]. Moreover, recalling the *Principle of Gentle Explosion* (cf. Definition 3.1), we note that the presence of $\bot$ in the logic validates the inference rule *ex falsum sequitur quodlibet*: $\bot \vdash_{\mathbf{mbC}} \beta$, which in the paraconsistent paradigm is clearly distinct from *ex contradictione sequitur quodlibet*, see e.g. [14].

---

[6]Some strong extensions of **mbC**, such as the logic **Cie** (and its extensions), do not distinguish between inconsistency and contradiction as a consequence of their axiomatic presentation, and some may even allow for the reduction of double negations (this is, however, contrary to what happens in most other **LFI**s; see [8] for a discussion). We will thus restrict ourselves to relatively weak systems starting from the minimal (weakest) Logic of Formal Inconsistency **mbC**.

mbC is the basis for a potentially infinite hierarchy of logics, going up to da Costa logics C_n, paraconsistent many-valued logics [10], paraconsistent modal logics [4, 5], and eventually reaching classical logic, all of them characterized by axiomatic systems extending **mbC** (see [8, Ch. 3–4] for a detailed discussion).

It has been proved that the logics in the da Costa hierarchy C_n (C_1 included) are hardly algebraizable. This is partly due to the non-validity of a replacement meta-theorem which would establish the validity of intersubstitutivity of provable equivalents (IpE) for such logics. Indeed, Theorem 3.51 in [15] shows that IpE cannot hold in any paraconsistent extension of the logic **Ci** (or, for that matter, in any **LFI**) in which $(\neg \alpha \vee \neg \beta) \vdash \neg(\alpha \wedge \beta)$ holds or $\neg(\alpha \wedge \beta) \vdash (\neg \alpha \vee \neg \beta)$ holds.

We will sketch the main lines of **RmbC**, an extension of **mbC** (as mentioned, a minimal **LFI** extending **CPL⁺**) which satisfies the *replacement property*, a meta-property that grants that if $\alpha \leftrightarrow \beta$ is a theorem then $\gamma[p/\alpha] \leftrightarrow \gamma[p/\beta]$ is a theorem, for every formula $\gamma(p)$. As mentioned, **mbC** does not generally satisfy replacement for sentences containing $\neg$ and $\circ$. By adding replacement for $\neg$ and $\circ$ as new *global* inference rules, full replacement can be recovered, in the sense that if $\alpha \leftrightarrow \beta$ is a theorem then $\neg \alpha \leftrightarrow \neg \beta$ is also a theorem, and if $\alpha \leftrightarrow \beta$ is a theorem then $\circ \alpha \leftrightarrow \circ \beta$ also is. As discussed in [9], this makes **RmbC** and its extensions fully algebraizable in the standard Lindenbaum-Tarski sense, which enables their combination with other similarly algebraizable logics by means of *algebraic fibring* [6, 7]. This is very important for us, since we want to be able to combine **RmbC** with other modal logics (e.g., extending **K**) for the sake of formally reconstructing Gödel's arguments.

Definition 4.4. *The logic* **RmbC**, *defined over the language containing* $\{\wedge, \vee, \rightarrow, \neg, \circ\}$ *is obtained from* **mbC** *by adding the following inference rules:*

$$\frac{\alpha \leftrightarrow \beta}{\neg \alpha \leftrightarrow \neg \beta} \quad (R_\neg) \qquad\qquad \frac{\alpha \leftrightarrow \beta}{\circ \alpha \leftrightarrow \circ \beta} \quad (R_\circ)$$

As observed in [9], where **RmbC** was introduced and from where we borrow its presentation, the rules that grant replacement are *global* instead of *local* rules; this means that in order to apply each global rule the corresponding premise must be a theorem. This is similar to what happens with the *necessitation rule* in modal logics. Observe that adding global rules of this kind requires special care with the definition of derivation in the logic.

Definition 4.5 (Derivation in **RmbC**).

- A derivation *of a formula φ in* **RmbC** *is a finite sequence of formulas $\varphi_1 \ldots \varphi_n$ such that φ_n is φ and, for every $1 \leq i \leq n$, either φ_i is an instance of an axiom of* **RmbC***, or φ_i is the consequence of some inference rule of* **RmbC** *whose premises appear in the sequence $\varphi_1 \ldots \varphi_{i-1}$.*

- *We say that a formula φ is* derivable *in (or a theorem of)* **RmbC***, denoted by $\vdash_{\textbf{RmbC}} \varphi$, if there exists a derivation of φ in* **RmbC***.*

Definition 4.6 (Derivation from assumptions in **RmbC**). *Let $\Gamma \cup \{\varphi\}$ be a set of formulas in* **RmbC***. We say that φ is derivable in* **RmbC** *from Γ, and we write $\Gamma \vdash_{\textbf{RmbC}} \varphi$, if either φ is derivable in* **RmbC***, or there exists a finite, non-empty subset $\{\gamma_1, \ldots, \gamma_n\}$ of Γ such that the formula $(\gamma_1 \wedge \gamma_2 \wedge \ldots \wedge \gamma_n) \to \varphi$ is derivable in* **RmbC***.*

Remark 4.7. *The* deduction meta-theorem *is not generally valid in* **RmbC** *when considering the (global) notion of derivation in Definition 4.5. However, it is easy to see (by the properties of $\wedge$ and $\to$ inherited from* **CPL**$^+$*) that the* deduction meta-theorem *does hold when considering the (local) notion of derivation from assumptions in Definition 4.6. Note also that* **RmbC** *is a Tarskian and finitary logic (see [9] for details).*

As presented in [9], a sound and complete semantics for **RmbC** can be given by means of a suitable class of *Boolean algebras with LFI operators* (BALFIs), a (non-additive) generalization of the standard *Boolean algebras with operators* (BAOs) used in algebraic semantics for (normal) modal logics. It is important to highlight that the possibility of such a semantic characterization for paraconsistent logic **RmbC** opens the door to its combination with other logics (e.g., modal logics) by means of *algebraic fibring* [6, Ch. 3], as well as its use in many other application areas for algebraic methods in logic. Moreover, a neighborhood semantics characterizing **RmbC** (and its extensions) has been introduced in [9], where it was shown that **RmbC** can be defined within the minimal bimodal non-normal logic **E** (cf. *minimal models* in [16]). In this respect, **RmbC** can indeed be considered as a (non-normal) modal logic.

We have exploited this fact in our *Isabelle/HOL* reconstruction [13], where we conduct automated reasoning with combinations of **RmbC** and other modal logics, by employing the technique of *shallow semantical embeddings* (SSE) [2].

Before finishing this section, it is convenient to emphasize some properties of the consistency operator and of the paraconsistent negation in the logic **RmbC**. Note that these properties concern particularly the notions of consistency and inconsistency:

Theorem 4.8. *The following properties hold in* **RmbC**:

(1) $\bot \vdash_{\mathbf{RmbC}} \beta$;

(2) $\circ\alpha, \neg\alpha \vdash_{\mathbf{RmbC}} \alpha \to \bot$;

(3) $\alpha \land \neg\alpha \vdash_{\mathbf{RmbC}} \neg\circ\alpha$ *but* $\neg\circ\alpha \nvdash_{\mathbf{RmbC}} \alpha \land \neg\alpha$;

(4) $\circ\alpha \vdash_{\mathbf{RmbC}} \neg(\alpha \land \neg\alpha)$ *but* $\neg(\alpha \land \neg\alpha) \nvdash_{\mathbf{RmbC}} \circ\alpha$;

(5) $\neg\alpha \to \beta \vdash_{\mathbf{RmbC}} \alpha \lor \beta$ *but* $\alpha \lor \beta \nvdash_{\mathbf{RmbC}} \neg\alpha \to \beta$;

(6) $\circ\alpha, \alpha \lor \beta \vdash_{\mathbf{RmbC}} \neg\alpha \to \beta$;

(7) $\alpha \to \beta \nvdash_{\mathbf{RmbC}} \neg\beta \to \neg\alpha$ *but* $\circ\beta, \alpha \to \beta \vdash_{\mathbf{RmbC}} \neg\beta \to \neg\alpha$;

(8) $\alpha \to \neg\beta \nvdash_{\mathbf{RmbC}} \beta \to \neg\alpha$ *but* $\circ\beta, \alpha \to \neg\beta \vdash_{\mathbf{RmbC}} \beta \to \neg\alpha$;

(9) $\neg\alpha \to \beta \nvdash_{\mathbf{RmbC}} \neg\beta \to \alpha$ *but* $\circ\beta, \neg\alpha \to \beta \vdash_{\mathbf{RmbC}} \neg\beta \to \alpha$;

(10) $\neg\alpha \to \neg\beta \nvdash_{\mathbf{RmbC}} \beta \to \alpha$ *but* $\circ\beta, \neg\alpha \to \neg\beta \vdash_{\mathbf{RmbC}} \beta \to \alpha$.

Proof. The first two properties can be easily derived from Definition 4.3 above and the *Principle of Gentle Explosion*. The other proofs can also be easily adapted from [8], Chapter 2, Propositions 2.3.3, 2.3.4, and 2.3.5. □

The above properties help us to understand the connections between consistency, non-consistency, contradictions, and non-contradictions, as well as to anticipate some of their effects on Gödelian arguments. Item **(1)** corresponds to the principle *ex falso sequitur quodlibet* (interpreting 'falso' as *falsum*, i.e., $\bot$). Item **(2)** shows that negating a consistent formula implies its classical negation. Item **(3)** shows that contradiction implies non-consistency, but not vice-versa. Item **(4)** shows that consistency implies non-contradiction, but not vice-versa. Items **(5)** and **(6)** show that disjunction cannot be fully recovered from negation and implication, as in the classical case (however, this can be done when some parts are consistent). Items **(7)** to **(10)** show that contraposition rules for implication do not hold when the paraconsistent negation is considered, but will hold under the guarantee of consistency for the consequent (disregarding negation) in the original conditional form.

The discussion in this section has been aimed at explaining our choice of a paraconsistent scenario for dealing with the plethora of variants of Gödel's theorems that emerge by weakening the Principle of Explosion and, in particular, when consistency is liberated from being defined as non-contradiction. The following sections introduce a logic combination featuring both **LFI** operators and a (normal) modal operator $\Box$ aimed at capturing the notion of provability in formal systems like Peano Arithmetic, drawing upon systems of modal logic extending **K**. We also discuss some notable (though not exhaustive) conclusions achieved through the reconstruction of proofs for paraconsistent variants of Gödel's theorems (in joint work with the proof assistant *Isabelle/HOL*).

5 A paraconsistent logic of provability

We have so far introduced the logic **RmbC** as a candidate logic for the formalization of the Gödelian proofs. However, the most important component for such a logical system is still missing, namely, a *provability operator*, since we want to encode the notion of being derivable/provable directly in our object language. Following the tradition of provability logic [30, 3], we will employ the modal operator $\Box$ for this purpose. But first of all, what is this provability logic? Generally speaking, every time we apply modal logic to the study of formal provability (in some given expressive system) it becomes provability logic. However, not every system of modal logic is appropriate for modeling the notion of derivability in a system including (Peano) arithmetic (e.g., some common modal axioms like T or D are unqualified). Hence, the logics that usually come into consideration when it comes to provability are normal modal logics extending the well-known system **K** with either the axiom 4: $\vdash \Box\phi \to \Box\Box\phi$ (logic **K4**) and/or the inference rule $LR: \vdash \Box\phi \to \phi \implies \vdash \phi$ (logic **K(4)LR**), or the axiom $L: \vdash \Box(\Box\phi \to \phi) \to \Box\phi$ (logic **GL**).

To get an idea of how these modal logics relate to provability, let us consider a formal system **F** that includes Peano Arithmetic. We write $\vdash_F \phi$ to indicate that ϕ is a theorem of **F**. If ϕ is an expression of the language of **F** (i.e., an F-formula), we shall let $\lceil \phi \rceil$ denote the corresponding numeral for the Gödel number of ϕ (which we henceforth call the *Gödel numeral* of ϕ).[7] Given the F-formula $Pf_F(y,x)$, stating that there is a proof in **F** with Gödel numeral y for the formula with Gödel numeral x, we can construct the formula $Pr_F(x) := \exists y. Pf_F(y,x)$. Hence $Pr_F(\lceil \psi \rceil)$ expresses that $\lceil \psi \rceil$ is the Gödel numeral of a sentence ψ that is provable in **F**.

The link between modal logic and provability in **F** (both sharing the primitive logical connectives $\bot$ and $\to$)[8] becomes explicit by considering the following notion (cf. [3, Ch. 3]).

Definition 5.1 (Realization). *A realization $r(p)$ is a function that assigns to each sentence letter p a sentence of the language of **F**. A realization r induces a translation $(\cdot)^r$ such that:*

1. $(p)^r = r(p)$
2. $(\bot)^r = \bot$
3. $(\phi \to \psi)^r = (\phi)^r \to (\psi)^r$
4. $(\Box \phi)^r = Pr_F(\lceil (\phi)^r \rceil)$

Remark 5.2. Since we aim at obtaining a paraconsistent provability logic, we need to add the following additional items to Definition 5.1 above:

5. $(\neg \phi)^r = \neg (\phi)^r$ *(for **LFI**s only)*
6. $(\circ \phi)^r = \circ (\phi)^r$ *(for **LFI**s only)*

Observe that, in adding the last two items, we assume the existence of counterparts for $\neg$ and $\circ$ in the language of **F**. Since **F** is also assumed to contain (Peano) arithmetic, we can, at least in principle, articulate arithmetic formulas featuring $\neg$ or $\circ$. This *prima facie* ability to employ a

---

[7] Recall that we can establish an injection between the set of F-formulas (and also their sequences) and the set of natural numbers, in such a way that each natural number is recursively associated with at most one formula according to Gödel's arithmetization procedure. Also recall that a *numeral* corresponding to some natural number n is the F-formula consisting of the symbol **0** preceded by n occurrences of the symbol **S**.

[8] Recall that $\sim \phi$ can be defined as $\phi \to \bot$. Other connectives can be defined by employing $\sim$ and $\to$ as usual.

paraconsistent negation, as well as a consistency operator, in arithmetic formulas has interesting philosophical repercussions. Their analysis is, however, beyond the scope of this paper. An interesting discussion can be found in [24].

The link between theoremhood in both systems, i.e., between a system **F** containing Peano Arithmetic and some extensions of **K**, is given by the three results below.

Proposition 5.3. $\vdash_{GL} \phi$ *if and only if for every realization* r, $\vdash_F (\phi)^r$.

Proposition 5.4. *If* $\vdash_{K4} \phi$, *then for every realization* r, $\vdash_F (\phi)^r$.

Corollary 5.5. *If* $\vdash_K \phi$, *then for every realization* r, $\vdash_F (\phi)^r$.

Proof. Proposition 5.3 is Solovay's arithmetical completeness theorem for the logic **GL** [28]. Proposition 5.4 (arithmetical soundness theorem for **K4**) and its Corollary 5.5 (arithmetical soundness for **K**) are earlier results; see [30] and [3] for discussion. □

We now recall the well-known derivability conditions for $Pr_F(x)$ (drawing on the ones introduced by M.H. Löb [21]) and their counterpart axioms in the modal logic **K4** (note that the logic **GL** is an extension of **K4** [3]).

Proposition 5.6 (Derivability conditions). *Let* **F** *be a formal system containing Peano Arithmetic; we have:*

1. *If* $\vdash_F \phi$ *then* $\vdash_F Pr_F(\lceil \phi \rceil)$
2. $\vdash_F Pr_F(\lceil \phi \rceil) \wedge Pr_F(\lceil \phi \to \psi \rceil) \to Pr_F(\lceil \psi \rceil)$
3. $\vdash_F Pr_F(\lceil \phi \rceil) \to Pr_F(\lceil Pr_F(\lceil \phi \rceil) \rceil)$

Proof. Consult, e.g., [21] or [3, Ch. 2]. □

Remark 5.7. Observe that the previous results apply in general to every system **F** which contains (classical) Peano Arithmetic. Hence it also applies for arithmetic systems featuring **LFI** operators ¬ and ∘; recall from Definition 4.3 that ⊥ (and in consequence ∼) is definable in the **LFI**s.

Definition 5.8. *The conditions above are encoded in* **K** *as follows:*

1. *Necessitation rule:* $\vdash \phi \implies \vdash \Box\phi$
2. *Axiom K:* $\vdash \Box(\phi \to \psi) \to (\Box\phi \to \Box\psi)$

Moreover, we have in modal logic **K4**:

3. *Axiom 4:* $\vdash \Box\phi \to \Box\Box\phi$

Furthermore, we can enrich our logic with another postulate further restricting the behavior of the provability operator $\Box$. This postulate draws upon the following result:

Proposition 5.9 (Löb's Theorem). *Let ϕ be any sentence of the language of* **F**. *We have that*

$$\text{if} \quad \vdash_F Pr_F(\ulcorner\phi\urcorner) \to \phi \quad \text{then} \quad \vdash_F \phi.$$

Proof. Consult M.H. Löb's original result [21] (see also [3]). □

Definition 5.10. *Löb's theorem can be encoded in (extensions of) modal logic* **K** *as follows:*

- *Löb's rule LR:* $\vdash \Box\phi \to \phi \implies \vdash \phi$
- *Löb's axiom L:* $\vdash \Box(\Box\phi \to \phi) \to \Box\phi$

The modal logic **KLR** is obtained by extending **K** with the inference rule *LR* above; similarly, the logic **K4LR** is obtained by extending **K4** with rule *LR*. The modal logic **GL** (Gödel–Löb logic, or provability logic in the strict sense) is obtained by extending **K** with axiom *L*. Note that axiom 4 follows from *L*, and also that logics **K4LR** and **GL** validate the same formulas (consult [3] for this and other interesting results).

Concerning our present purposes, we define a paraconsistent logic of provability **RmbC⊕K** by means of algebraic fibring (which generalizes *fusion* of normal modal logics) [6, Ch. 3] between the logic **RmbC** and the logic **K**, sharing the connectives $\{\land, \lor, \to, \bot\}$.[9] It is evident that **RmbC⊕K** is an **LFI** (Def. 3.1), as well as its extensions featuring axioms 4 and *L*.

---

[9]Recall that the classical negation $\sim$ and the bottom particle $\bot$ are definable in logic **RmbC**, see Definition 4.3. Observe that we can combine both logics by means of algebraic fibring, since, as recently shown in [9], **RmbC** can be seen as a fragment of a (non-normal) modal logic with an algebraic semantics based on Boolean algebras extended with additional operations. Further modal axioms such as 4 and *L* can be employed to extend this logic combination (as we do in Section 6). Results concerning the preservation of meta-properties (soundness, completeness, interpolation, etc.) for combinations of logics employing the algebraic fibring approach can be consulted in [6, Ch. 2–3].

6 Which assumptions lie behind paraconsistent incompleteness theorems?

We now put the logical machinery developed in the previous sections to work. We provide a formulation of Gödel's incompleteness theorems, utilizing our paraconsistent provability logic **RmbC⊕K** (and its extensions), and reconstruct their proofs. It is important to observe that, in the proofs below, we reason by applying *global* inference rules to theorems (Definition 4.5). This also includes applying rules like *modus ponens* and derived rules like those in Theorem 4.8. Note that this is different than carrying out derivations from assumptions (Definition 4.6), where the use of global rules is not allowed (unless the assumptions are theorems), but, on the flip side, the *deduction meta-theorem* is recovered (recall the corresponding discussion in Section 4). Not only the correctness of these proofs has been mechanically verified using the proof assistant *Isabelle/HOL*, but Isabelle also helped to fill numerous gaps and provided relevant counter-examples: our recognition to the *dolce guida*, the sweet guide as the Beatrice of Dante. The corresponding source files are available online [13].

6.1 First incompleteness theorem (G_1^{par})

The (paraconsistent) formulation of Gödel's theorem presented in Theorem 6.1 below, as well as its corresponding proof, draws upon the analysis by Smoryński [26, §2], Raatikainen [23], and Epstein & Carnielli [18, Ch. 24]. It is worth mentioning that these three variants (as well as many others) are indeed very similar, up to an additional premise (which we will discuss below).

Theorem 6.1 (Gödel's first incompleteness theorem G_1^{par}). *Assume $\boldsymbol{F}$ is a* consistent *formal system which contains Peano Arithmetic. Let G_F be a formula that satisfies $\vdash_F\ G_F \leftrightarrow \neg Pr_F(\lceil G_F \rceil)$, we have:*

- G_F *exists;*   *(existence lemma)*
- $\nvdash_F\ G_F$;   *(non-provable lemma)*
- *under an* additional premise, $\nvdash_F\ \neg G_F$.   *(non-refutable lemma)*

Proof. As remarked previously, the first condition (*existence*), drawing on Gödel's arithmetization procedure, is simply assumed in this analysis. Recalling the discussion in Section 5, we provide a semi-formal proof for the other two lemmas, as encoded in the logic **RmbC⊕K** (conveniently using $\vdash$ as shorthand for $\vdash_{\mathbf{RmbC \oplus K}}$). The proof rests on an (intuitionistically valid) use of the method of *reductio ad absurdum*.

The proof for lemma *non-provable* is simple. Assume that G_F is a theorem of **F**; this becomes formulated in our provability logic as: $\vdash G_F$. Thus, applying *necessitation* (Definition 5.8), we obtain $\vdash \Box G_F$. Moreover, since G_F satisfies the fixed-point lemma $\vdash G_F \leftrightarrow \neg\Box G_F$, we obtain $\vdash \neg G_F$ **reasoning by contraposition** under the premise $\vdash {\circ}\Box G_F$ (cf. Theorem 4.8). Next, from $\vdash G_F$ and $\vdash \neg G_F$ we derive an 'inconsistent' (*absurdum*) statement $\vdash \bot$ by drawing on the **definition of** $\bot$ (Definition 4.3), under the premise $\vdash {\circ}G_F$. Finally, $\nvdash G_F$ follows by *reductio ad absurdum*.

As for lemma *non-refutable*, we draw on the proof presented by Smoryński in [26, §2], which operates under "an additional assumption" (in the author's own words), and which we interpret here in modal terms. The additional premise is '$\vdash_F Pr_F(\lceil \varphi \rceil)$ implies $\vdash_F \varphi$' (for any φ), which becomes formalized using our provability logic as the rule: $\vdash \Box\varphi \implies \vdash \varphi$. The proof goes as follows:

Assume that G_F is refutable in **F**, i.e., that $\vdash \neg G_F$; thus, employing (the right-to-left direction of) the fixed-point lemma and **reasoning by contraposition** under the premise $\vdash {\circ}G_F$, we obtain $\vdash \Box G_F$. We further obtain $\vdash G_F$, by instantiating φ as G_F in the additional premise previously discussed and applying it as a rule, thus obtaining a contradiction. As in the previous case, this contradiction leads to $\vdash \bot$ (*absurdum*) under the premise $\vdash {\circ}G_F$ (recall the **definition of** $\bot$ in Definition 4.3). Finally, this gives us $\nvdash \neg G_F$ by *reductio ad absurdum*. □

Remark 6.2. Observe that in the previous proof we have made salient (using **boldface**) the inferences which, though classically valid, require further additional assumptions from the point of view of the **LFI**s. They correspond to contraposition and '$\bot$-introduction' steps. Remember from Theorem 4.8 that contraposition requires the (**LFI**) consistency of the formula (disregarding negation) in the consequent. That is, we require $\vdash {\circ}\Box G_F$ and $\vdash {\circ}G_F$ as further premises in order to validate the corresponding contraposition steps in the proofs for lemmas *non-provable* and *non-refutable* respectively, as remarked above. Also note that these two assumptions are necessary and sufficient, in the context of **LFI**s, to validate the respective (final) '$\bot$-introduction' steps; recall from Definition 4.3 that $\bot := {\circ}\varphi \wedge \varphi \wedge \neg\varphi$ (for an arbitrary φ) acts as the bottom particle in **LFI**s.

6.2 Second incompleteness theorem (G_2^{par})

We formalize two (paraconsistent) variants of the proof for G_2^{par}; the first one draws from Smoryński [26, §2], and the second one draws from Boolos [3, Ch. 3]. Both variants presuppose the existence of an object-logical formula $Cons_F$ which faithfully encodes the consistency of a system **F** (i.e., $\nvdash_F \bot$) in the arithmetic. We will employ the formula $\neg\Box\bot$ as a working formalization for $Cons_F$ in provability logic, following the established practice (see, e.g., [3]).

Theorem 6.3 (Gödel's second incompleteness theorem G_2^{par}). *Assume **F** is a consistent formal system which contains Peano Arithmetic. Let $Cons_F$ be a formula of **F** representing the consistency of the system. We also assume the derivability conditions on Pr_F from Proposition 5.6. We have two variants:*

*(i) Let G_F be some formula of **F** that satisfies the fixed-point lemma: $\vdash_F G_F \leftrightarrow \neg Pr_F(\ulcorner G_F \urcorner)$. We show that $\vdash_F G_F \leftrightarrow Cons_F$. Hence G_2^{par} (i.e., $\nvdash_F Cons_F$) follows as an immediate corollary from G_1^{par} (namely from one half: $\nvdash_F G_F$) by replacement.*

(ii) From Löb's theorem (Proposition 5.9), it follows that $\nvdash_F Cons_F$.

Proof. We now provide a semi-formal proof for (i), drawing on the proof presented in [26, §2], paraphrased using the logic **RmbC⊕K4**, i.e., we have *axiom 4*: $\vdash \Box\varphi \to \Box\Box\varphi$ (for any φ) as a further premise.

The proof for the left-to-right direction of (i) is relatively simple. We start with $\vdash \bot \to G_F$ (since $\bot$ implies anything). Applying *necessitation* we obtain $\vdash \Box(\bot \to G_F)$, and then $\vdash \Box\bot \to \Box G_F$ by instantiating modal axiom K (with $\bot$ and G_F) and applying *modus ponens*. Now, we obtain $\vdash \Box G_F \to \neg G_F$ from the (left-to-right) fixed-point lemma **by contraposition**, under the premise $\vdash \circ\Box G_F$ (cf. Theorem 4.8). Chaining the two previous formulas together we get $\vdash \Box\bot \to \neg G_F$, and, reasoning **by contraposition** under the premise $\vdash \circ G_F$, we obtain $\vdash G_F \to \neg\Box\bot$ as desired.

The proof for the right-to-left direction is more elaborate. We start with the left-to-right direction of the fixed-point lemma $\vdash G_F \to \neg\Box G_F$. **Applying contraposition**, under the premise $\vdash \circ\Box G_F$, we obtain

$\vdash \Box G_F \to \neg G_F$, which becomes $\vdash \Box\Box G_F \to \Box\neg G_F$ after applying *necessitation* followed by *modus ponens* (after instantiating axiom K with $\Box G_F$ and $\neg G_F$). Instantiating axiom 4 with G_F, which gives us $\vdash \Box G_F \to \Box\Box G_F$, and applying the chain rule, we get $\vdash \Box G_F \to \Box\neg G_F$, which can be trivially expanded to $\vdash \Box G_F \to (\Box G_F \wedge \Box\neg G_F)$. Now, assuming $\vdash \circ G_F$ as a further premise, we obtain $\vdash \Box\circ G_F$ by necessitation, and thus $\vdash \Box G_F \to \Box\circ G_F$ (since implication is classical). Both previously derived formulas give us: $\vdash \Box G_F \to (\Box G_F \wedge \Box\neg G_F \wedge \Box\circ G_F)$. Moreover, as is well known, normal propositional modal logics satisfy $\vdash \Box(\alpha_1 \wedge \ldots \wedge \alpha_n) \leftrightarrow (\Box\alpha_1 \wedge \ldots \wedge \Box\alpha_n)$ for finite n; this makes the consequent in the previous formula equivalent to $\Box(G_F \wedge \neg G_F \wedge \circ G_F)$, thus giving us $\vdash \Box G_F \to \Box(G_F \wedge \neg G_F \wedge \circ G_F)$. Now, recalling the **definition of** $\bot$ (Definition 4.3), we obtain $\vdash \Box G_F \to \Box\bot$ (using *replacement* of equivalents). Reasoning, again, **by contraposition**, under the premise $\vdash \circ\Box\bot$, we get $\vdash \neg\Box\bot \to \neg\Box G_F$. Finally, drawing on the fixed-point lemma, we obtain $\vdash \neg\Box\bot \to G_F$ by *replacement*.

We now provide a semi-formal proof for (ii). This proof is considerably simpler than the previous one; however, it relies on a stronger premise, namely Löb's theorem (Proposition 5.9). As the corresponding representation in provability logic we choose the Löb's rule variant $LR: \vdash \Box\varphi \to \varphi \implies \vdash \varphi$ (Definition 5.10).[10] Thus, our logic of formalization corresponds to the logic combination **RmbC⊕KLR**.

We start by assuming that $Cons_F$ is provable, i.e., $\vdash \neg\Box\bot$. We thus have $\vdash \Box\bot \to \bot$ by the **properties of negation**, under the additional premise $\vdash \circ\Box\bot$. Instantiating rule LR with $\bot$ ($\vdash \Box\bot \to \bot \implies \vdash \bot$), we obtain $\vdash \bot$; from where $\not\vdash \neg\Box\bot$ follows by *reductio ad absurdum*.
□

Remark 6.4. Similarly as we did for G_1^{par}, we have made salient (using **boldface**) the inferences which, though classically valid, require further additional assumptions from the point of view of the **LFI**s (corresponding in this case to contraposition and '$\bot$-introduction' inferences). As for the first step of the proof for (i), we note that *ex falso (qua falsum:* $\bot$*) sequitur quodlibet* is still valid in the **LFI**s. As regards the subsequent contraposition steps in that proof, we have remarked that $\vdash \circ\Box G_F$, $\vdash \circ G_F$ and $\vdash \circ\Box\bot$ are required as further premises for those steps to succeed; recall that contraposition requires the consistency of the formula in the consequent (disregarding negation), cf. Theorem 4.8.

---

[10] A proof along similar lines can be given if employing modal Löb's axiom L.

Also recall that $\bot$ is defined in the **LFI**s as $\circ\varphi \land \varphi \land \neg\varphi$ for arbitrary φ (Definition 4.3); hence $\circ G_F$ is also required for the step involving '$\bot$-introduction' to get off the ground. As regards the proof for (ii), we note that the inference $\vdash \neg\varphi$, $\vdash \circ\varphi \implies \vdash \varphi \to \bot$ is valid in **LFI**s, as the reader can easily verify. It is also worth noting that, in contrast to variant (i), neither $\vdash \circ\Box G_F$ nor $\vdash \circ G_F$ are required as further premises in this second variant.

7 A long story short: conclusions

Our results in Section 6 show that a condition for validating a paraconsistent version G_1^{par} of Gödel's first incompleteness theorem is to assume that both G_F and $\Box G_F$ (i.e., $Pr_F(G_F)$) are consistent in the **LFI** sense of 'contradiction-intolerant'. This is mainly due to the failure of explosion and contraposition in the logic **RmbC** (and generally in **LFI**s), unless the $\circ$-consistency of certain sentences is assumed.

Similarly to the proof reconstruction of G_1^{par}, the paraconsistent version G_2^{par} of Gödel's second incompleteness theorem requires some further premises in order for contraposition steps to succeed, namely, the $\circ$-consistency of G_F, $\Box G_F$ and $\Box\bot$ for the first presented variant (i). As regards the second variant (ii) of G_2^{par}, we have seen that only the $\circ$-consistency of $\Box\bot$ is required. Recall that, in contrast to the first variant (i), variant (ii) of G_2^{par} does not depend on G_1^{par}, but relies on Löb's theorem instead.

We may summarize, in a sketchy form, the ideas above as follows:

If $FP(G_F)$ and $\circ G_F$ and $\circ\Box G_F$ then G_1^{par};

where $FP(G_F)$ stands for 'G_F is a fixed-point for $\neg\Box(\cdot)$' and G_1^{par} stands for '$Consistency_F$ entails $Incompleteness_F$'. Reasoning by contraposition:[11]

If $FP(G_F)$ and not G_1^{par} then not $\circ G_F$ or not $\circ\Box G_F$.

A similar reasoning applies to G_2^{par}, as the reader can verify. Based on this, in an exercise of counterfactual imagination, we can envision that, if things had been different in the thirties (e.g., if logics like **LFI**s

---

[11] We can do this, since our reasoning (meta-logic) is taken to be classical.

already existed), the Gödelian results could have been presented along the following lines:

Theorem 7.1 (Gödel's Existence Theorem). *For every consistent ($\neq$ non-contradictory) and complete formal system $\boldsymbol{F}$, which includes Peano Arithmetic, a sentence G_F can be constructed such that G_F or $Pr_F(G_F)$ is a non-consistent statement (i.e., it is 'contradiction-tolerant'). Moreover, if $\boldsymbol{F}$ can prove its own consistency, then $Pr_F(\bot)$ is also a non-consistent statement.*

A less technical, more conceptual conclusion is that, by adopting a lighter, more flexible negation (such as the paraconsistent negation featured in **RmbC**), we genuinely avoid the Gödelian objection, which is mistakenly taken to be universal (although Gödel himself never saw it this way). To be sure, Gödel's argumentation is still sound, it just becomes interpreted, more appropriately, as an existence claim.

Limitations of Gödel's arguments are totally understandable, specially if we take into account that they were aimed at challenging the foundations of mathematics, whose notion of negation is the classical one, with its brutal simplification, conflating negation, denial, subtraction, and falsity in just one idea. But a Gödelian objection cannot be readily directed against the subtle linguistic and pragmatic usage of negation, nor at its usage in contemporary areas like knowledge representation in computer science. The non-mathematical usage of negation needs to adhere to some additional postulates to fall prey to Gödel's arguments; we have shown that the consistency (in the sense of **LFI**s, namely, 'contradiction-intolerance') of the formulas G_F and $Pr_F(G_F)$ is among them. In this respect, an interesting question is whether there are any 'natural' mathematical statements (i.e., those not involving the numerical coding of logical notions) which could be shown to be undecidable in our basic paraconsistent systems, just as the celebrated Paris–Harrington theorem is a 'natural' undecidable combinatorial statement in the standard case. This and similar issues deserve further investigation.

Acknowledgements: The first author acknowledges support from the National Council for Scientific and Technological Development (CNPq), Brazil, under research grant 307376/2018-4. We are indebted to Marcelo Coniglio for early discussions on these ideas.

References

[1] L. Beklemishev and D. Shamkanov. Some abstract versions of Gödel's second incompleteness theorem based on non-classical logics. 2014.

[2] C. Benzmüller. Universal (meta-)logical reasoning: Recent successes. *Science of Computer Programming*, 172:48–62, 2019.

[3] G. Boolos. *The Logic of Provability*. Cambridge University Press, 1995.

[4] J. Bueno-Soler. Two semantical approaches to paraconsistent modalities. *Logica Universalis*, 4(1):137–160, 2010.

[5] J. Bueno-Soler. Models for anodic and cathodic multimodalities. *Logic Journal of IGPL*, 20(2):458–479, 2012.

[6] W. A. Carnielli, M. Coniglio, D. M. Gabbay, P. Gouveia, and C. Sernadas. *Analysis and Synthesis of Logics: how to cut and paste reasoning systems*, volume 35. Springer, 2008.

[7] W. A. Carnielli and M. E. Coniglio. Combining logics. In E. N. Zalta, editor, *The Stanford Encyclopedia of Philosophy*. Metaphysics Research Lab, Stanford University, winter 2016 edition, 2016.

[8] W. A. Carnielli and M. E. Coniglio. *Paraconsistent Logic: Consistency, Contradiction and Negation*. Springer, 2016.

[9] W. A. Carnielli, M. E. Coniglio, and D. Fuenmayor. Logics of formal inconsistency enriched with replacement: an algebraic and modal account. Submitted for publication, preprint available from https://www.cle.unicamp.br/eprints/index.php/CLE_e-Prints.

[10] W. A. Carnielli, M. E. Coniglio, and J. Marcos. Logics of formal inconsistency. In D. Gabbay and F. Guenthner, editors, *Handbook of Philosophical Logic*, volume 14, pages 1–93, Amsterdam, 2007. Springer.

[11] W. A. Carnielli, M. E. Coniglio, and A. Rodrigues. On formal aspects of the epistemic approach to paraconsistency. In M. Freund, M. F. de Castro, and M. Ruffino, editors, *Logic and Philosophy of*

Logic: Recent Trends in Latin America and Spain, pages 48–74. College Publications, 2018.

[12] W. A. Carnielli, M. E. Coniglio, and A. Rodrigues. Recovery operators, paraconsistency and duality. *Logic Journal of the IGPL*, 2019. First published online at https://doi.org/10.1093/jigpal/jzy054.

[13] W. A. Carnielli and D. Fuenmayor. Isabelle/HOL sources associated with this paper. Online available at Github: https://github.com/davfuenmayor/Goedel-Incompleteness-Isabelle, 2020.

[14] W. A. Carnielli and J. Marcos. Ex contradictione non sequitur quodlibet. *Bulletin of Advanced Reasoning and Knowledge*, 1(1):89–109, 2001. On line at http://www.advancedreasoningforum.org/bark01.

[15] W. A. Carnielli and J. Marcos. A taxonomy of **C**-systems. In W. A. Carnielli, M. E. Coniglio, and I. M. L. D'Ottaviano, editors, *Paraconsistency - the Logical Way to the Inconsistent*, volume 228 of *Lecture Notes in Pure and Applied Mathematics*, pages 1–94, New York, 2002. Marcel Dekker.

[16] B. F. Chellas. *Modal Logic: an Introduction*. Cambridge University Press, 1980.

[17] D. Dubois and H. Prade. Inconsistency management from the standpoint of possibilistic logic. *International Journal of Uncertainty, Fuzziness and Knowledge-Based Systems*, 23:15–30, 2015.

[18] R. L. Epstein and W. A. Carnielli. *Computability: Computable Functions, Logic, and the Foundations of Mathematics*. Advanced Reasoning Forum, 2008. 3rd Edition.

[19] T. Franzén. *Gödel's Theorem: An Incomplete Guide to its Use and Abuse*. AK Peters, Wellesley, 2005.

[20] D. Fuenmayor and C. Benzmüller. Automating emendations of the ontological argument in intensional higher-order modal logic. In G. Kern-Isberner, J. Fürnkranz, and M. Thimm, editors, *KI 2017: Advances in Artificial Intelligence*, volume 10505 of *LNAI*, pages 114–127. Springer, 2017.

[21] M. H. Löb. Solution of a problem of Leon Henkin 1. *The Journal of Symbolic Logic*, 20(2):115–118, 1955.

[22] R. McNaughton et al. Alfred Tarski, a decision method for elementary algebra and geometry. *Bulletin of the American Mathematical Society*, 59(1):91–93, 1953.

[23] P. Raatikainen. Gödel's incompleteness theorems. In E. N. Zalta, editor, *The Stanford Encyclopedia of Philosophy*. Metaphysics Research Lab, Stanford University, fall 2018 edition, 2018.

[24] S. Shapiro. Incompleteness and inconsistency. *Mind*, 111(444):817–832, 2002.

[25] P. Smith. *An Introduction to Gödel's Theorems*. Cambridge University Press, 2013.

[26] C. Smoryński. The incompleteness theorems. In J. Barwise, editor, *Handbook of Mathematical Logic*, volume 4, pages 821–865. North-Holland Publishing Company, 1977.

[27] R. M. Smullyan. *Gödel's incompleteness theorems*. Oxford University Press, 1992.

[28] R. M. Solovay. Provability interpretations of modal logic. *Israel Journal of Mathematics*, 25(3-4):287–304, 1976.

[29] A. Tarski. A decision method for elementary algebra and geometry. In *Quantifier elimination and cylindrical algebraic decomposition*, pages 24–84. Springer, 1998.

[30] R. Verbrugge. Provability logic. In E. N. Zalta, editor, *The Stanford Encyclopedia of Philosophy*. Metaphysics Research Lab, Stanford University, fall 2017 edition, 2017.

On existence in arithmetic

Edgar L. B. Almeida[∀]
Rodrigo A. Freire[∃]

[∀]Federal Institute of Brasília (IFB), Brazil
[∃]Department of Philosophy, University of Brasília, Brazil

Abstract

The aim of this paper is to develop a systematic analysis for the notion of existence towards a precise, conceptual definition of the intuitive notion of *set existence axiom* in the arithmetical context. This goal is achieved by the main definition, that is coherent with the reverse mathematics foundational program and also with a systematic analysis of the notion of existence in the context of the set theory. Additionally, from the existential analysis is extracted a criterion that separates formal systems that are framed from different mathematical practices even though they are considered equivalent in view of bi-interpretability.

1 Introduction

A precise definition of *set existence axiom* in the arithmetical context is missing, despite the widespread use of a corresponding undefined notion, and our goal is to fill this gap. This notion plays a key role in the foundational program known as reverse mathematics - explicitly acknowledged in the influential Simpson's book [10]. According to him, mathematics can be divided in two: set-theoretic mathematics and ordinary mathematics. The latter can be characterized by its independence from abstract set-theoretic concepts and it encompasses geometry, number theory, calculus, countable algebra and mathematical logic, for example. The former, on the contrary, relies heavily on set-theoretic concepts and it covers general topology, abstract functional analysis, and set theory itself.

Reverse mathematics is occupied with ordinary mathematics only, for the strength of set existence axioms for ordinary mathematics is, in a sense, much weaker than that required for set-theoretic mathematics. By doing that, reverse mathematics discharges ordinary mathematics from the set-theoretic heavyweight system.

In this scenario, reverse mathematics uses second order arithmetic to answer the following "*Main Question: Which set existence axioms are needed to prove the theorems of ordinary, non-set-theoretic mathematics?*" [10, p. 2]. Although the Main Question explicitly mentions *set existence axioms*, such a notion remains undefined and unaddressed throughout the investigation. This is not an apparent problem for Simpson's book - he proceeds by choosing and applying set existence axioms as needed, and the question of what does it mean to be an existence axiom is never raised. However, for foundational studies, it is desirable to have a conceptual definition.

In order to provide a conceptual definition of existence axiom, we consider that the relevant notion of existential import of a sentence parallels the notion of validity: while the validity of a sentence is determined by its truth-conditions, the existential import of a sentence is determined by its existential-conditions. The motivation for this approach is that it does not seem reasonable to associate the existential import of a sentence with its syntax. After all, from that association, it would follow that prenex sentences with existential quantifiers would have existential import, implying that each first-order sentence is logically equivalent to one with existential import. Since we want a notion of existential import which is stable under logical equivalence, that should not be our path.

Naturally, other approaches can be taken. For example, Corcoran and Massoud ([1] and [2]) pursue an entirely different path:

> "Let us say that a given universalized conditional *has existential import* if it implies the corresponding existentialized conjunction. It may seem awkward at first but we will also say that a given existentialized conjunction *has existential import* if it is implied by the corresponding universalized conditional". [1, p. 3].

According to this, a sentence of the form $\forall x(Px \to Qx)$ has existential import if it implies the sentence $\exists x(Px \wedge Qx)$, and the sentence $\exists x(Px \wedge Qx)$ has existential import if it is implied by $\forall x(Px \to Qx)$. Therefore, there would be a situation in which two logically equivalent sentences are different with respect to their existential import. Moreover, for every sentence φ there would be a logically equivalent sentence ψ with the same logical form and such that φ has existential import if and only if ψ does not have.

We think that stability under logical equivalence is a crucial feature for a definition of existential import, so the Corcoran and Massoud's approach is unsuitable in our view. In order to achieve a satisfactory definition, we have adapted the work of Freire [5], concerning set theory, to arithmetic. Accordingly, the key role is played by the notion of existence requirement degree. The degrees of existence requirement correspond to existential-conditions. There are natural existential-conditions for sentences in set theory, based on the cumulative hierarchy concept, hence arithmetic can inherit this previously explored environment once interpreted in set theory. The following assumption is a basic component of our plan. If we are given a standard bi-interpretation[1] $\mathcal{A}$ between two formal systems T and T', then the existential-conditions of a sentence φ in T can be identified with the existential-conditions of its image $\varphi^{\mathcal{A}}$.

For instance, suppose that T is a formal system for arithmetic and T' is a formal system for set theory; φ is a $L(T)$-sentence, $\mathcal{A}$ is a standard interpretation of T in T' and $\mathcal{B}$ is a standard interpretation of T' in T such that the composition of $\mathcal{A}$ with $\mathcal{B}$ is isomorphic to the identity and the composition of $\mathcal{B}$ with $\mathcal{A}$ is also isomorphic to the identity. Then, the existential-conditions of φ and $\varphi^{\mathcal{A}}$ are the same and, consequently, they have the same existential import (assuming the hypothesis that existential import is determined by the existential-conditions).[2]

From this, if T' is a formal system for set theory and $\varphi^{\mathcal{A}}$ is a $L(T')$-sentence, then the existential-conditions of $\varphi^{\mathcal{A}}$ is identified with the closure-conditions that a domain $\mathcal{D}$ must fulfill for the validity of $(\varphi^{\mathcal{A}})^{\mathcal{D}}$ in T'.

The method of analysis can be illustrated by the following diagram:

---

[1] For a precise presentation of the notion of interpretation here considered see [9, §4.7]. The theories T and T' are bi-interpretable if there is an interpretation $\mathcal{A}$ of T in T' and an interpretation $\mathcal{B}$ of T' in T such that their compositions are isomorphic to the identity. By an abuse of terminology, we say that $\mathcal{A}$ is an interpretation of T in T' when, strictly speaking, $\mathcal{A}$ is an interpretation of T in an appropriate conservative extension of T'.

[2] Since the existential analysis of arithmetic featured in this paper considers a specific pair of interpretations $\mathcal{A}$ and $\mathcal{B}$, the results are, in principle, relative to the given pair of interpretations. So we should develop the existential analysis of the arithmetic sentence φ by first working out the analysis of the arithmetic equivalent $\varphi^{(\mathcal{A},\mathcal{B})}$ and then transposing it back to φ. However, for the sake of cleaner notation, the interpretation $\mathcal{B}$ will be often dropped. The existence of bi-interpretations between systems for arithmetic and set theory which are not trivial variations of the standard bi-interpretations that will be presented in the next two sections is an open issue, but we will have something more to say about the dependence on the choice of $\mathcal{A}$ and $\mathcal{B}$.

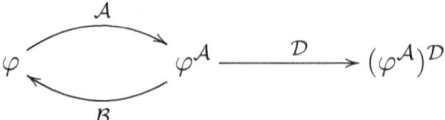

What are the closure-conditions on $\mathcal{D}$ to guarantee the validity of $(\varphi^{\mathcal{A}})^{\mathcal{D}}$ in T'?

Keeping the above diagram in mind, we stipulate the following:

Definition 1. *Let Z be a formal system for arithmetic, T a theory of sets and $\mathcal{A}$ an interpretation of $L(Z)$ in T. Under these conditions, $\mathcal{D}$ is said to be an $\in$-arithmetical interpretation of the language of T if and only if the composition of $\mathcal{D}$ and $\mathcal{A}$ is an interpretation of the language of Z in T satisfying the following conditions:*

- *If A is the domain of $\mathcal{A}$, then $(Ax)^{\mathcal{D}}$ is Ax.*

- *If $\mathfrak{s}$ is a symbol in the signature of the language of Z, then $(\mathfrak{s}^{\mathcal{A}})^{\mathcal{D}}$ is $\mathfrak{s}^{\mathcal{A}}$.*

2 Second order arithmetic

Our presentation of the formal system Z_2 for second order arithmetic separates and highlights the relational, functional and existential components of each axiom. The language of Z_2 is $\{A, P, S, M, Z, N, C, \in\}$, and the formulas $A(x, y, z)$ and $P(x, y, z)$ are read, respectively, *z is the sum of x with y* and *z is the product of x and y*. The formulas $S(x,y)$, $M(x,y)$ and Zx express that *the successor of x is y*, *x is less than y* and *x is zero*, respectively. Now, Nx means that *x is a number* and Cx expresses that *x is a set of numbers*. The usual defined symbols $\subseteq$, ω, etc will be used without been explicitly introduced. They must be understood as the abbreviation of standard formulas defining them, and their explicit formulation can be viewed, for example, in Drake [3]. The axioms of Z_2 are distributed in eight groups.

1. Relational axioms: the universal closure of each formula below is a relational axiom.

    (a) $Nx \wedge Ny \wedge Nz \wedge Zy \wedge S(x,z) \to y \neq z$.

    (b) $Nx \wedge Ny \wedge Nz \wedge S(y,x) \wedge S(z,x) \to y = z$.

(c) $Nx \wedge Ny \wedge Zy \to A(x,y,x)$.
(d) $Nx \wedge Ny \wedge Nz \wedge Nv \wedge Nu \wedge S(y,u) \wedge S(z,v) \wedge A(x,y,z) \to A(x,u,v)$.
(e) $Nx \wedge Ny \wedge Zy \to P(x,y,y)$.
(f) $Nx \wedge Ny \wedge Nz \wedge Nv \wedge Nu \wedge S(y,u) \wedge A(z,x,v) \wedge P(x,y,z) \to P(x,u,v)$.
(g) $Nx \wedge Ny \wedge Zy \to \neg M(x,y)$.
(h) $Nx \wedge Ny \wedge Nz \wedge S(y,z) \to (M(x,z) \leftrightarrow M(x,y) \vee x = y)$.

2. Closure axioms:

   (a) $\exists x \, (Nx \wedge Zx)$.
   (b) $\forall x \, (Nx \to \exists y \, (Ny \wedge S(x,y)))$.
   (c) $\forall x \forall y \, (Nx \wedge Ny \to \exists z \, (Nz \wedge A(x,y,z)))$.
   (d) $\forall x \forall y \, (Nx \wedge Ny \to \exists z \, (Nz \wedge P(x,y,z)))$.

3. Axioms of functionality: the universal closure of each formula below is an axiom of functionality.

   (a) $Nx \wedge Ny \wedge Zx \wedge Zy \to x = y$.
   (b) $Nx \wedge Nu \wedge Nv \wedge S(x,u) \wedge S(x,v) \to u = v$.
   (c) $Nx \wedge Ny \wedge Nu \wedge Nv \wedge A(x,y,u) \wedge A(x,y,v) \to u = v$.
   (d) $Nx \wedge Ny \wedge Nu \wedge Nv \wedge P(x,y,u) \wedge P(x,y,v) \to u = v$.

4. Axioms for numbers and sets:

   (a) $\forall x \, (Zx \to Nx)$.
   (b) $\forall x \, [(Nx \vee Cx) \wedge \neg(Nx \wedge Cx)]$.
   (c) $\forall x \forall y \, (Nx \wedge S(x,y) \to Ny)$.

5. Membership axiom: $\forall x \forall y \, (x \in y \to Nx \wedge Cy)$.

6. Axiom of Induction:

   $\forall x \, [Cx \wedge (Zy \to y \in x) \wedge \forall z \forall u \, (Nz \wedge z \in x \wedge S(z,u) \to u \in x) \to \forall z \, (Nz \to z \in x)]$.

7. Extensionality axiom:

   $\forall x \forall y \, [Cx \wedge Cy \wedge \forall z \, (z \in x \leftrightarrow z \in y) \to x = y]$.

8. Totality axioms:

   (a) $\forall x \forall y \, (Cx \wedge S(x,y) \to Zy)$.
   (b) $\forall x \forall y \forall z \, ((Cx \vee Cy) \wedge A(x,y,z) \to Zz)$.
   (c) $\forall x \forall y \forall z \, ((Cx \vee Cy) \wedge P(x,y,z) \to Zz)$.
   (d) $\forall x \forall y \, (Cx \vee Cy \to \neg M(x,y))$.

9. Axiom of Comprehension: for each $L(Z_2)$-formula φ whose variables are $x, z_1, ..., z_n$ (assuming that y is not among those variables), the following formula is a comprehension axiom:

$$\forall z_1 \cdots \forall z_n \exists y \, (Cy \wedge \forall x \, (x \in y \leftrightarrow Nx \wedge \varphi)).$$

The closure axioms guarantee that the relation symbols Z, S, A and P are total, i.e., for each point in the domain of the corresponding relation there exists a point in the range such that these points are related. Sometimes we will call these formulas *the existential part* of the corresponding relational symbol. The axioms of functionality express that those relations are functions: for each point in a relation's domain there is just one related point in the range. With this axiomatization the usual successor operation can be set up by the conjunction of axioms 1(a), 1(b), 2(b) and 3(b). Similarly to the operations of sum and multiplication and the constant zero. The other axioms are standard.

We will introduce a set theory, denoted by ZFe, that is bi-interpretable with Z_2. The language of ZFe has only one predicate symbol - the membership symbol $\in$ - and its axioms are:

- Some of the axioms of ZF: the axiom of extensionality; the axiom of regularity; the axiom of separation; the axioms of comprehension; the axiom of pair; the axiom of union; the axiom of replacement and the axiom of infinity.

- Axiom of finite power set:

$$\forall x \, (\text{Finite}(x) \to \exists y \, \forall z (z \in y \leftrightarrow z \subseteq x)).$$

- Axiom of enumerability: $\forall x \, (\text{Enum}(x))$, where

$$\text{Enum}(x) \leftrightarrow \exists f \, (func(f) \wedge bij(f) \wedge img(f) \subseteq \omega \wedge dom(f) = x)$$

and $func(f)$, $bij(f)$ and $img(f)$ mean, respectively, that f is *function*, f is *bijective* and *the image of* f.

A natural strategy to construct an interpretation I of Z_2 in ZFe is based on the association between natural numbers and ordinals and between sets of natural numbers and sets of ordinals. Unfortunately, the association will not work without some adjustment, because an ordinal is a set of ordinals, so that the natural number 2 and the set of natural numbers $\{0, 1\}$ are associated with the same ordinal. We can circumvent this problem by the usual trick for obtaining disjoint interpretations of N and C. The disjoint union of the adjusted interpretations of N and C gives us the domain U_I of I.

$$N^I = \{\langle 0, x\rangle : x \in \omega\}; \quad C^I = \{\langle 1, x\rangle : x \subseteq \omega\}; \quad U_I = N^I \cup C^I.$$

The predicate symbols Z and M are interpreted by the relations

$$Z^I = \{\langle 0, 0\rangle : 0 \in \omega\}.$$

$$M^I = \{\langle\langle 0, x\rangle, \langle 0, y\rangle\rangle : x \in y \in \omega\}.$$

The interpretation $\in^I$ is defined from the predicate $\in_2$ in the following:

$$\langle i, x\rangle \in_2 \langle j, y\rangle \leftrightarrow x \in y,$$

$$\in^I = \{\langle x, y\rangle : N^I x \wedge C^I y \wedge x \in_2 y\}.$$

In ZFe the interpretation of the arithmetical operations are defined from successor, sum and product of ordinals, which are denoted respectively by x^+, $x \oplus y$ and $x \odot y$.

$$\dagger^* : N^I \to N^I \text{ such that } \langle 0, x\rangle \mapsto \langle 0, x^+\rangle,$$
$$+^* : N^I \times N^I \to N^I \text{ such that } \langle\langle 0, x\rangle, \langle 0, y\rangle\rangle \mapsto \langle 0, x \oplus y\rangle,$$
$$\cdot^* : N^I \times N^I \to N^I \text{ such that } \langle\langle 0, x\rangle, \langle 0, y\rangle\rangle \mapsto \langle 0, x \odot y\rangle.$$

The last step is to define the interpretations of S (successor), A (sum) and P (product):

$$S^I = \{\langle x, y\rangle : (x \in N^I \wedge y = x^{\dagger^*}) \vee (x \in C^I \wedge y = \langle 0, 0\rangle)\},$$

$$A^I = \{\langle x, y, z\rangle : (x \in N^I \wedge y \in N^I \wedge z = x+^*y) \vee ((x \in C^I \vee y \in C^I) \wedge z = \langle 0, 0\rangle)\},$$

$$P^I = \{\langle x, y, z\rangle : (x \in N^I \wedge y \in N^I \wedge z = x\cdot^*y) \vee ((x \in C^I \vee y \in C^I) \wedge z = \langle 0, 0\rangle)\}.$$

Following these definitions, it is a routine exercise to verify that

$$\langle U_I, N^I, C^I, M^I, Z^I, S^I, A^I, P^I\rangle$$

is a model for Z_2. Such a model is a definable class in ZFe, hence can be taken as an interpretation of Z_2 in ZFe.[3]

The interpretation from ZFe back to Z_2 involves a massive amount of codification of sets in numbers and sets of numbers; it will not be presented here because, for our purposes, it is enough to know that there is an interpretation J of ZFe in Z_2 such that the pair (I, J) is a bi-interpretation between Z_2 and ZFe. A rigorous and complete presentation of an appropriate J can be seen in Simpson [10, VII.3] and in McLarty [7].

The final step before our main definition is provided by the natural closure conditions under which the existential import analysis will take place. Those conditions reflect the key role played by the second coordinate in the set-theoretic constitution of numbers and sets of numbers under the above interpretation.

Definition 2. *Let D a set of ordered pairs in which the first coordinate is an ordinal.*

- *D is a 2-nonempty domain iff $\langle 0, \alpha \rangle \in D$ for some ordinal α.*

- *D is a 2-transitive domain iff D is a 2-nonempty domain and $\langle 0, y \rangle \in D$ whenever $\langle \alpha, x \rangle \in D$ and $y \in x$, for some ordinal α.*

- *D is a 2-supertransitive domain iff D is a 2-nonempty domain and $\langle 1, y \rangle \in D$ whenever $\langle \alpha, x \rangle \in D$ and $y \subseteq x$, for some ordinal α.*

- *D is a 2-level domain iff D is a 2-nonempty domain and $D = \{\langle k, x \rangle \mid k \in \{0, 1\} \wedge x \subseteq \beta \wedge \beta \in \alpha)\}$, where α is an ordinal.*

- *D is a 2-limit level domain iff D is a 2-nonempty domain and $D = \{\langle k, x \rangle \mid k \in \{0, 1\} \wedge x \subseteq \beta \wedge \beta \in \lambda)\}$, where λ is a limit ordinal.*

The next definition coordinates the notions of existential import and interpretation and plays a key role in our analysis.

---

[3] This close relation between the model-theoretic discourse and the interpretation-theoretic discourse is stated in Shoenfield's passage: *"We have so far discussed structures in English. We could, of course, translate the entire discussion into any language in which there is sufficient set-theoretic notation to discuss functions, predicates, etc."* [9, p. 61].

Definition 3. *Let us fix a standard bi-interpretation between Z_2 and ZFe, that is, a standard interpretation $\mathcal{A}$ of Z_2 in ZFe and a standard interpretation $\mathcal{B}$ of ZFe in Z_2 such that their compositions are isomorphic to the identity. A $L(Z_2)$-sentence φ:*

- Admits degree 0 of existence requirement *iff for every 2-nonempty $\in$-arithmetical interpretation $\mathcal{D}$ of the language of ZFe in ZFe, the formula $(\varphi^{\mathcal{A}})^{\mathcal{D}}$ is valid in ZFe.*

- Admits degree 1 of existence requirement *iff for every nonempty 2-transitive $\in$-arithmetical interpretation $\mathcal{D}$ of the language of ZFe in ZFe, the formula $(\varphi^{\mathcal{A}})^{\mathcal{D}}$ is valid in ZFe.*

- Admits degree 2 of existence requirement *iff for every nonempty 2-supertransitive $\in$-arithmetical interpretation $\mathcal{D}$ of the language of ZFe in ZFe, the formula $(\varphi^{\mathcal{A}})^{\mathcal{D}}$ is valid in ZFe.*

- Admits degree 3 of existential requirement *iff for every ordinal α the corresponding 2-level domain $\mathcal{D}$ is an $\in$-arithmetical interpretation of the language of ZFe in ZFe, and $Ord(\alpha) \to (\varphi^{\mathcal{A}})^{\mathcal{D}}$ is valid in ZFe.*

- Admits degree 4 of existence requirement *iff for every limit ordinal λ the corresponding 2-limit level $\mathcal{D}$ is an $\in$-arithmetical interpretation of the language of ZFe in ZFe, and $Lim(\lambda) \to (\varphi^{\mathcal{A}})^{V(\lambda)}$ is valid in ZFe.*

- Admits degree 5 of existence requirement *iff the identity interpretation $\mathbf{V} = \{x : x = x\}$ is an $\in$-arithmetical interpretation of the language of ZFe in ZFe and $(\varphi^{\mathcal{A}})^{\mathbf{V}}$ is valid in ZFe.*

If $\mathcal{D}$ is an $\in$-arithmetical interpretation, then we will write $\varphi^{\mathcal{AD}}$ instead of $(\varphi^{\mathcal{A}})^{\mathcal{D}}$. Each clause in the above definition coordinates a degree of existence requirement that φ may admit with a closure condition on domains fulfilling the existential-conditions of φ. We assume that $\mathbf{V}$ is a limit level, so that all closure conditions obtain in $\mathbf{V}$. One must picture the least degree of existence requirement admitted by φ as the minimal closure condition on the universe $\mathbf{V}$ fulfilling the existential demands of $\varphi^{\mathcal{A}}$ in ZFe, that is, as the existential import of φ in Z_2.

We have that the sentences which are valid in Z_2 are the only sentences in Z_2 admitting a degree of existence requirement. Indeed, if $Z_2 \vdash \varphi$ then ZFe $\vdash \varphi^{\mathcal{A}}$; therefore ZFe $\vdash \varphi^{\mathcal{AV}}$ and we have that φ admits a degree of existence requirement. But if φ is not valid then it is not the case that ZFe $\vdash \varphi^{A\mathbf{V}}$, hence it is nonsense to ask for the identification of the exact closure condition on $\mathbf{V}$ fulfilling the existential demands of φ in Z_2.

Furthermore, if a sentence admits some existence requirement degree d, then it admits all the existence requirement degrees greater than d. In fact, if φ admits degree 0 of existence requirement then $\varphi^{A\mathcal{D}}$ is valid in every 2-nonempty $\mathcal{D}$ and, in particular, $\varphi^{A\mathcal{D}}$ is valid in every 2-transitive domain, etc and hence φ admits degrees 1, 2, 3, 4 and 5 also. This remark legitimates the association of a unique degree of existence requirement to each valid arithmetical sentence.

Definition 4. *The* degree of existence requirement *of a $L(Z_2)$-sentence φ (relative to a bi-interpretation) is the least degree of existence requirement that φ admits (under that bi-interpretation). If d is the degree of existence requirement of φ, we say that φ has degree d of existence requirement.*

The next result concerns the stability of the degrees of existence requirement under logical consequence: the degree of a logical consequence is bounded by the degrees of the premises.

Theorem 1 (Following [5, Proposition 7]). *Let φ and ψ be $L(Z_2)$-sentences. If φ has degree of existence requirement d and ψ is a logical consequence of φ, then ψ has degree of existence requirement no greater than d.*

Proof. If ψ is a logical consequence of φ, then ψ^A is a logical consequence of φ^A; consequently, $\varphi^A \to \psi^A$ is logically valid. Let $\mathcal{D}$ be any $\in$-arithmetical interpretation satisfying the closure condition corresponding to degree d. Since φ has degree d of existence requirement, then ZFe $\vdash \varphi^{A\mathcal{D}}$. From this and the logical validity of $\varphi^A \to \psi^A$, we have that ψ^A is valid in $\mathcal{D}$, hence ψ admits degree d of existence requirement. So, ψ has degree $d' \leq d$ of existence requirement. □

A straightforward consequence of this result is that two logically equivalent arithmetical sentences have the same degree of existence requirement. Moreover, two sentences which are logically equivalent in the inclusive logic have exactly the same quantitative and qualitative existential demands, and, in this sense, our existential analysis is stable under logical variations. This theorem, together with definition 3, guarantee that a theorem has no more existential import than its premisses. In other words, logical consequences do not demand more ontological commitments than their hypotheses.

It is clear that our analysis of the existential import of a sentence is relative to a specific pair of fixed interpretations. Therefore, we will not make explicit reference to that pair and we will use the short expression *degree of existence requirement of a sentence*, instead of *degree of existence requirement of a sentence with respect to such-and-such bi-interpretation*. If there are two bi-interpretations between the theories under analysis, then it is natural to inquire about the stability of the degree of existence requirement under a change of bi-interpretation. The next definition points to a first answer.

Definition 5. *Let T_1 and T_2 be formal systems playing a role similar to that of Z_2 and ZFe in definition 3. Let $\mathcal{A}$ and $\mathcal{A}^\star$ be interpretations of T_1 in T_2, and let $\mathcal{B}$ and $\mathcal{B}^\star$ be interpretations of T_2 in T_1 such that the pairs $(\mathcal{A}, \mathcal{B})$ and $(\mathcal{A}^\star, \mathcal{B}^\star)$ are bi-interpretations. Under these conditions, the bi-interpretations $(\mathcal{A}, \mathcal{B})$ and $(\mathcal{A}^\star, \mathcal{B}^\star)$ are equivalent for the evaluation of the existence requirement iff for every axiom φ of T_1 the degree of existence requirement of $\varphi^{\mathcal{A}} \leftrightarrow \varphi^{\mathcal{A}^\star}$ is less than or equal to the minimum of the degrees of existence requirement of $\varphi^{\mathcal{A}}$ and $\varphi^{\mathcal{A}^\star}$.*

The next result is a partial answer to the question about the stability of the degree of existence requirement with respect to the bi-interpretation: if two bi-interpretations are equivalent for the evaluation of existence requirement, then the respective degrees of existence requirement are equal.

Proposition 1. *If $(\mathcal{A}, \mathcal{B})$ and $(\mathcal{A}^\star, \mathcal{B}^\star)$ are bi-interpretations of T_1 in T_2 which are equivalent for the evaluation of existence requirement, then for every axiom φ of T_1 the degree of existence requirement of φ with respect to $\mathcal{A}$ is equal to the degree of existence requirement of φ with respect to $\mathcal{A}^\star$.*

Proof. Assume, without loss of generality, that the degree of existence requirement of $\varphi^{\mathcal{A}}$ is less than or equal to the degree of $\varphi^{\mathcal{A}^\star}$. If the degree of existence requirement of $\varphi^{\mathcal{A}}$ is d, from the hypothesis that the bi-interpretations are equivalent, we have that $\varphi^{\mathcal{A}} \leftrightarrow \varphi^{\mathcal{A}^\star}$ admits degree d of existence requirement. So, $\varphi^{\mathcal{A}^\star}$ admits degree d of existence requirement. Therefore, the degree of existence requirement of φ^{A^*} is d also. □

We can derive here two important qualitative notions. The first one is the notion of *productivity*, which is directly connected to definition 3. The other notion gives the distinction between *conditional* and *unconditional* assertions, which is very natural in the purely set-theoretic context and has its place here too. For example the axiom of the empty set in set theory is an unconditional and nonproductive assertion - it unconditionally affirms that there is a set which is empty and it admits degree 0 of existence requirement. Something analogous happens with arithmetic: the claim that there exists the number zero is an unconditional assertion. In opposition, the axiom of union is a productive and conditional assertion: it does not admit degree 0 and it conditionally affirms that if there is a set, then there is a set which is the union of the given set. Something analogous can be said about arithmetical assertions: the existence of the sum of two numbers is conditioned to the existence of the summands.

The notion of productivity is easily defined: an assertion is productive if it does not admit degree 0. The distinction between conditional and unconditional assertions can become accurate with the help of the empty structures. The basic idea is that unconditional existence assertions are not valid in the empty structure, while conditional assertions are valid in this structure. However, standard first order logic presupposes that all domains of interpretation are nonempty, and for this reason the underlying logic of this work is the inclusive logic. The inclusive logic is described by Mendelson [8, §2.6], to which we add axioms of equality and identity.[4]

Definition 6. *Let φ be a $L(Z_2)$-sentence.*

- *φ is a nonproductive assertion iff φ admits degree 0 of existence requirement.*

- *φ is a productive assertion iff φ does not admit degree 0 of existence requirement.*

---

[4] Axioms of equality are the universal closure of formulas with the form $x_1 = x_2 \to \cdots \to x_n = y_n \to fx_1...x_n = fy_1...y_n$ and $x_1 = x_2 \to \cdots \to x_m = y_m \to Px_1...x_m \to Py_1...y_m$ in which f and P are, respectively, an n-ary function symbol and an m-ary relation symbol. An axiom of identity is a sentence of the form $\forall x(x = x)$.

- φ is a conditional assertion *iff* φ is valid in the empty interpretation.

- φ is an unconditional assertion *iff* φ is not valid in the empty interpretation.

A productive assertion of degree d is a productive assertion whose degree of existence requirement is d.

We can now evaluate the axioms of second order arithmetic.

Theorem 2. *The axioms of Z_2 are classified as follows*[5]:

1. *Relational axioms, axioms of functionality, membership axiom, axioms for numbers and sets and axioms of totality: conditional unproductive assertions.*

2. *Extensionality axiom and axiom of induction: conditional productive assertion with degree 1.*

3. *Axiom of comprehension: conditional productive assertion with degree 5.*

4. *First closure axiom: unconditional productive assertion with degree 1.*

5. *The remaining closure axioms: conditional productive assertions with degree 4.*

Proof. With the exception of the first closure axiom, all other axioms are universal formulas and, consequently, they are conditional assertions. The first closure axiom is not valid in the empty structure, so it is an unconditional assertion. The degrees of the axioms are directly determined from the definitions, and we will be provide the details just for some paradigmatic cases. In the following, $\mathcal{D}$ denotes $\in$-arithmetical interpretations with domain D.

RELATIONAL AXIOMS. The interpretation of the second relational axiom

$$\forall x \forall y \forall z \, (Nx \wedge Ny \wedge Nz \wedge Zx \wedge S(y,z) \to x \neq z)$$

by interpretation I is equivalent to

---
[5]The degree of an axiom schema is defined as the greatest degree of its instances.

$$\forall x \forall y \forall z \, (N^I x \wedge N^I y \wedge N^I z \wedge Z^I x \wedge S^I(y,z) \to x \neq z).$$

The $\mathcal{D}$-interpretation of this is

$$\forall x \forall y \forall z \, [Dx \wedge Dy \wedge Dz \wedge$$
$$\wedge (N^I x)^\mathcal{D} \wedge (N^I y)^\mathcal{D} \wedge (N^I z)^\mathcal{D} \wedge (Z^I x)^\mathcal{D} \wedge (S^I yz)^\mathcal{D} \to (x \neq z)^\mathcal{D} \,].$$

From the definition of $\in$-arithmetic interpretation, this is

$$\forall x \forall y \forall z \, [Dx \wedge Dy \wedge Dz \wedge N^I x \wedge N^I y \wedge N^I z \wedge Z^I x \wedge S^I yz \to x \neq z].$$

The validity of this last sentence in any 2-nonempty domain D follows immediately from the definitions of N^I, Z^I and S^I, and the axiom has degree 0.

In exactly the same way, the other relational axioms, axioms of functionality, axioms for numbers and sets, membership axiom and totality axioms have degree 0.

CLOSURE AXIOMS. Let us begin with the first axiom of this group. When it is interpreted in ZFe successively by I and then by $\mathcal{D}$ we obtain

$$\exists x (Dx \wedge N^I x \wedge Z^I x).$$

If D is a 2-transitive domain (so that there is some $\langle 0, \alpha \rangle \in D$), then, from the 2-transitivity of D, we conclude that $\langle 0, 0 \rangle \in D$. So, the axiom admits degree 1 but not degree 0. Indeed, the sentence above is not valid in $D = \{\langle 0, 1 \rangle\}$.

Consider now the second axiom of this group and, as in the previous cases, interpreting the axiom in ZFe and then in $\mathcal{D}$ gives

$$\forall x \, (Dx \wedge N^I x \to \exists y \, (Dy \wedge N^I y \wedge S^I xy)).$$

Consider that D is the 2-level domain $\{\langle k, z \rangle \mid k \in \{0,1\} \wedge z \subseteq \beta \wedge \beta \in \alpha\}$. If α is a finite successor ordinal $n+1$, then $x = \langle 0, n \rangle$ is in D and the antecedent of the sentence is valid in D. Suppose that the consequent is valid in D too. So, there is $y \in D$ such that $S^I xy \in D$, i.e., $y = \langle 0, n+1 \rangle = \langle 0, \alpha \rangle \in D$, absurd. So, the axiom does not admit degree 3. If D is a 2-limit level, it is immediate that if the antecedent of the sentence above is valid in D, then the consequent is also; consequently the axiom has degree 4. The other axioms of this group are analysed in the exactly same way.

AXIOM OF INDUCTION. Interpreting this axiom gives
$$\forall x \left[Dx \wedge C^I x \wedge \forall y \left(Dy \wedge N^I y \wedge Z^I y \to y \in^I x \wedge \right. \right.$$
$$\wedge \forall z \forall w \left(Dz \wedge Dw \wedge N^I z \wedge S^I zw \wedge z \in^I x \to w \in^I x \right) \right) \to$$
$$\left. \to \forall y \left(Dy \wedge N^I y \to y \in^I x \right) \right].$$

Assume that the antecedent holds, *i.e.*, assume that (i) x is a set of numbers in the sense of I in D, (ii) if y is a zero in the sense of I in D, then it belongs to x and (iii) if z is a number in D which belongs to x in the sense of I and w is a successor of z in the sense of I that belongs to D in the sense of I, then w belongs to x in the sense of I too. Moreover, assume that $y \notin^I x$. From the antecedent y is a number in I, so there is some y_0 such that $y = \langle 0, y_0 \rangle \notin_2 \langle 1, x_0 \rangle = x$. Since $\in_2$ is well-founded, if we assume that D is 2-transitive then there is an $\in_2$-minimal element $u = \langle 0, u_0 \rangle$ such that $u \in D$ and $u_0 \in \omega$ and $\langle 0, u_0 \rangle \notin_2 x$. From $u \notin_2 x$ and $\langle 0, 0 \rangle \in x$, we conclude that $u_0 \neq \emptyset$. Since $u_0 \neq \emptyset$ and $u_0 \in \omega$, there is a $t_0 \in \omega$ such that $u_0 = t_0^+$. From the fact that D is 2-transitive and $\langle 0, u_0 \rangle \in D$ and $t_0 \in u_0$ we conclude that $t = \langle 0, t_0 \rangle \in D$ and $u \in^I x$, absurd. So, the axiom admits degree 1.

With $D = \{\langle 0, 1 \rangle, \langle 1, 0 \rangle\}$ we see that the axiom of induction does not admit degree 0. Indeed, the antecedent of the axiom is vacuously true in D but if we take $y = \langle 0, 1 \rangle$ and $x = \langle 0, 1 \rangle$, then it is not the case that $y \in^I x$.

EXTENSIONALITY AXIOM. Interpreting the axiom in ZFe by I and then by $\mathcal{D}$ gives
$$\forall x \forall y \forall z (\psi \to \gamma \to x = y),$$
where ψ is $Dx \wedge Dy \wedge C^I y \wedge C^I y \wedge Dz \wedge Uz$ and γ is $z \in^I x \leftrightarrow z \in^I y$. Assume ψ and $x \neq y$. We have to conclude $\neg \gamma$.

We can take $y = \langle 1, y_0 \rangle$ and $x = \langle 1, x_0 \rangle$. From the fact that $x \neq y$, we have $\langle 1, x_0 \rangle \neq \langle 1, y_0 \rangle$. Therefore, there is a z_0 in the symmetric difference $x_0 \Delta y_0$ and, without lost of generality, we may assume $z_0 \in x_0$. From the assumption that $\langle 1, x_0 \rangle \in D$, if D is 2-transitive then $\langle 0, z_0 \rangle \in D$. We know that $z_0 \notin y_0$, therefore $z = \langle 0, z_0 \rangle \notin^I \langle 1, y_0 \rangle = y$. Therefore, extensionality admits degree 1.

If we take $z = \langle 0, 0 \rangle$, $x = \langle 1, 1 \rangle$, $y = \langle 1, 2 \rangle$ and $D = \{x, y, z\}$, then the antecedent $\psi \to \gamma$ is valid in D but $x \neq y$. So, the axiom does not admit degree 0.

AXIOM OF COMPREHENSION. Let $\bar{z}$ denote the sequence of variables $z_1, \ldots, z_n$ that occur in φ and let $D\bar{z}$ denote $Dz_1 \wedge \ldots \wedge Dz_n$. The interpretation of the axiom by I and $\mathcal{D}$ results in

$$\forall \bar{z} \exists y \forall x \left(D\bar{z} \wedge Dx \wedge Dy \wedge U_I \bar{z} \wedge C^I y \wedge U_I x \to \left(x \in^I y \leftrightarrow N^I x \wedge \varphi^I(\bar{z}, x) \right) \right).$$

If D is the 2-limit level domain determined by the ordinal ω, then the sentence above is not valid. In fact, there is no y containing all of N^I, which corresponds to the instance in which φ is a tautology. Therefore, comprehension has degree 5 of existence requirement. □

We have evaluated the productivity of the axioms.

3 Conclusions

We are now in position to achieve the main goal of this work, a precise definition of the notion of set existence assertion in arithmetic.

Definition 7 (Main Definition). *A $L(Z_2)$-sentence φ is an assertion of existence iff φ is an unconditional assertion or a productive assertion and φ is a nonexistence assertion iff φ is an unproductive conditional assertion. A sentence has existential import iff it is an assertion of existence. A $L(Z_2)$-sentence φ is an axiom of existence iff φ is an axiom of Z_2 and φ has existential import.*

This definition is worthy of our attention because it is directly related to the *notion of existence axiom* present in the reverse mathematics program. As an evidence of the uniformity between this definition and the reverse mathematics program can be seen again in Simpson's book. There the Main Question is always answered considering restrictions of either the induction axiom, or the comprehension axiom, or both. For example, Weyl's predicativism is formalized by the formal system ACA_0, which is obtained from Z_2 with the restriction that the axiom of comprehension is valid for arithmetical sentences only. Feferman's predicativism is identified with the formal system $\Pi_1^1 - CA_0$, framed by restricting the axiom of comprehension to Π_1^1-formulas. Bishop's constructivism is associated to the theory RCA_0, obtained restricting the axiom of comprehension to Δ_1^0-formulas and restricting the axiom of induction to Σ_1^0-formulas. These examples make a lot of sense according to our analysis: comprehension is the strongest existence axiom in the context of second order arithmetic.

Additionally, when the existential analysis of Z_2-axioms, summarized in theorem 2, is compared to the existential features of ZFC-axioms presented in [5, Proposition 41], we can say in a precise way that the informal forecast made by Simpson is sound:

> (...) "that the set existence axioms which are needed for set-theoretic mathematics are likely to be much stronger than those which are needed for ordinary mathematics". [10, p. 2].

An interesting byproduct concerns the relational formulation of arithmetic adopted here and the closure axioms. These axioms are assertions of existence according to the Main Definition but they do not play a relevant role in the existential analysis developed in reverse mathematics. There are at least two ways to understand this omission.

The first one is very straightforward: the existential analysis developed in this work throws light on existential aspects of sentences that were in the shadows until now. In particular, the existential character of functional closure axioms was hidden by the use of function symbols, therefore, not included in the intuitions that motivates the existential analysis in reverse mathematics. According to this point of view, the analysis presented in this work reveals existential nuances that are usually ignored by reverse mathematics.

The second one is that this omission is justified because the corresponding existential import comes from the use of function symbols in the formal framework formalizing second-order arithmetic. It is easy to eliminate that kind of existential import. In order to achieve this we need just to move to a relational framework and eliminate the closure axioms. According to this perspective the existential import given by the closure axioms is not unavoidable, hence it is not due to the arithmetical practice. They emerge from a choice in the process of formalizing arithmetic.

The analysis developed to elucidate the notion of set existence axiom in Z_2 can be applied to make subtle distinctions between formal systems forged from distinct practices. The idea that two bi-interpretable formal systems are equivalent is well disseminated in the foundational studies. This idea displays some plausibility. Indeed, if two systems are bi-interpretable, their languages are inter-definable, every theorem of one formal system is interpreted as a theorem of the other one and every deduction developed in a formal system can be recursively converted into a deduction of the interpreted sentence. From a deductive-theoretical perspective, the differences between the formal systems involved seem to be irrelevant.

A concrete example of this attitude is seen in Kaye and Wong's work [6]. They affirm that the purpose of their work is to investigate the following

> **Folklore Result** First-order theories Peano arithmetic and ZF set theory with the axiom of infinity negated are equivalent, in the sense that each is interpretable in the other and the interpretations are inverse to each other. [6, p. 497].

They have shown that this result is wrong - the theory that is bi-interpretable with Peano arithmetic is not ZF set theory with the axiom of infinity negated - but they have not questioned the idea that bi-interpretable theories are equivalent. Their attitude motivate us to ask this *Question:*

> In which sense, if any, the axiomatizations of a formal system for arithmetic T and a bi-interpretable set theory T^* could be inequivalent?

Nothing prevents the application of the strategy developed in this work to answer this general question. After all, if a formal system T is bi-interpretable with T^* and T^* is a set theory, the existential analysis realized in ZFe for Z_2 can be, *mutatis mutandis*, applied to T in T^*. If the existential classification of the axioms of both theories is the same, then the axiomatic systems will be equivalent in a stronger sense. Otherwise, the existential classification of the selected axioms provides a criterion for distinguishing the axiomatic systems. This criterion conceptually separates the axioms of second order arithmetic Z_2 and ZFe set theory. Those systems are usually taken as indistinguishable from a model-theoretic point of view, as remarked by Enayat in this passage:

> "[T]he recent discussion about Woodin's reference to the well-known fact that the standard model for second order number theory is "essentially the same" as the model (M, epsilon), where M is the set of hereditary countable sets. As suggested out by Bill Tait, the above two structures are intimately related at the interpretability level, i.e., they are *bi-interpretable*. Note that this is stronger than saying that they are mutually interpretable". [4]

Following the same ideas given in the previous sections and considering nonempty domains, nonempty transitive domains, and so on, instead of 2-nonempty domains, 2-transitive domains, and so on, in definition 3, from the results presented in [5] (in particular from the proofs of propositions 12, 13 and remark 34), it follows that the axioms of ZFe can be classified:

- The axiom of regularity is an unproductive conditional assertion.

- The axiom of extensionality is a productive conditional assertion with degree 1.

- The axiom of comprehension is a productive conditional assertion with degree 2.

- The axiom of union is a productive conditional assertion with degree 3.

- The axioms of finite power set, pair and enumerability are productive conditional assertions with degree 4.

- The axiom of replacement is a productive conditional assertion with degree 5

- The axiom of infinity is a productive unconditional assertion with degree 5

- The axiom of empty set is an unproductive unconditional assertion.

The comparison of the results above and theorem 2 reveals that the formal systems of the structures mentioned in Enayat's quote above are, in a precise way, not equivalent: *their axioms have distinct ontological demands.* Such differences are an indication that these systems reflect different mathematical practices.

A natural further direction in this work is to understand the existential import of axioms of other theories which are bi-interpretable with set theories. In particular Peano Arithmetic is specially interesting, since it is bi-interpretable with the set theory described in Kaye and Wong's work.

References

[1] John Corcoran and Hassan Masoud. Existential-Import Mathematics. *The Bulletin of Symbolic Logic*, 21(1):1–14, 2015.

[2] John Corcoran and Hassan Massoud. Existential Import Today: New Metatheorems; Historical, Philosophical, and Pedagogical Misconceptions. *History and Philosophy of Logic*, 36(1):39–61, 2015.

[3] Frank R Drake. *Set Theory, an introduction to large cardinals*, volume 76 of *Studies in Logic and the Foundations of Mathematics*. North-Holland, Amsterdam, 1974.

[4] Ali Enayat. (2010, January 18). *Woodin's pair of articles on CH*. FOM mail listl. https://cs.nyu.edu/pipermail/fom/2010-January/014318.html, accessed February 12, 2020.

[5] Rodrigo A. Freire. On Existence in Set Theory. *Notre Dame Jornal of Formal Logic*, 53(4):525–547, 2012.

[6] Richard Kaye and Tin Lok Wong. On Interpretations of Arithmetic and Set Theory. *Notre Dame Journal of Formal Logic*, 48(4):497–510, 2007.

[7] Colin McLarty. Interpreting Set Theory in Higher Order Arithmetic. *ArXiv e-prints*, jul 2012.

[8] Elliot Mendelson. *Introduction to Mathematical Logic*. Textbooks in Mathematics. CRC Press, New York, 6th edition, 2015.

[9] Joseph R. Shoenfield. *Mathematical Logic*. Association for Simbolic Logic, Massachusetts, 2nd edition, 1967.

[10] Stephen G. Simpson. *Subsystems of Second Order Arithmetic*. Perspectives in Logic. Association for Symbolic Logic, Cambridge University Press, New York, 2nd edition, 2009.

Part 3

Non-classical inferences

A NOTE ON SEMI-IMPLICATION WITH NEGATION

Arnon Avron

School of Computer Science, Tel-Aviv University, Israel

Abstract

In [4] we introduced a general notion of *semi-implication*, which is mainly based on the relevant deduction property (RDP) — a weak form of the classical-intuitionistic deduction theorem which has motivated the design of the intensional fragments ($\mathbf{R}_\to$ and $\mathbf{R}_{\neg\to}$) of the relevance logic $\mathbf{R}$. We showed there that with exactly one exception, in the pure language of $\to$ the connective $\to$ has in a finitary logic $\mathbf{L}$ the RDP iff $\mathbf{L}$ has a strongly sound and complete Hilbert-type system which is an extension by axiom schemas of $HR_\to$ (the standard axiomatization of $\mathbf{R}_\to$). Another result proved in [4] is that if a logic $\mathbf{L}$ has a conjunction or a disjunction, then it cannot have a semi-implication for which the basic relevant property of *variable sharing* obtained. The goal of this note is to show that in contrast, the inclusion of a (relevant) negation does not affect the results of [4] in an essential way.

In [4] we introduced a general notion of *semi-implication* which generalizes the implications used in classical logic, intuitionistic logic and its extensions, and relevance logics. It is mainly based on the relevant deduction property (RDP) — a weak form of the classical-intuitionistic deduction theorem which has motivated the design of the intensional fragments ($\mathbf{R}_\to$ and $\mathbf{R}_{\neg\to}$) of the relevance logic $\mathbf{R}$. (See [1, 6].) We showed there that with one exception, in the pure language of $\to$ the connective $\to$ has in a finitary logic $\mathbf{L}$ the RDP iff $\mathbf{L}$ has a strongly sound and complete Hilbert-type system which is an extension by axiom schemas of $HR_\to$ (the standard axiomatization of $\mathbf{R}_\to$). The only exception is $\mathbf{CL}_\leftrightarrow$, the pure equivalential fragment of classical logic (where we denote the biconditional by $\to$). Another result proved in [4] is that if a logic $\mathbf{L}$ has a conjunction or a disjunction, then it cannot have a semi-implication for which the *variable sharing property* obtained. (The latter is the most basic property of any relevance logic, including $HR_\to$.) The goal of this note is to show that in contrast, the inclusion of a (relevant) negation does not affect the results of [4] in an essential way.

In the sequel $\mathcal{L}$ denotes a propositional language. φ, ψ vary over the formulas of $\mathcal{L}$. p, q vary over the atomic formulas of $\mathcal{L}$. $\mathcal{T}$ varies over theories of $\mathcal{L}$ (where by a 'theory' we simply mean here a set of formulas of $\mathcal{L}$).

Definition 1. A (propositional) *logic* is a pair $\mathbf{L} = \langle \mathcal{L}, \vdash_{\mathbf{L}} \rangle$, where $\mathcal{L}$ is a propositional language, and $\vdash_{\mathbf{L}}$ is a structural and non-trivial Tarskian consequence relation for $\mathcal{L}$. A logic $\mathbf{L} = \langle \mathcal{L}, \vdash_{\mathbf{L}} \rangle$ is *finitary* if so is $\vdash_{\mathbf{L}}$.[1]

Definition 2. Let $\mathbf{L} = \langle \mathcal{L}, \vdash_{\mathbf{L}} \rangle$ be a propositional logic, and let $\to$ be a (primitive or defined) connective of $\mathcal{L}$.

1. $\to$ has in $\mathbf{L}$ the *relevant deduction property* (RDP) if it satisfies the following condition: $\mathcal{T}, \varphi \vdash_{\mathbf{L}} \psi$ iff either $\mathcal{T} \vdash_{\mathbf{L}} \psi$ or $\mathcal{T} \vdash_{\mathbf{L}} \varphi \to \psi$.

2. $\to$ is called a *semi-implication for* $\mathbf{L}$ if $\to$ has in $\mathbf{L}$ the RDP, and in addition there are formulas φ and ψ such that $\vdash_{\mathbf{L}} \varphi \to \psi$ but $\nvdash_{\mathbf{L}} \psi \to \varphi$.

3. $\to$ has in $\mathbf{L}$ the *variable sharing property* if $\mathsf{Atoms}(\varphi) \cap \mathsf{Atoms}(\psi) \neq \emptyset$ whenever $\vdash_{\mathbf{L}} \varphi \to \psi$. (Here $\mathsf{Atoms}(\varphi)$ is the set of atomic formulas in φ.)

Next we turn to the negation connective. Our starting point is the notion of a *negation associated with a semi-implication* $\to$ that was introduced and analyzed in Chapter 11 of [5]. What we need here from that analysis is one fact that immediately follows from Proposition 11.76 of [5] and Theorem 3.5 of [4]: If $\mathbf{L}$ is a logic in a language which contains $\to$ and $\neg$, $\to$ has in $\mathbf{L}$ the RDP, and $\neg$ is a negation associated with $\to$, then $\mathbf{L}$ contains the system $\mathbf{LL}_{\overset{\neg}{\to}}$ below.

Definition 3. $HLL_{\overset{\neg}{\to}}$ is the Hilbert-type proof system in $\{\to, \neg\}$ which is presented in Figure 1.

Note 4. Let $\mathbf{LL}_{\overset{\neg}{\to}}$ be the pure implication-negation fragment of linear logic ([7]). $HLL_{\overset{\neg}{\to}}$ is the axiomatization of this fragment given in [3]. An axiomatization $HLL_{\to}$ for $\mathbf{LL}_{\to}$ (the pure implicational fragment of linear logic) is obtained by omitting from $HLL_{\overset{\neg}{\to}}$ the negation axioms [N1] and [N2].

---

[1] This is the notion of propositional logic which has been used in [4], as well as throughout in [5]. See either for the definitions of the notions involved.

Axioms:

[Id] $\varphi \to \varphi$ (Identity)

[Tr] $(\varphi \to \psi) \to ((\psi \to \theta) \to (\varphi \to \theta))$ (Transitivity)

[Pe] $(\varphi \to (\psi \to \theta)) \to (\psi \to (\varphi \to \theta))$ (Permutation)

[N1] $(\varphi \to \neg\psi) \to (\psi \to \neg\varphi)$ (Contraposition)

[N2] $\neg\neg\varphi \to \varphi$ (Double negation)

Rule of inference:

[MP] $\dfrac{\varphi \quad \varphi \to \psi}{\psi}$

Figure 1: The proof system $HLL_{\to}$

Definition 5. Let $\mathcal{L} = \{\to, \neg\}$.

1. $HR_{\to}$ is the extension of $HLL_{\to}$ by the following axiom:

   [Ct] $(\varphi \to (\varphi \to \psi)) \to (\varphi \to \psi)$ (Contraction)

2. $HCL_{\leftrightarrow}$ is the extension of $HLL_{\to}$ by the following axiom:

   [Eq] $(\varphi \to (\varphi \to \psi)) \to \psi$ (Equivalence)

Theorem 6. *Let* **L** *be a finitary extension of* $\mathbf{LL_{\to}}$. $\to$ *has in* **L** *the RDP iff* **L** *has a strongly sound and complete Hilbert-type system which is an extension by axiom schemas of either* $HR_{\to}$ *or* $HCL_{\leftrightarrow}$.

Proof. Immediate from Corollary 4.4 and Theorem 4.5 of [4]. □

Theorem 7. *A finitary extension of* $\mathbf{LL_{\to}}$ *has a semi-implication connective* $\to$ *iff it has a strongly sound and complete Hilbert-type system which is an extension by axiom schemas of* $HR_{\to}$.

Proof. Using Theorem 6, the proof is practically identical to the proof of Theorem 6.4 in [4]. □

Note 8. *The last theorem means that* $\mathbf{R_{\to}}$ *is the minimal logic which has a semi-implication together with an associated negation.*

Corollary 9. *Suppose $\to$ has in a finitary logic* **L** *the VSP, the RDP, and an associated negation. Then* **L** *is an axiomatic extension of* $\mathbf{R}_{\to,\neg}$, *and $\to$ is a semi-implication for it.*

Proof. Similar to the proof of Corollary 6.6 [4]. □

Note 10. In [2] one can find an infinite family of logics in the language of $R_{\to,\neg}$, all of which has the VSP and are extensions of $R_\to$ by axiom schemas.

Let us turn now to the question how many logics are there that have a connective $\to$ which has both the RDP and an associated negation, but is not a semi-implication. By the results above, this is equivalent to the question how many non-trivial axiomatic extensions $HCL_{\leftrightarrow,\neg}$ has.

Note 11. The two main facts about $HCL_\leftrightarrow$ (the pure implicational fragment of $HCL_{\leftrightarrow,\neg}$) that were shown in [4] were:

1. $HCL_\leftrightarrow$ is strongly sound and complete for $\mathbf{CL}_\leftrightarrow$.

2. $HCL_\leftrightarrow$ is strongly Post-complete: it has no proper extension in its language.[2]

In what follows we show that $HCL_{\leftrightarrow,\neg}$ has similar, though weaker, properties.

Definition 12. $\mathbf{CL}_{\leftrightarrow,\neg}$ is the equivalence-negation fragment of classical logic.

The most important fact that is known about $HCL_{\leftrightarrow,\neg}$ is that it is *weakly* complete for $\mathbf{CL}_{\leftrightarrow,\neg}$, that is: $\vdash_{HCL_{\leftrightarrow,\neg}} \varphi$ iff $\vdash_{\mathbf{CL}_{\leftrightarrow,\neg}} \varphi$. ([8].[3] See also a review by Bennet of this paper in JSL, Vol. 2, P. 173, 1937.) It is also easily proved that $HCL_{\leftrightarrow,\neg}$ is *strongly* sound for $\mathbf{CL}_{\leftrightarrow,\neg}$, that is: if $\mathcal{T} \vdash_{HCL_{\leftrightarrow,\neg}} \varphi$ then $\mathcal{T} \vdash_{\mathbf{CL}_{\leftrightarrow,\neg}} \varphi$. However, in [4] it was shown that $HCL_\to$ (the pure implicational fragment of $HCL_{\to,\neg}$) is not only strongly sound and weakly complete for $\mathbf{CL}_\to$, but is also *strongly complete* for it. Unfortunately, the corresponding proposition is not true for $HCL_{\leftrightarrow,\neg}$.

Proposition 13. $\neg p, p \vdash_{\mathbf{CL}_{\leftrightarrow,\neg}} q$, *but* $\neg p, p \nvdash_{HCL_{\leftrightarrow,\neg}} q$.

---
[2] Already in [9] (P. 307) it was shown that $\mathbf{CL}_\leftrightarrow$ is Post-complete in the sense that one cannot add any new axiom to it in its language.
[3] The system used in [8] is easily seen to be equivalent to $HCL_{\leftrightarrow,\neg}$.

Proof. The first part is obvious. The second one easily follows from the fact that $HCL_{\neg\leftrightarrow}$ has the RDP (Theorem 6). □

Corollary 14. *$HCL_{\neg\leftrightarrow}$ is not strongly sound and complete for $\mathbf{CL}_{\neg\leftrightarrow}$.*

Definition 15. *$HCL^i_{\neg\leftrightarrow}$ is the system obtained from $HCL_{\neg\leftrightarrow}$ by adding to it $\varphi \to \neg\varphi$ as an axiom.*

Theorem 16. *$HCL^i_{\neg\leftrightarrow}$ is the sole non-trivial proper axiomatic extension of $HCL_{\neg\leftrightarrow}$.*

Proof. Let HL be a non-trivial proper axiomatic extension of $HCL_{\neg\leftrightarrow}$. Let φ be a formula such that $\vdash_{HL} \varphi$, but $\nvdash_{HCL_{\neg\leftrightarrow}} \varphi$. By the weak completeness of $HCL_{\neg\leftrightarrow}$ for $\mathbf{CL}_{\neg\leftrightarrow}$, φ is not valid in classical logic. Let ν be a classical valuation such that $\nu(\varphi) = f$. Let φ' be obtained from φ by substituting $p \to p$ for every atomic formula q such that $\nu(q) = t$, and $\neg(p \to p)$ for every atomic formula q such that $\nu(q) = f$. Then $\vdash_{HL} \varphi'$. On the other hand, $v(\varphi') = f$ for every valuation v, and so $\neg\varphi'$ is a tautology. By weak completeness again, it follows that $\vdash_{HCL_{\neg\leftrightarrow}} \neg\varphi'$. Hence both φ' and its negation are theorems of HL. Since $\neg\varphi' \to (\varphi' \to (\psi \to \neg\psi))$ is a tautology (where $\to$ is interpreted as the biconditional), we get that $\vdash_{HL} \psi \to \neg\psi$ for every ψ. Hence HL is an extension of $HCL^i_{\neg\leftrightarrow}$. Suppose that it is a proper extension of $HCL^i_{\neg\leftrightarrow}$. Then there is a formula ψ such that $\vdash_{HL} \psi$, while $\nvdash_{HCL^i_{\neg\leftrightarrow}} \psi$. Let $\psi^\star$ be a formula in which $\neg$ does not occur, and is equivalent to ψ in $HCL^i_{\neg\leftrightarrow}$. (It is easy to show that such a formula exists.) Then $\vdash_{HL} \psi^\star$, while $\nvdash_{HCL^i_{\neg\leftrightarrow}} \psi^\star$. Hence $\nvdash_{HCL_\to} \psi^\star$ too. Since $HCL_\to$ is Post-complete (Note 11), this implies that every formula in $\to$ is provable in HL. In particular $\vdash_{HL} p$ when p is an atom, and so HL is trivial. A contradiction. Hence HL and $HCL^i_{\neg\leftrightarrow}$ are equivalent. □

Note 17. By Theorem 16, $HCL_{\neg\leftrightarrow}$ is not even Post-complete.[4] However, the difference from $HCL_\leftrightarrow$ is small: $HCL_{\neg\leftrightarrow}$ has just one proper axiomatic extension.

---

[4] That $\mathbf{CL}_{\neg\leftrightarrow}$ has no Post-complete axiomatization has already been noted in Section 8 of [9].

References

[1] A. R. Anderson and N. D. Belnap. *Entailment: The Logic of Relevance and Necessity, Vol.I.* Princton University Press, 1975.

[2] A. Avron. Relevant entailment - semantics and formal systems. *Journal of Symbolic Logic*, 49:334–342, 1984.

[3] A. Avron. The semantics and proof theory of linear logic. *Theoretical Computer Science*, 57:161–184, 1988.

[4] A. Avron. Semi-implication: A chapter in universal logic. In A. Koslow and A. Buchsbaum, editors, *The Road to Universal Logic, Volume I.*, Studies in Universal Logic, pages 59–72. Birkhűser, 2015.

[5] A. Avron, O. Arieli, and A. Zamansky. *Theory of Effective Propositional Paraconsitent Logics*, volume 75 of *Studies in Logic (Mathematical Logic and Foundations)*. College Publications, 2018.

[6] J. M. Dunn and G. Restall. Relevance logic. In *Handbook of Philosophical Logic*, volume 6, pages 1–136. Kluwer, 2002. Second edition.

[7] J. E. Girard. Linear logic. *Theoretical Computer Science*, 50:1–102, 1987.

[8] E. Gh. Mihailescu. Recherches sur la negation el l'équivalence dans le calcul des proposition. *Annales scientiftques de l'Université de Jas*, pages 369–408, 1937.

[9] A. N. Prior. *Formal Logic*. Oxford, 1962. second edition.

A roadmap of paraconsistent hybrid logics

Diana Costa$^\diamond$
Manuel A. Martins$^\square$

$^\diamond$ University College London, United Kingdom
$^\square$ CIDMA – Center for R&D in Mathematics and Applications, Department of Mathematics, University of Aveiro, Portugal

Abstract

Paraconsistent logics allow inconsistencies without the collapse of systems and their development has been driven not only by theoretical interest, but also by genuine problems in different scientific domains, such as Computer Science, Medicine or Robotics. On the other hand, the description of relational structures is easily formalized by resorting to hybrid logics. The addition of nominals and a satisfaction operator to modal logic makes it possible to specify what happens at a particular state as well as to specify equalities and transitions between states. This roadmap explores two recent new logics which combine both of the above mentioned. In a first instance, we allow inconsistent propositional variables, and later we extend paraconsistency to transitions. In the latter, modal operators are no longer dual.

From modal to hybrid logics

The study of modal logics begins by considering an extension of classical propositional logic that incorporates modalities [9]. The traditional alethic modalities regard possibility, necessity and impossibility but there are other modalities that have been formalized such as temporal, epistemic and deontic ones. The use of modal logics is widespread as they provide a simple formalism for working with relational structures (or multigraphs). The basic modal logic incorporates the modal operator $\square$, which expresses necessity, and its dual $\lozenge$, that captures the notion of possibility; thus the formula $\square\varphi$ represents that φ is the case in every possible circumstance and the formula $\lozenge\varphi$ that φ is the case in at least one possible circumstance. These two modal operators are dual in a similar way as quantifiers $\forall$ and $\exists$ are, namely $\square\varphi \equiv \neg\lozenge\neg\varphi$.

In 1959 Saul Kripke introduced what is now the standard semantics for modal logics. Truth in a model is relative to points in a set, the domain, which are usually taken to represent possible worlds, times, epistemic states, states in a computer, or something else. Therefore a propositional variable may have different truth values relative to different points. A model also includes an accessibility relation between worlds and a valuation that assigns to each propositional variable the set of worlds where it holds. Modal formulas are evaluated as follows: $\Diamond\varphi$ is true in a world w if and only if there is a world accessible from w where φ is true; similarly $\Box\varphi$ is true in a world w if and only if φ is true in all worlds accessible from w.

Unfortunately, modal logics lack in mechanisms for naming possible worlds, asserting equalities and describing accessibility relations between them. These limitations are overcome with hybrid logics, an extension of modal logic but whose history begins with Arthur Prior's work in tense logic in the 50s, [8]. In its most basic form we find a new class of atomic formulas, called nominals, which are true at exactly one state. We can say that nominals act as names for the unique world they are true at. Propositional hybrid logic also includes the so called satisfaction operator @, such that if i is a nominal and φ is an arbitrary formula, then $@_i\varphi$ is a new formula, called a satisfaction statement. This machinery allows us to express what happens at a specific state, thus can be viewed as a *jump* operator:

$$\begin{array}{ccc} @_i\varphi \text{ is true} & & \varphi \text{ is true} \\ \text{relatively to a state } w & \Longleftrightarrow & \text{in the state named by } i \end{array}$$

Note that the satisfaction operator @ shifts the state of evaluation w of a satisfaction statement $@_i\varphi$ to the state that is named by i, thus the state where we evaluate a satisfaction statement is irrelevant. Observe therefore that either $@_i\varphi$ is true at all worlds, or false at all worlds, depending on whether φ is true or false at the world named by i.

In particular we can express that two states are identical:

$$\begin{array}{ccc} @_i j \text{ is true} & \Longleftrightarrow & j \text{ is true} \\ \text{relatively to a state } w & & \text{in the state named by } i \\ & \Longleftrightarrow & i \text{ and } j \text{ name the same state} \end{array}$$

Observe that a world can be named by more than one nominal but a nominal names a single world.

Another special case is that of the accessibility between states:

$@_i \Diamond j$ is true
relatively to a state w $\iff$ the state named by j is reachable in one step from the state named by i

In a standard modal logic there are properties of the underlying transition structure which are simply not definable but that are easily expressed in a hybrid extension, such as irreflexivity, asymmetry or antisymmetry. Nonetheless, although being strictly more expressive than its modal fragment, the basic hybrid logic (where only nominals and the satisfaction operator are added) does not increase the complexity of the problem of determining whether a formula is valid or not, which is still decidable. However, in the strong Priorean logic, where quantification over world variables is possible, the complexity of that problem increases.

We can see nominal-like features and glimpses of the hybrid machinery in the work of Arthur Prior. In what he called the I-(later U-)calculus, propositions of the tense calculus are treated as predicates expressing properties of dates, represented by variables. He established that the formula px should be read as "p at x" and considered a binary relation I over dates, where xIy should be read as "y is later than x", [15]. By representing the time of utterance by means of an arbitrary date x, Fp represents "it is now the case that it will be the case that p happens" and it is equated with $\exists y(x\text{I}y \wedge py)$. Similarly for the past, Pp is equated with $\exists y(y\text{I}x \wedge py)$. In order to obtain a Kripke model for the logic of time, possible worlds become moments in time and the accessibility relation is taken as an ordering relation between moments in time. Thus the formula Fp is true at moment w if for some moment v in the future, *i.e.*, a moment such that $w < v$, p is true at v; analogously the formula Pp is true at moment w if for some moment v in the past, *i.e.*, a moment such that $v < w$, p is true at v.

In the 80s, the Bulgarian school of logic revived the interest in hybrid logic. It started with the proof that the union of two accessibility relations is definable in the basic modal language, *i.e.*, that the formula $\langle \text{T} \rangle p \leftrightarrow \langle \text{U} \rangle p \vee \langle \text{S} \rangle p$ is valid on a frame precisely if R_T, the relation that interprets the modality T, is the union of R_U and R_S, the relations that interpret modalities U and S, respectively. Yet, and it came as a surprise, the intersection of two accessibility relations does not work in the same way [29]. Gargov, Passy and Tinchev showed that the intersection can be defined using nominals, stating that $\langle \text{T} \rangle i \leftrightarrow \langle \text{U} \rangle i \wedge \langle \text{S} \rangle i$ [28]. The same occurs for complementation: although there is no formula of the basic modal logic that is valid on a frame where R_U is the complement of R_S, there is such a formula when nominals are added to the language, namely $\langle U \rangle i \leftrightarrow \neg \langle S \rangle i$.

Another interesting result is that named models, *i.e.*, models in which each world is named by at least one nominal, can be completely described by a set of formulas of the form $@_i p, @_i j, @_i \Diamond j$, known as the diagram of a model [7].

Tableau [12], Gentzen, and natural deduction [13] style proof-theory for hybrid logic work very well compared to ordinary modal logic. Usually, when a modal tableau, Gentzen, or natural deduction system is given, it is for one particular modal logic and it has turned out to be problematic to formulate such systems for modal logics in a uniform way without introducing metalinguistic machinery. This can be remedied with hybridization which enables the formulation of uniform tableau, Gentzen, and natural deduction systems for wide classes of logics.

Hybrid logics have been an opulent source of inspiration for many researchers in many areas and have been studied under the scope of feature logic, model theory, proof theory and natural language. It is worth mentioning some of the most relevant works in the field of hybrid logic, amongst which those of Patrick Blackburn, Maarten Marx, Carlos Areces, Balder ten Cate and Torben Braüner with studies on interpolation and complexity of hybrid logics [4, 2, 3], on bisimulation in hybrid logics, results on Hilbert axiomatizations for some extensions of basic hybrid logic [11], and the development of first-order hybrid logic [16], intuitionistic hybrid logic [17] and many-valued hybrid logic [30]. More recently hybrid logics took another dimension and we can find works on hierarchical hybrid logic [33] and hybridization [35, 32, 36] by Alexandre Madeira, Manuel Martins and Luís Barbosa. The group constituted by María Manzano, Antónia Huertas, Carlos Areces, Manuel Martins and Patrick Blackburn have been actively working on hybrid type theory [1, 34] and on the concept of intensionality [10].

However, there is a combination barely explored: we are talking about a crossing between hybrid logic and paraconsistency.

Notes on paraconsistency

Paraconsistent logics were created with the purpose of allowing inconsistencies without producing the collapse of systems. They do so by excluding the *Principle of Non-Contradiction* which states that from contradictory premises any formula can be derived. The prefix *"para"*, of Greek origin, has three synonyms: (1) "against", as in *"paradox"*, for "against the common sense"; (2) "beyond", as in *"paranormal"*; and finally (3) "very similar", "connected" or "nearby" as in *"parallel"* and *"parabola"*. Łukasiewicz was a pioneer in discussing the possibility of violating this principle. His disciple, Stanisław Jaśkowski, constructed the first system of propositional paraconsistent logic and for the last sixty years many philosophers, logicians and mathematicians have become involved in this area, with the Brazilian logic school (Newton da Costa, Walter Carnielli, Jean-Yves Béziau, João Marcos, Alexandre Costa-Leite) taking a prominent role.

Discussed for almost a century, paraconsistency is a growing topic of interest and many paraconsistent logics have been developed over the years, either to meet different aims or to target genuine problems in different scientific domains, such as Computer Science, Medicine or Robotics. In Computer Science, subdomains like requirements engineering ([25]), artificial intelligence ([26]) and automated reasoning within information processing knowledgebases ([24]), are amongst the most relevant areas in which paraconsistent logic can address difficulties raised by inconsistent data.

Traditionally, the consensus amongst the computer science community is that inconsistencies are undesirable. Many believe that databases, knowledgebases, and software specifications should be completely free of inconsistencies and thus try to eradicate them by any means possible. However, if contradictory information is the norm rather than the exception in the real world, it should be formalized and used to our advantage. Furthermore, contradictory information does not always mean wrong information, it can be part of a fraudulent operation thus detecting it would be a major step, while resolving it would result in the loss of valuable information.

Comparing heterogeneous sources often involves comparing conflicts and there are situations of our daily lives where we apply a paraconsistent reasoning. The simple gathering of the opinions of a group of people about a certain subject (when formalized roughly known as discussive logic) is a major source of contradictions. For some common examples, suppose that we are dealing with a group of clinicians giving advice to a patient, a group of witnesses of an incident or a set of newspaper reports covering some event – in all of these situations, some degree of inconsistency is expected. Therefore inconsistencies are no longer seen purely as anomalies and paraconsistent logics are now viewed as flexible logical systems able to handle heterogeneous and complex data as they accommodate inconsistency in a sensible manner that treats contradictions as informative.

In what follows we will make on overview of the cases when hybrid logic meets paraconsistency.

Hybrid logics meet paraconsistency (and paracompleteness)

We can roughly say that there are three components of hybrid logics that are prone to inconsistencies: (a) propositional variables, (b) accessibility relations and (c) nominals. Versions of hybrid logic where propositional variables are allowed to be inconsistent without trivializing the whole system can be found in [14] where intuitionistic hybrid logic and Nelson's logic N4 meet and [21] where a restriction of classical hybrid logic to formulas in negation normal form allows inconsistent and incomplete information on propositional variables. We will focus our attention in the latter.

Quasi-hybrid (QH) logic [21] is inspired by Besnard and Hunter's quasi-classical (QC) logic introduced in [6]. Whilst their work considered only formulas in conjunctive normal form, the hybrid version considers only formulas in negation normal form. This is still a huge limitation. Nonetheless, inconsistent and incomplete information about propositional variables is allowed by simply splitting the usual valuation, V, into two valuations, V^+, V^- such that $V^+(p)$ is the set of worlds where the propositional variable p holds and $V^-(p)$ as the set of worlds where $\neg p$ holds. We have thus guaranteed a way in which propositional variables may take one of four values at each world: both true and false at w if $w \in V^+(p) \cap V^-(p)$, only true if $w \in V^+(p), w \notin V^-(p)$, only false if $w \notin V^+(p), w \in V^-(p)$, or neither true nor false if $w \notin V^+(p) \cup V^-(p)$. The key thing to remember is that $w \vDash \neg p \nLeftrightarrow w \nvDash p$.

The semantics for disjunction, as in QC logic, resorts to the classical notion of disjunctive syllogism. This will preserve the link between a disjunct and its negation; for a disjunction $\varphi \vee \psi$ to hold when one of the disjuncts and its negation both hold, for example φ and $\neg \varphi$, the other disjunct, in this case ψ, must hold as well. Observe that in QH logic, $\neg \varphi$ is replaced with $\sim \varphi$ which is the formula that results from putting $\neg\varphi$ in negation normal form. A detailed discussion about the use of the disjunctive syllogism can be found in [5].

As in classical hybrid logic, the new structures where propositional variables may be inconsistent or incomplete can be described by a set of atomic formulas. Loosely speaking, a diagram is constituted by all *evidence* of what happens at specific states (at the most basic level – that of propositional variables), all *evidence* about the presence of transitions, and finally all *evidence* about equalities between states. Classically, if a formula is part of a diagram, then it is satisfied in the structure the diagram represents; on the other hand, a formula missing from the diagram is such that it is not satisfied in the structure, therefore its negation is. In QH logic there is not a direct connection between the satisfaction of a propositional variable and its negation, since their interpretations resort to distinct valuations. Therefore, in order to describe what happens at specific states we need both formulas of the form $@_i p$ and $@_i \neg p$. Evidence about transitions and equalities between states are represented by formulas of the form $@_i \Diamond j$ and $@_i j$; since the behaviour of these formulas remains classical, either they are part of a diagram, meaning that they are satisfied in the structure we are describing, or they are missing from the diagram, which means that their negation is satisfied in the structure. Therefore, a diagram is composed of sets of formulas of the form $@_i p, @_i \neg p, @_i j, @_i \Diamond j$ for $p \in \text{Prop}, i, j \in \text{Nom}$. Here is a simple example:

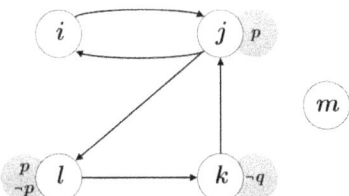

Figure 1: A structure with local inconsistencies.

The diagram of the structure represented in Figure 1 is as follows:

$\{@_i i, @_j j, @_k k, @_l l, @_m m,$ // nominal equalities
$@_j p, @_k \neg q, @_l p, @_l \neg p,$ // local properties
$@_i \Diamond j, @_j \Diamond i, @_j \Diamond l, @_k \Diamond j, @_l \Diamond k\}$ // transitions

A sound and complete tableau system and a decision procedure to check if a formula is a consequence of a set of formulas can be found in [18]. The construction of a tableau follows the usual steps, with restrictions to avoid repetition of formulas, the application of the same rule and entering loops. However, since inconsistencies are allowed, a branch will not close when arbitrary formulas φ and $\neg\varphi$ occur, as for example the case $\varphi = p$ is acceptable and does not lead to explosion. A branch will close when formulas φ and φ^* occur, where φ^* holds if and only if φ does not. The occurrence of a formula and its starred version corresponds to a *real* inconsistency. Observe that since a nominals' behaviour remains classical, the formulas i^* and $\neg i$ are equivalent. A formula φ is a consequence of a set of formulas Δ whenever φ holds in all the structures that satisfy all formulas in Δ, *i.e.* are models of Δ. A tableau-based decision procedure to check this consists of the construction of a tableau with root Δ, φ^*. A formula φ is a consequence of Δ if and only if the tableau closes. Moreover, the tableau construction algorithm can be used to obtain syntactic representations of models for a set of formulas. Several measures of inconsistency for models and databases were introduced in [21], allowing for comparisons and ultimately leading to choosing the *best*, as in least inconsistent, model or database. There is also a notion of bisimulation that preservers satisfiability in QH logic. The definition is a straightforward extension of the classical one and can be found in [21].

Let us now move on to our second topic: versions of hybrid logic with paraconsistency at the level of accessibility relations. Even though we can find modal versions where that is the case, such as Modal bilattice logic MBL in [38], the work of Wansing and Odintsov with BK^{FS} logic in [37] and Many-valued modal logic by Fitting in [27], the only hybrid variants are found under the form of Many-valued hybrid logic [30] and more recently Double-belnapian hybrid logic (DBHL) [23]. We move on to discussing the latter, and how it compares with the previous ones.

DBHL is a four-valued logic at the level of both propositional variables and accessibility relations whose main characteristic is the fact that the modal operators $\Box$ and $\Diamond$ do not act as duals.[1] The argument invoked to sustain this choice is that when it is not possible φ, formally represented as $\neg\Diamond\varphi$, it should not be assumed that the negation of φ is necessarily the case, $\Box\neg\varphi$. If the duality was kept, the usual semantics for modal operators would make it so that in a structure that satis-

---

[1] For the sake of simplicity, this paper deals with the case with a single modality, which is then ommitted. In [23], the multimodal case is considered.

fies the formulas $@_i\Diamond p$ and $\neg @_i\Diamond p$, negation would be carried towards the propositional variable and the latter would be equivalent to $@_i\Box\neg p$. Thus we would conclude that p was inconsistent in a world accessible from i. In this scenario, negation shifts from the perspective of the relation to the propositional variable and that is precisely what one wants to avoid.

In DBHL, $@_i\Diamond\varphi$ and $@_i\neg\Diamond\varphi$ shall be interpreted as "there is evidence of a transition from the world named by i to a world where φ holds" and "there is evidence that all transitions from i to worlds where φ holds are *missing*", respectively. The latter is not compatible with the interpretation of $@_i\Box\neg\varphi$ which is that "there is evidence that all transitions from the world named by i lead to a world where $\neg\varphi$ holds".[2] This alone, is already enough reason to claim that DBHL is neither an extension of pre-existing paraconsistent modal logics with hybrid logic features (namely nominals and the satisfaction operator), nor can it be captured by MVHL. For a more detailed comparison, check [23]. Additionally, in the semantics for disjunction, the classical notion of disjunctive syllogism is used, as in [21].

The structures underlying this system will incorporate two valuations in order to deal with contradictions at the level of propositional variables, V^+ and V^-, and will, analogously, consider two accessibility relations, R^+ and R^- in order to deal with contradictions at the level of the accessibility relations. The semantics for nominals is the usual one: each nominal holds at a unique state. There is now no restriction to the language used, all the usual hybrid formulas are evaluated.

We update the tableau system for QH logic in order to incorporate rules for the new formulas allowed and adjust the rules for modal operators. The new system is still sound, complete and terminating. Once again, resorting to a tableau-based decision procedure it is possible to check if a formula follows from a set of formulas and it is also possible to build models for a certain database, *i.e.*, given a set of hybrid formulas Δ a tableau whose root consists of all formulas in Δ allows us to extract a syntactic representation of a structure where all the initial formulas

---

[2]We have recently submitted an extended version of [23] where the semantics for $\neg\Box\varphi$ is slightly different. In what we called 4HL, $\neg\Box\varphi$ holds at a world w if and only if there exists a world w' such that wR^-w' (in other words, there is no evidence of the lack of transition between w and w') and φ does not hold at w'; in DBHL the last condition is that $\neg\varphi$ holds at w'. This does not change the interpretation of $\neg\Box i$ since nominals behave classically, thus diagrams (more on this in what follows) represent the same structure in both DBHL and 4HL.

hold. The syntactic representation referred is none other than the diagram of the structure. For structures where local inconsistencies and inconsistencies on transitions are allowed, the representation in DBHL resorts to sets of formulas of the form $@_i p, @_i \neg p$, to represent the former, formulas of the form $@_i \Diamond j, @_i \neg \Diamond j$ to represent the latter, and $@_{ij}$ to represent equalities between nominals. Note that the non-classical behaviour that was exclusive of propositional variables in QH logic, is, in DBHL, also shared by modal formulas, which is the reason why, not only do we need formulas of the form $@_i p$ and $@_i \neg p$, but also of the form $@_i \Diamond j$ and $@_i \neg \Diamond j$. Additionally, observe that, analogously to what happens in QH logic, formulas of the form $@_i \neg j$ are simply unnecessary. An example is as follows:

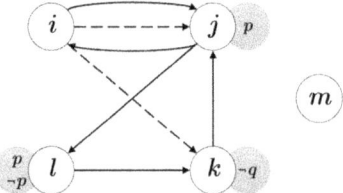

Figure 2: A structure with local inconsistencies and inconsistent transitions.
A full line indicates evidence of a transition and a dashed line indicates evidence of the lack of a transition.

The diagram of the structure represented in Figure 2 is as follows:

$\{@_i i, @_j j, @_k k, @_l l, @_m m,$ // nominal equalities
$@_j p, @_k \neg q, @_l p, @_l \neg p,$ // local properties
$@_i \Diamond j, @_i \neg \Diamond j, @_i \neg \Diamond k, @_j \Diamond i,$ // transitions
$@_j \Diamond l, @_k \Diamond j, @_l \Diamond k\}$

The measures of inconsistency for models and databases can be applied to these new structures, bearing in mind that they would only count the number of inconsistencies at the level of propositional variables. These measures can of course be extended so that they count the number of inconsistent transitions as well. Both absolute and relative measures (which come as a ratio between the number of actual inconsistencies and the number of possible inconsistencies) can be found in [18]. Curiously enough, a suitable notion of bisimulation in DBHL that would preserve satisfiability was not as simple to find as that for QH logic. In fact, the usual construction did not preserve satisfiability. Nonetheless, a new notion of bisimulation for 4HL, a marginally different logic from DBHL briefly discussed in Footnote 2, has been proved invariant.

At this moment the reader may wonder about what happens when nominals join the party. The third topic we introduced in the beginning of the section is still unexplored. Allowing inconsistencies at the level of nominals has several nuances. First of all, it should be defined when is it that a nominal is inconsistent: is it the case that the same nominal holds in more than one world, making it closer in behaviour to propositional variables, or shall we keep the restriction that a nominal holds only in a single world and assume that at that world its negation may hold as well? This choice and the constructions carried by it and how it fits with previous work, especially when intertwined with inconsistencies at the level of accessibility relations, will certainly constitute a nice challenge.

A wrap-up of the present and future of paraconsistent hybrid logics

Have you ever hear of the story of the mountaineers lost in the Alps who rescued themselves after a member of the group found a map of the Pyrenees? Even though they were holding a wrong map, the discussion created around it was enough to propel the mountaineers in the right direction towards safety.[3] What the story illustrates is how powerful discussing the information we have at hands can be. It is clearly also a way for us to advocate in favour of paraconsistency. Gathering information is rarely immune to gluts and gaps, but when we embrace these imperfect bits of information and make them part of a bigger system, we can reach meaningful conclusions.

This roadmap intended to give a brief flavour on variants of hybrid logic where contradictions are allowed, in a first instance at the level of propositional variables, and latter at the level of accessibility relations as well. Even though the combination of hybrid logic with paraconsistency is still understudied, it certainly holds a lot of potential. Hybrid logic's machinery, in special nominals and the satisfaction operator, have been widely used and almost surely in situations where some degree of inconsistency is expected. That is why we strongly believe that a thor-

---

[3]The story is attributed to the Hungarian biochemist and Nobel laureate Albert Szent-Gyorgi and dates back to the 1930s.

ough study of this topic must be carried. Brief incursions in the fields of robotics and healthcare where paraconsistent hybrid logics are suggested as a means to formalize systems can be found in [20] and [19, 22], respectively.

DBHL makes the description of inconsistent maps an easy task; building a map from inconsistent information is equally simple. Some of our next steps include the development of graph algorithms where the underlying graph is replaced with one where transitions and local data may contain inconsistencies. We aim to explore problems as the travelling salesman one where the map given is either incomplete or inconsistent and answer the question "is there a way to travel from point A to point B?" with a level of (un)certainty.

Talking about levels of uncertainty, it is often the case that propositions are neither totally true nor totally false, but rather somewhat true and somewhat false. The study of situations when positive and negative evidence of the occurrence of a proposition do not add up to 1 is another topic for future research. There are several real-life situations where this may occur, namely we may get conflicting results when a property and its complement are being independently evaluated using different methods.

In the near future we hope to come up with a suitable notion of composition of four-valued relations. The classical case is described with dynamic logic [31] and we reckon that some interesting behaviours will emerge when belnapian relations are considered.

Acknowledgements. Diana Costa acknowledges the support from the UK's Engineering and Physical Sciences Research Council - EPSRC (Grant number EP/R0068 65/1). Both authors acknowledge the support given by The Center for Research and Development in Mathematics and Applications (CIDMA) through the Portuguese Foundation for Science and Technology (FCT - Fundação para a Ciência e a Tecnologia), references UIDB/04106/2020 and UIDP/04106/2020.

References

[1] Carlos Areces, Patrick Blackburn, Antonia Huertas, and María Manzano. Completeness in hybrid type theory. *Journal of Philosophical Logic*, 43(2):209–238, Jun 2014.

[2] Carlos Areces, Patrick Blackburn, and Maarten Marx. A road-map on complexity for hybrid logics. In J. Flum and Mario Rodríguez-Artalejo, editors, *Computer Science Logic*, Lecture Notes in Computer Science, pages 307–321. Springer, 1999.

[3] Carlos Areces, Patrick Blackburn, and Maarten Marx. Hybrid logics: Characterization, interpolation and complexity. *Journal of Symbolic Logic*, 66(3):977–1010, 2001.

[4] Carlos Areces, Patrick Blackburn, and Maarten Marx. Repairing the interpolation theorem in quantified modal logic. *Annals of Pure and Applied Logic*, 124(1-3):287–299, 2003.

[5] Ofer Arieli. On the application of the disjunctive syllogism in paraconsistent logics based on four states of information. In *Proceedings of the Twelfth International Conference on Principles of Knowledge Representation and Reasoning*, KR'10, pages 302–309. AAAI Press, 2010.

[6] Philippe Besnard and Anthony Hunter. Quasi-classical logic: Non-trivializable classical reasoning from inconsistent information. In Christine Froidevaux and Jãrg Kohlas, editors, *Symbolic and Quantitative Approaches to Reasoning and Uncertainty*, volume 946 of *Lecture Notes in Computer Science*, pages 44–51. Springer Berlin Heidelberg, 1995.

[7] Patrick Blackburn. Internalizing labelled deduction. *Journal of Logic and Computation*, 10(1):137–168, 2000.

[8] Patrick Blackburn. Representation, reasoning, and relational structures: A hybrid logic manifesto. *Logic Journal of the IGPL*, 8(3):339–365, 2000.

[9] Patrick Blackburn, Maarten de Rijke, and Yde Venema. *Modal logic*. Cambridge University Press, 2001.

[10] Patrick Blackburn, Manuel A. Martins, María Manzano, and Antonia Huertas. Rigid first-order hybrid logic. In Rosalie Iemhoff, Michael Moortgat, and Ruy J. G. B. de Queiroz, editors, *Logic, Language, Information, and Computation - 26th International Workshop, WoLLIC 2019, Utrecht, The Netherlands, July 2-5, 2019, Proceedings*, volume 11541 of *Lecture Notes in Computer Science*, pages 53–69. Springer, 2019.

[11] Patrick Blackburn and Balder ten Cate. Pure extensions, proof rules, and hybrid axiomatics. *Studia Logica*, 84(2):277–322, 2006.

[12] Thomas Bolander and Patrick Blackburn. Termination for hybrid tableaus. *Journal of Logic and Computation*, 17(3):517–554, 2007.

[13] Torben Braüner. Natural Deduction for Hybrid Logic. *Journal of Logic and Computation*, 14(3):329–353, 06 2004.

[14] Torben Braüner. Axioms for classical, intuitionistic, and paraconsistent hybrid logic. *Journal of Logic, Language and Information*, 15(3):179–194, 2006.

[15] Torben Braüner. Arthur Prior's temporal logic and the origin of contemporary hybrid logic. In Pedro Schmechtig and Gerhard Schönrich, editors, *Persistenz - Indexikalität - Zeiterfahrung*, Philosophische Analyse / Philosophical Analysis, pages 301–336. Ontos Verlag, 2011.

[16] Torben Braüner. First-order hybrid logic: introduction and survey. *Logic Journal of the IGPL*, 22(1):155–165, 2014.

[17] Torben Braüner and Valeria de Paiva. Intuitionistic hybrid logic. *Journal of Applied Logic*, 4(3):231–255, 2006.

[18] Diana Costa. *Hybrid logics with paraconsistency*. PhD thesis, University of Aveiro, 2019. Available at https://ria.ua.pt/handle/10773/28287.

[19] Diana Costa and M. A. Martins. Inconsistencies in health care knowledge. In *2014 IEEE 16th International Conference on e-Health Networking, Applications and Services (Healthcom)*, pages 37–42, October 2014.

[20] Diana Costa and Manuel A. Martins. Intelligent-based robot to deal with contradictions. In *2016 International Conference on Autonomous Robot Systems and Competitions, ICARSC 2016, Bragança, Portugal, May 4-6, 2016*, pages 199–204, 2016.

[21] Diana Costa and Manuel A. Martins. Paraconsistency in hybrid logic. *Journal of Logic and Computation*, 27(6):1825–1852, 2017.

[22] Diana Costa and Manuel A. Martins. Measuring inconsistent diagnoses. In *20th IEEE International Conference on e-Health Networking, Applications and Services, Healthcom 2018, Ostrava, Czech Republic, September 17-20, 2018*, pages 1–4. IEEE, 2018.

[23] Diana Costa and Manuel A. Martins. A four-valued hybrid logic with non-dual modal operators. In Luís Soares Barbosa and Alexandru Baltag, editors, *Dynamic Logic. New Trends and Applications - Second International Workshop, DaLí 2019, Porto, Portugal, October 7-11, 2019, Proceedings*, volume 12005 of *Lecture Notes in Computer Science*, pages 88–103. Springer, 2019.

[24] Sandra de Amo and Mônica Pais. A paraconsistent logic programming approach for querying inconsistent databases. *International Journal of Approximate Reasoning*, 46(2):366–386, 2007.

[25] Neil Ernst, Alexander Borgida, Ivan Jureta, and John Mylopoulos. Agile requirements engineering via paraconsistent reasoning. *Information Systems*, 43(0):100 – 116, 2014.

[26] J. Filho, Germano Lambert-Torres, and Jair Abe. *Uncertainty Treatment Using Paraconsistent Logic: Introducing Paraconsistent Artificial Neural Networks*. IOS Press, Amsterdam, The Netherlands, 2010.

[27] Melvin Fitting. Many-valued modal logics. *Fundam. Inform.*, 15(3-4):235–254, 1991.

[28] George Gargov, Solomon Passy, and Tinko Tinchev. Modal environment for boolean speculations. In DimiterG. Skordev, editor, *Mathematical Logic and Its Applications*, pages 253–263. Springer US, 1987.

[29] R. Goldblatt and S. Thomason. Axiomatic classes in propositional modal logic. In JohnNewsome Crossley, editor, *Algebra and Logic*, volume 450 of *Lecture Notes in Mathematics*, pages 163–173. Springer Berlin Heidelberg, 1975.

[30] Jens Hansen, Thomas Bolander, and Torben Braüner. Many-valued hybrid logic. In *Advances in Modal Logic*, pages 111–132, 2008.

[31] David Harel, Dexter Kozen, and Jerzy Tiuryn. *Dynamic logic.* Cambridge, MA: MIT Press, 2000.

[32] Alexandre Madeira, José M. Faria, Manuel A. Martins, and Luís Soares Barbosa. Hybrid specification of reactive systems: An institutional approach. In *Software Engineering and Formal Methods*, pages 269–285, 2011.

[33] Alexandre Madeira, Renato Neves, Manuel A. Martins, and Luís S. Barbosa. Hierarchical hybrid logic. *Electronic Notes in Theoretical Computer Science*, 338:167 – 184, 2018. The 12th Workshop on Logical and Semantic Frameworks, with Applications (LSFA 2017).

[34] María Manzano, Manuel A. Martins, and Antonia Huertas. Completeness in equational hybrid propositional type theory. *Stud Logica*, 107(6):1159–1198, 2019.

[35] Manuel A. Martins, Alexandre Madeira, Razvan Diaconescu, and Luís Soares Barbosa. Hybridization of institutions. In *Conference on Algebra and Coalgebra in Computer Science*, pages 283–297, 2011.

[36] Renato Neves, Alexandre Madeira, Manuel A. Martins, and Luís Soares Barbosa. Hybridisation at work. In *Conference on Algebra and Coalgebra in Computer Science*, pages 340–345, 2013.

[37] Sergei P. Odintsov and Heinrich Wansing. Disentangling FDE-based paraconsistent modal logics. *Studia Logica*, 105(6):1221–1254, Dec 2017.

[38] Umberto Rivieccio, Achim Jung, and Ramon Jansana. Four-valued modal logic: Kripke semantics and duality. *Journal of Logic and Computation*, 27(1):155–199, 2017.

A RELATIONAL MODEL FOR THE LOGIC OF DEDUCTION

Hércules de Araujo Feitosa
Angela Pereira Rodrigues Moreira
Marcelo Reicher Soares

Department of Mathematics, São Paulo State University (UNESP), Brazil

Abstract

The logic of deduction originated from Tarski consequence operators. While Tarski with your operator highlighted important characteristics of deduction and described them as a function, the logic of deduction put those characteristics in a logical environment. The first proof of adequacy for the logic of deduction was proposed via algebraic models, with TK-algebras. Naturally any set with a Tarski operator is a case of a TK-algebra and this condition permits the adequacy. Posteriorly some new models for the logic of deduction were proposed. In this paper, we introduce another one completely connected with a very popular definition of consequence relation. This is a model in Kripke style, but it is not exactly a Kripke model, given that the logic of deduction is not a normal modal logic.

Introduction

The logic of deduction or logic **TK** was proposed as a counterpart of Tarski consequence operators. In order to clean the way in which we develop this paper, we start with three very short sections where we remember the definition of Tarski consequence operator, show that TK-algebras are completely connected with consequence operators, and present the logic **TK**. In other section we present consequence as a relation and compare this definition with the definition of consequence operator. These two definitions are cases of Tarski spaces. As an original result, motivated by the consequence relation defined in this paper, we introduce a new relational semantic for the logic **TK**. This model is relational and it is in the Kripke style, but it is not exactly a Kripke model because the logic of deduction is not a normal modal logic. In [5] a relational model for **TK** was introduced, but motivated by modal properties of this logic, what generates a neighbourhood model.

1 Consequence operator

In this section we present the definition of Tarski consequence operator and Tarski spaces.

Definition 1.1. A consequence operator on E is a function $^- : \mathcal{P}(E) \to \mathcal{P}(E)$ such that, for every $A, B \subseteq E$:
 (i) $A \subseteq \overline{A}$
 (ii) $A \subseteq B \Rightarrow \overline{A} \subseteq \overline{B}$
 (iii) $\overline{\overline{A}} \subseteq \overline{A}$.

Of course, from (i) and (iii), for every $A \subseteq E$, it holds $\overline{\overline{A}} = \overline{A}$.

Definition 1.2. A consequence operator $^- : \mathcal{P}(E) \to \mathcal{P}(E)$ is finitary when, for every $A \subseteq E$:
 (iv) $\overline{A} = \cup\{\overline{A_f} : A_f \text{ is a finite subset of } A\}$.

Definition 1.3. A Tarski space (Tarski deductive system or closure space) is a pair $(E, ^-)$ such that E is a non-empty set and $^-$ is a consequence operator on E.

Definition 1.4. The set A is closed in a Tarski space $(E, ^-)$, if $\overline{A} = A$, and A is open if its complement relative to E, denoted by A^C, is closed in $(E, ^-)$.

Since, for all $A \subseteq E$, it follows that $\overline{\overline{A}} = \overline{A}$, then $\overline{A}$ is closed in $(E, ^-)$.

Proposition 1.5. *If* $(E, ^-)$ *is a Tarski space, then:*
 (i) $\overline{A}$ *is the least closed set that includes* A
 (ii) the set E *is closed*
 (iii) the set $\emptyset$ *is open.*

So $\overline{\emptyset}$ and E correspond to the least and the greatest closed sets, respectively, associated to the consequence operator $^-$.

Definition 1.6. The set $\overline{A}$ is the closure of A.

Proposition 1.7. If $(E,^-)$ is a Tarski space, then:
  (i) every intersection of closed sets is closed set in $(E,^-)$
  (ii) $\overline{A} = \cap \{X : A \subseteq X \text{ and } X = \overline{X}\}$.

Proof: (i) If $\{A_i\}$ is a collection of closed sets, then $\cap_i A_i \subseteq \overline{\cap_i A_i} \subseteq \cap_i \overline{A_i} = \cap_i A_i$. Hence, $\cap_i A_i = \overline{\cap_i A_i}$. ∎

Definition 1.8. The interior of A is the set: $\mathring{A} = \cup \{X \subseteq E : X \subseteq A \text{ and } X \text{ is open}\}$

Proposition 1.9. If $(E,^-)$ is a Tarski space, then for every $A, B \subseteq E$:
  (i) $\mathring{A} \subseteq A \subseteq \overline{A}$
  (ii) $\mathring{\mathring{A}} \subseteq \mathring{A}$
  (iii) $\mathring{\emptyset} \subseteq \emptyset$
  (iv) $A \subseteq B \Rightarrow \mathring{A} \subseteq \mathring{B}$.

2 TK-algebras

TK-algebras (cf. [4]) were introduced motivated by the concept of Tarski consequence operator. Now the definition of TK-algebra.

Definition 2.1. A TK-algebra is a 6-tuple $\mathcal{A} = (A, 0, 1, \vee, \sim, \bullet)$ such that $(A, 0, 1, \vee, \sim)$ is a Boolean algebra and $\bullet$ is a new operator, called operator of Tarski, such that:
  (i) $a \vee \bullet a = \bullet a$
  (ii) $\bullet a \vee \bullet (a \vee b) = \bullet (a \vee b)$
  (iii) $\bullet (\bullet a) = \bullet a$.

Since we are working with Boolean algebras, the item (i) of the previous definition asserts that, for every $a \in A$, $a \leq \bullet a$.

Examples:
(a) The space of sets $\mathcal{P}(A)$ with $A \neq \emptyset$ and $\bullet a = a$, for all $a \in A$, is a TK-algebra.
(b) The space of sets $\mathcal{P}(\mathbb{R})$ with $\bullet X = X \cup \{0\}$ is a TK-algebra.
(c) The space of sets $\mathcal{P}(\mathbb{R})$ with $\bullet X = \cap \{I : I \text{ is an interval and } X \subseteq I\}$ is a TK-algebra.

Proposition 2.2. If $\mathcal{A} = (A, 0, 1, \vee, \sim, \bullet)$ is a TK-algebra, then it holds:
  (i) $\sim \bullet a \leq \sim a \leq \bullet \sim a$
  (ii) $a \leq b \Rightarrow \bullet a \leq \bullet b$
  (iii) $\bullet (a \wedge b) \leq \bullet a \wedge \bullet b$
  (iv) $\bullet a \vee \bullet b \leq \bullet (a \vee b)$.

3 The logic TK

The Logic **TK** (cf. [3]) was introduced from the TK-algebras.

The propositional logic **TK** is constructed over the propositional language $L = \{\neg, \vee, \rightarrow, \blacklozenge, p_1, p_2, p_3, ...\}$ with the following axioms and rules:

(CPC) $\quad\quad\quad\quad \varphi$, if φ is a tautology

(TK_1) $\quad\quad\quad\quad \varphi \rightarrow \blacklozenge\varphi$

(TK_2) $\quad\quad\quad\quad \blacklozenge\blacklozenge\varphi \rightarrow \blacklozenge\varphi$

(MP) $\quad\quad\quad\quad \dfrac{\varphi \rightarrow \psi, \varphi}{\psi}$

($RM^\blacklozenge$) $\quad\quad\quad\quad \dfrac{\vdash \varphi \rightarrow \psi}{\vdash \blacklozenge\varphi \rightarrow \blacklozenge\psi}.$

As usual, we write $\vdash \varphi$ to indicate that φ is a theorem. If $\Gamma \cup \{\varphi\}$ is a set of formulas, then Γ deduces φ, what is denoted by $\Gamma \vdash \varphi$, if there is a finite sequence of formulas $\varphi_1, ..., \varphi_n$ such that $\varphi_n = \varphi$ and, for every $\varphi_i, 1 \leq i \leq n$:

φ_i is an axiom, or

$\varphi_i \in \Gamma$, or

φ_i is obtained from previous formulas of the sequence by some of the deduction rules.

The TK-algebras are algebraic models for the logic **TK**, as it is shown in [3].

4 Consequence relation

Now we see the consequence as a relation.

Definition 4.1. Let E be a non-empty set. A consequence relation on E is a relation $\vdash \, \subseteq \mathcal{P}(E) \times E$ such that, for every $A \cup B \cup \{x, y\} \subseteq E$:

(i) $x \in A \Rightarrow A \vdash x$

(ii) $A \vdash x$ and $A \subseteq B \Rightarrow B \vdash x$

(iii) $A \vdash x$ and $B \cup \{x\} \vdash y \Rightarrow A \cup B \vdash y$.

Proposition 4.2. *The conditions (i) and (iii) implies (ii).*
Proof: If $A \vdash x$ and $A \subseteq B$, using (i), $B \cup \{x\} \vdash x$ and, by (iii), $A \cup B \vdash x$. As $A \subseteq B$, then $B \vdash x$. ∎

Proposition 4.3. *The conditions (iii) implies (iv)* $A \vdash x$ *and* $A \cup \{x\} \vdash y \Rightarrow A \vdash y$.

Proposition 4.4. *If $\vdash$ is a consequence relation on E and $^- : \mathcal{P}(E) \to \mathcal{P}(E)$ is defined for each $A \subseteq E$ by $\overline{A} = \{x \in E : A \vdash x\}$, then $^-$ is a consequence operator.*

Proposition 4.5. *If $^- : \mathcal{P}(E) \to \mathcal{P}(E)$ is a Tarski consequence operator, then the induced relation $A \vdash x \Leftrightarrow x \in \overline{A}$ is a consequence relation.*

So we have a bridge connecting consequence relations and consequence operators.

5 A new model for TK

Now we define a relational model for **TK**.

Definition 5.1. A frame for **TK** is a structure $\langle W, \vdash \rangle$ in which W is a non-empty set of possible worlds and $\vdash$ is a relation on $\mathcal{P}(W) \times W$ such that:
 (i) $x \in A \Rightarrow A \vdash x$
 (ii) $A \vdash x$ and $B \cup \{x\} \vdash y \Rightarrow A \cup B \vdash y$.

Of course, $\langle W, \vdash \rangle$ is a consequence relation.

Definition 5.2. A valuation v on W is a function from the set of atomic formulas of **TK** to $\mathcal{P}(W)$.

Definition 5.3. The valuation v must be extended to the set of all formulas by:
 (i) $v(\neg \varphi) = v(\varphi)^C$
 (ii) $v(\varphi \vee \psi) = v(\varphi) \cup v(\psi)$
 (iii) $v(\varphi \to \psi) = v(\varphi)^C \cup v(\psi)$
 (iv) $v(\blacklozenge \varphi) = \{y \in W : v(\varphi) \vdash y\}$.

So, if $\top$ and $\bot$ represent respectively any formula **TK**-valid and **TK**-invalid, then $v(\top) = W$ and $v(\bot) = \emptyset$.

Definition 5.4. A model for **TK** is a pair $\mathfrak{M} = \langle \mathfrak{F}, v \rangle$ or a triple $\mathfrak{M} = \langle W, \vdash, v \rangle$ such that $\mathfrak{F} = \langle W, \vdash \rangle$ is a frame for **TK** and v a valuation on W.

We denote that a model $\mathfrak{M}$ in a world x satisfies a formula ψ or that ψ is true in the world x of $\mathfrak{M}$ by $(\mathfrak{M}, x) \vDash \psi$.

Definition 5.5. The satisfaction of ψ is inductively defined by:
(i) if ψ is a propositional variable p, then $(\mathfrak{M}, x) \vDash p \Leftrightarrow x \in v(p)$
(ii) $(\mathfrak{M}, x) \vDash \neg \varphi \Leftrightarrow (\mathfrak{M}, x) \nvDash \varphi$
(iii) $(\mathfrak{M}, x) \vDash \varphi \vee \sigma \Leftrightarrow (\mathfrak{M}, x) \vDash \varphi$ or $(\mathfrak{M}, x) \vDash \sigma$
(iv) $(\mathfrak{M}, x) \vDash \varphi \to \sigma \Leftrightarrow (\mathfrak{M}, x) \nvDash \varphi$ or $(\mathfrak{M}, x) \vDash \sigma$
(v) $(\mathfrak{M}, x) \vDash \blacklozenge \varphi \Leftrightarrow$ if there is $y \in W$ such that $v(\varphi) \vdash x$ and $v(\varphi) \cup \{x\} \vdash y$, then $(\mathfrak{M}, y) \vDash \varphi$.

The condition (v) uses the bridge between consequence operator and consequence relation.

Definition 5.6. A formula φ is valid in the model $\mathfrak{M} = \langle W, \vdash, v \rangle$ if it is true in every world $x \in W$. The formula φ is valid if it is true in every model $\mathfrak{M}$.

We denote that φ is valid in $\mathfrak{M}$ by $\mathfrak{M} \vDash \varphi$, and that φ is valid by $\vDash \varphi$.

If Γ is a set of formulas and $\mathfrak{M}$ a model, then we write $\mathfrak{M} \vDash \Gamma$ if, and only if, $\mathfrak{M} \vDash \varphi$, for every $\varphi \in \Gamma$.

Definition 5.7. Let $\Gamma \cup \{\varphi\}$ be a set of formulas. The set of formulas Γ implies φ when, for every model $\mathfrak{M}$, if $\mathfrak{M} \vDash \Gamma$, then $\mathfrak{M} \vDash \varphi$.

We denote that Γ implies φ by $\Gamma \vDash \varphi$.

Acknowledgements

This work has been sponsored by FAPESP and CNPq.

References

[1] Chellas, B. (1980) *Modal Logic: an introduction.* Cambridge: Cambridge University Press.

[2] Feitosa, H. A.; Nascimento, M. C. Logic of deduction: models of pre-order and maximal theories. *South American Journal of Logic*, v. 1, p. 283-297, 2015.

[3] Feitosa, H. A.; Nascimento, M. C.; Grácio, M. C. C. Logic TK: algebraic notions from Tarki's consequence operator. *Principia*, v. 14, p. 47-70, 2010.

[4] Nascimento, M. C.; Feitosa, H. A. As álgebras dos operadores de consequência. *Revista de Matemática e Estatística*, v. 23, n. 1, p. 19-30, 2005.

[5] Mortari, C. A.; Feitosa, H. A. A neighbourhood semantic for the Logic TK. *Principia*, v. 15, p. 287-302, 2011.

[6] Rasiowa, H. *An algebraic approach to non-classical logics*. Amsterdam: North-Holland, 1974.

FROM PRAGMATIC TRUTHS TO EMOTIONAL TRUTHS

Andrew Schumann

University of Information Technology and Management in Rzeszow, Poland

Abstract

In this paper, I propose an extension of partial models introduced first by Irene Mikenberg, Newton C. A. Da Costa, and Rolando Chuaqui. The partial models were made to explicate the notion of pragmatic truths. In my extension of partial models we can define partial relations for explicating the notion of emotional truths.

1 Introduction

Recall that in classical logic a well-formed proposition is evaluated as either true or false and in conventional logics as a degree of truth (the latter could be however very different, e.g. it could run the unit interval $[0, 1]$ as in fuzzy logics, trees of some data as in spatial logics and behavior logics, sets of truth values as in higher-order fuzzy logics and some paraconsistent logics, etc.) [2]. However, the question raised by Ronald de Sousa [5], [6], [7] how we should evaluate emotional (performative) propositions that contain speaker's pragmatic values such as 'I fear this one,' 'I order you to leave the room,' 'I am thinking about' is open still.

In [5], there was introduced the rule of evaluating emotional propositions such as 'Fear that p':

'Fear that p' is satisfied iff p is true, but it is successful iff p is actually dangerous.

Let $E(p)$ be a performative proposition, where E is an emotion and p is a proposition. Then the rule of R. de Sousa is formulated as follows:

$E(p)$ is satisfied iff p is true $E(p)$ is successful iff p actually fits E's formal object.

We can assume that a performative proposition $E(p)$ cannot be successful if it is not satisfied:

$E(p)$ is successful iff p is true and p actually fits E's formal object.

So, in everyday speeches a non-defective performative proposition is evaluated as *either successful or unsuccessful* in the given context of utterance (notice that within the more detailed consideration in the same way as in non-classical logics, the meaning of non-defective simple performative proposition could be evaluated as a degree of successfulness). It seems as though we can consider performative propositions just within informal logic and never within symbolic logic taking into account the fact that the successfulness of performative propositions depends on human actions and utterance contexts and cannot be completely formalized.

The point is that the classical conception of semantical truth was developed by Alfred Tarski [14]. According to him, the semantical model for the class of n-ary predicates $\{P_i^n\}_i$ as atomic propositions is defied as an ordered system $\mathfrak{D} = \langle D, \{R_j^n\}_j, \{P_i^n\}_i, V, \{\top, \bot\}\rangle$, where D is a non-empty set, R_j^n are n-ary relations on this set, V is an evaluation function: $(\{P_i^n\}_i \times \{R_j^n\}_j) \mapsto \{\top, \bot\}$, where $\top$ is the meaning to be true and $\bot$ is the meaning to be false. So, $V(P_i^n, R_j^n) = t$ is to read P_i^n has the value t (either true, $\top$, or false, $\bot$) on the relation R_j^n.

> An atomic n-ary proposition P_i^n is true in $\mathfrak{D}$ (symbolically $\mathfrak{D} \models P_i^n$) iff there exists a relation R_j^n in $\mathfrak{D}$ such that the evaluation function V gives the meaning $\top$ for P_i^n on R_j^n.

Within this approach, we assume that the reality is completely given for us in the form, first, of all real objects (the set D) and, second, of all n-ary relations on D which can realize our n-ary predicates as atomic propositions. As we see, this assumption puts forward a context-free interpretation of atomic propositions. And, as a consequence, within it we cannot consider a kind of semantics for performative propositions.

The main disadvantage of the Tarskian approach is that we suppose to have a complete knowledge about reality. But, as we guess, it is absolutely impossible. In order to avoid this disadvantage, Irene Mikenberg, Newton C. A. Da Costa, and Rolando Chuaqui [9] proposed to replace the set of relation $\{R_j^n\}_j$ from the Tarskian model by the set of partial relations $\{\tilde{R}_j^n\}_j$.

Let us recall that each usual relation R_j^n is regarded as a subset of the Cartesian product $\underbrace{D \times D \times \cdots \times D}_{n} = D^n$, so that we can define the complement of R_j^n in D^n denoted by $\neg R_j^n$ in the way: $\neg R_j^n = D^n \backslash R_j^n$. So, for each relation R_j^n we have the rule of excluded middle $R_j^n \cup \neg R_j^n = D^n$ and the rule of contradiction $R_j^n \cap \neg R_j^n = \emptyset$.

The partial relation $\tilde{R}_j^n$ is said to be a relation which does not cover the whole D^n. It means that $\tilde{R}_j^n$ consists of all n-tuples that we know that belong to $\tilde{R}_j^n$. There are also n-tuples that we know that do not belong to $\tilde{R}_j^n$. Let us denote them by $\neg \tilde{R}_j^n$. It means that for $\tilde{R}_j^n$ we have the rule of contradiction: $\tilde{R}_j^n \cap \neg \tilde{R}_j^n = \emptyset$, but we do not have the rule of excluded middle: $\tilde{R}_j^n \cup \neg \tilde{R}_j^n \subset D^n$, i.e. $\tilde{R}_j^n \cup \neg \tilde{R}_j^n \neq D^n$. As a result, $D^n \backslash (\tilde{R}_j^n \cup \neg \tilde{R}_j^n)$ is the set of n-tuples for which it is not defined whether they belong or not to $\tilde{R}_j^n$.

Thus, within this new approach proposed in [9], we deal with an ordered system $\tilde{\mathfrak{D}} = \langle D, \{\tilde{R}_j^n\}_j, \{P_i^n\}_i, V, \{\top, \bot\} \rangle$, where $\tilde{R}_j^n$ are partial relations. This new structure is called pragmatic or partial and the new notion of truth is called pragmatic truth or quasi-truth. These notions give a weaker conception, which is more appropriate for the 'partialness' and the 'openness' of our reality. The idea of quasi-truth was further developed by Otávio Bueno and Edelcio G. de Souza in [3], where partial models are investigated, as well as by Newton C. A. da Costa, Otávio Bueno and Steven French in [4], where a new formulation of a coherence theory of truth is provided by using the resources of the partial structures approach, and in many other papers. The true meaning of atomic propositions is defined there as follows:

> An atomic n-ary proposition P_i^n is quasi-true in $\tilde{\mathfrak{D}}$ (symbolically $\tilde{\mathfrak{D}} \models P_i^n$) iff there exists a relation $\tilde{R}_j^n$ in $\tilde{\mathfrak{D}}$ such that the evaluation function V gives the meaning $\top$ for P_i^n on $\tilde{R}_j^n$.

Let $\mathfrak{D}$ and $\tilde{\mathfrak{D}}$ be structures of the same signature and each R_j^n in $\mathfrak{D}$ be an extension of $\tilde{R}_j^n$ to be defined for every n-tuples of D. Then between the Tarskian model $\mathfrak{D}$ and the pragmatic model $\tilde{\mathfrak{D}}$ we have the following inequality:

> If an atomic n-ary proposition P_i^n is true in $\mathfrak{D}$, then P_i^n is quasi-true in $\tilde{\mathfrak{D}}$, but not vice versa.

Emotional truths cannot be explicated within the Tarskian models $\mathfrak{D}$, but we can try to explicate them within the pragmatic (or partial) models $\tilde{\mathfrak{D}}$ as follows:

An atomic n-ary proposition P_i^n is emotionally true in $\tilde{\mathfrak{D}}$, if there exists a pragmatic model $\tilde{\mathfrak{D}}$, where P_i^n is quasi-true, and there are no Tarskian models $\mathfrak{D}$, where P_i^n is true.

In this paper, I try to sketch this definition formally.

2 Self-reference and indexicals

In conventional logic (it contains the classical one and the majority of non-classical logics), atomic propositions are being considered as claiming about the world presented as D of $\mathfrak{D}$ and these claims are true just in case the world is as it is claimed to be – the function V gives $\top$ for the atomic proposition P_i^n on $R_j^n \subseteq D^n$. We assume that each fact $R_j^n \subseteq D^n$ can be expressed by a proposition, i.e. there is a reference (correspondence) relation between facts and propositions (see Figure 1). This Tarskian approach to semantics was inspired by Aristotle and Bertrand Russell.

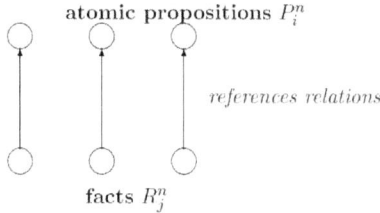

Figure 1: Tarskian semantics

However, very often we face propositions that refer to facts and it seems as though they express something about the world. Consider one of these propositions: 'I am just lying.' There are two options: either the proposition is true or it is not. Assume that it is true and then what it says is the case. As a result, the proposition is not true. Suppose, on the other hand, that it is not true, then this is what it says. Hence, the proposition is true. In either case it is both true and untrue. Such propositions are called semantic paradoxes (see Figure 2). They are self-referent.

Another example is to regard *performative* or *emotional propositions*. All propositions divide into informative and performative ones [11], [12],

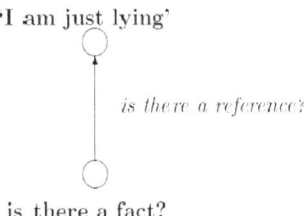

Figure 2: The Liar paradox

[13]. The first are built up by using an informative verb that describes an action which I can observe as process (e.g. 'walk,' 'jump,' 'run,' etc.), the second are built up by using a performative verb that describes an action which I cannot observe (e.g. 'think,' 'like,' 'hate,' etc.).

Let us consider the performative proposition 'You are looking good'. Let an appropriate fact say that a girl I am saying about is pretty indeed. In this case it is a true statement. However, we could assume that she is upset right now, e.g. her relative has just dead. Can I state that she is looking good now? Or she has got very tired, or she is very busy and so on. Further, assume that she is not very pretty, but she is very happy right now. Cannot I state that she is looking good?

Notice that 'be looking good' is a performative verb and its meaning depends on a context of utterance, not on facts. An appropriate statement is true if my utterance is accepted by a hearer I am talking to. Performative propositions such as 'you are looking good' are not true or false, they are successful or unsuccessful, i.e. their content is evaluated as either successful or unsuccessful in the given context of utterance (see Figure 3).

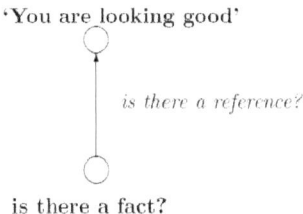

Figure 3: The performative statement

Let us assume that someone utters 'I am thinking' or shows this by gestures. Can I verify if (s)he is thinking in fact? No way, I cannot. The same case is if someone utters 'I like you' or shows this by gestures. The matter is that such propositions are more actions, than words. They have no references, because they are self-referent. I am thinking just in case I show it before a hearer. I like somebody just in case I let her/him know about it. The meaning of performative proposition is a fact of its successful utterance before a hearer, nothing more. Hence, meanings of performative propositions directly depend on our actions, not on the world.

Performative propositions (semantic paradoxes too) necessarily contain *indexicals*, expressions whose content, on a given occasion of use, depends systematically upon features of the context in which it is uttered. Indexicals include many types of words: 'I,' 'you,' 'he,' 'she,' etc. (the speaker of the context), 'here,' 'this,' 'that,' etc. (the context itself), 'now,' 'a day ago,' etc. (the time of the context), 'actually' (the world of the context), 'you' (the audience of the context), etc.

According to Saul Kripke, some propositions involving indexicals on a given occasion of use might express a content which seems to be knowable a priori such as 'I am the speaker of the context,' 'I am here now,' 'Stick S is one meter long at t_0.' Each of these examples, relative to our imagined contexts of utterance, expresses a *contingent a priori* [8].

Sometimes contingent a priori truths are said to be *analytic a posteriori* [1]. Notice that analytic a priori truths express logical relations (the latter may be verified in all possible worlds), synthetic a priori truths express semantic relations, more precisely rigid semantic connections (the latter may be verified just in some possible worlds), contingent a posteriori (analytic a posteriori) express pragmatic relations which hold everywhere, although they cannot be verified in possible worlds at all, they depend on human actions and possible worlds are not sufficient for verifying them.

3 Semantical models for atomic performative propositions

Let D be a set and I an infinite set of indices. Then the family D^I is the set of all functions $f \colon I \mapsto D$. A filter $\mathcal{F}$ on I is a family of sets $\mathcal{F} \subset 2^I$ for which: (1) $A \in \mathcal{F}$, $A \subset B \to B \in \mathcal{F}$; (2) $A_1,\ldots,A_n \in \mathcal{F} \to \bigcap_{k=1}^{n} A_k \in \mathcal{F}$ for any $n \geq 1$; (3) $\emptyset \notin \mathcal{F}$. The set of all complements for finite subsets of I is a filter and it is called a Fréchet filter on I and it is denoted by $\mathcal{U}_I$.

Let us define a relation $\backsim$ on the set D^I by $f \backsim g \equiv \{\alpha \in I \colon f(\alpha) = g(\alpha)\} \in \mathcal{U}_I$. It is easily be proved that the relation $\backsim$ is an equivalence. For each $f \in D^I$ let $[f]$ denote the equivalence class of f under $\backsim$. The ultrapower $D^I/\mathcal{U}_I$ is then defined to be the set of all equivalence classes $[f]$ as f ranges over D^I : $D^I/\mathcal{U}_I := \{[f] \colon f \in D^I\}$.

The ultrapower $D^I/\mathcal{U}_I$ is said to be a proper nonstandard extension of D and it is denoted by $^*D$, see [10]. Recall that each element of $^*D$ is an equivalence class $[f]$ where $f \colon I \to D$. There exist two groups of members of $^*D$: (1) equivalence classes of constant functions, e.g. $f(\alpha) = m \in D$ for all $\alpha \in I$. Such equivalence class is denoted by $^*m$ or $[f = m]$, (2) equivalence classes of functions that aren't constant.

The set $^\sigma D = \{^*m \colon m \in D\}$ is called a standard set. The members of $^\sigma D$ are called standard. It is readily seen that $^\sigma D$ and D are isomorphic: $^\sigma D \simeq D$.

Let D be a set of all possible facts as it is supposed in the Tarskian model $\mathfrak{D}$. Now let us define the standard extension of D denoted by $^\sigma D$ and its non-standard extension denoted by $^*D$. These $^\sigma D$ and $^*D$ can be considered as sets of infinite streams consisting of the members of D, but $^\sigma D$ and D are isomorphic. Let $^*D$ be the set of all situations, i.e. the set of all contexts of utterances, and $^\sigma D$ be the set of all facts.

Definition 1. *Assume that L is any first-order language containing n-ary predicates $\{P_i^n\}$ such that $\{P_i^n\} = \{^{inf}P_i^n\} \cup \{^{per}P_i^n\}$, where $\{^{inf}P_i^n\}$ is the set of informative atomic propositions and $\{^{per}P_i^n\}$ is the set of performative atomic propositions. Then the ordered system $^*\mathfrak{D} = \langle ^*D^n \setminus {}^\sigma D^n, {}^\sigma D, \{^*R_j^n\}_j, \{R_j^n\}_j, \{\tilde{R}_j^n\}_j, \{P_i^n\}_i, V, {}^*\{\top, \bot\}\rangle$ is called a situation model for informative and performative atomic propositions, where*

- $^*D^n \setminus {}^\sigma D^n$ *is the set of n-tuples of situations, ${}^\sigma D^n$ is the set of n-tuples of facts;*

- *each $^*R_j^n \subseteq {}^*D^n \setminus {}^\sigma D^n$ and each $R_j^n \subseteq {}^\sigma D^n$;*

- *each $\tilde{R}_j^n$ is a partial relation of ${}^\sigma D^n$;*

- V is a valuation from the set of atomic propositions $\{P_i^n\}$ such that
    - for any informative atomic proposition $^{inf}P_i^n$, $V(^{inf}P_i^n, R_j^n) = \{t\}$, where $\{t\}$ is the singleton such that $t \in {}^\sigma\{\bot, \top\}$;
    - for any performative atomic proposition $^{per}P_i^n$, $V(^{per}P_i^n, {}^*R_j^n) = {}^*T$, where $^*T \subseteq {}^*\{\bot, \top\} \setminus {}^\sigma\{\bot, \top\}$.

In the way of [3], we can consider partial structures $\tilde{\mathfrak{D}} = \langle {}^\sigma D, \{R_j^n\}_j, \{\tilde{R}_j^n\}_j, \{P_i^n\}_i, V, \{\top, \bot\}\rangle$, where for each $\tilde{R}_j^n$ there exists its extension R_j^n. Then an atomic proposition P_i^n is quasi-true in $\tilde{\mathfrak{D}}$ if P_i^n is true in $\tilde{\mathfrak{D}}$ in the Tarskian sense. If P_i^n is not quasi-true in $\tilde{\mathfrak{D}}$, we say that P_i^n is quasi-false.

Let us show how we can interpret semantic paradoxes in $^*\mathfrak{D}$ formulated as an atomic proposition. So, the Liar paradox is not addressed to some outside objects. It means that it is false on all non-empty sets, i.e. its truth meaning is equal to the infinite stream $^*\bot = \langle \bot, \bot, \ldots \rangle$. But it can be presented as an infinite stream of evaluations on the empty set:

$(V(\text{'this proposition is not true'}, \emptyset) = \bot) \longrightarrow (V(\text{'this proposition is not true'}, \emptyset) = \top) \longrightarrow (V(\text{'this proposition is not true'}, \emptyset) = \bot) \longrightarrow \ldots$

Thus, a pragmatic value of semantic paradox is considered as an infinite stream $^*t = \langle t, \neg t, t, \neg t, t, \ldots \rangle$, where $t \in \{\bot, \top\}$. This stream can be defined by the following two mutual recursions: $a = \langle t, a''\rangle$ and $a'' = \langle \neg t, a\rangle$. So, in fact we have the set consisting of two infinite streams $^*T = \{\langle \top, \bot, \top, \bot, \ldots\rangle, \langle \bot, \top, \bot, \top, \ldots\rangle\}$ and $^*T \subset {}^*\{\bot, \top\} \setminus {}^\sigma\{\bot, \top\}$.

Now, we can define $V(\text{'this proposition is not true'}, \emptyset) = \{\langle \top, \bot, \top, \bot, \ldots\rangle, \langle \bot, \top, \bot, \top, \ldots\rangle\}$ and for any $^*R^n \subseteq {}^*D$ we have $V(\text{'this proposition is not true'}, {}^*R^n) = \{^*\bot\}$.

Let us consider how emotional values of performative propositions are defined in $^*\mathfrak{D}$. A relation $^*R_j^n \cup \neg^*R_j^n$ for a performative atomic proposition P_i^n is said to be a type of situations to be successful talk (to be fulfilled on $^*R_j^n$) or unsuccessful talk (to be fulfilled on $\neg^*R_j^n$). A performative proposition P_i^n is successful if it is uttered under conditions of an appropriate successful talk. Then it is followed by an expected (e.g. desirable) action. A performative proposition P_i^n is unsuccessful if it is uttered in an unsuccessful talk, i.e. conditions of executing an appropriate performance are not kept. Then it will not be followed by the expected action. Out of the type of situation (relation) $^*R_j^n \cup \neg^*R_j^n$ with uttering P_i^n there are sets which are inappropriate for the meaning of P_i^n. The union of these inappropriate sets is denoted by $^*D^n \setminus (^*R_j^n \cup \neg^*R_j^n)$. Hence, as we see, the following proposition is obvious:

Proposition 1. *Let $\mathfrak{D}$ be a Tarskian submodel of $^*\mathfrak{D} = \langle ^*D^n \setminus {}^\sigma D^n, {}^\sigma D, \{^*R_j^n\}_j, \{R_j^n\}_j, \{\tilde{R}_j^n\}_j, \{P_i^n\}_i, V, ^*\{\top, \bot\}\rangle$, i.e. $\mathfrak{D} = \langle {}^\sigma D, \{R_j^n\}_j, \{P_i^n\}_i, V, ^*\{\top, \bot\}\rangle$, where each R_j^n is an extension of $\tilde{R}_j^n$ from $^*\mathfrak{D}$ and each $R_j^n \subseteq {}^\sigma D$. Then there are no performative atomic propositions realized in $\mathfrak{D}$. In other words, each $^*R_j^n$, where an appropriate performative atomic proposition is successful or unsuccessful, is partial.*

From this proposition it follows that in $^*\mathfrak{D}$ we can explicate emotional truths indeed.

Take the atomic proposition $P^n := $ 'You are looking good'. Now let us try to find its type of situation $^*R_j^n \cup \neg^*R_j^n$ combined its successful and unsuccessful talks. Assume the following set of the process alphabet: {'successful talk', 'unsuccessful talk', 'you are looking good', 'cheering up by means of compliment', 'not cheering up by means of compliment'}. We find out there one performative atomic proposition 'you are looking good' that is a label of the process 'my dialogue with the girl' and the four states of the process: {'successful talk', 'unsuccessful talk', 'cheering up by means of compliment', 'not cheering up by means of compliment'}. The process is defined as follows:

> 'my dialogue with the girl' = ('successful talk' ⟶ 'cheering up by means of compliment' | 'unsuccessful talk' ⟶ 'not cheering up by means of compliment')
>
> 'cheering up by means of compliment' = ('you are looking good today' ⟶ 'cheering up by means of compliment' | 'unsuccessful talk' ⟶ 'not cheering up by means of compliment')
>
> 'not cheering up by means of compliment' = ('you are looking good today' ⟶ 'not cheering up by means of compliment' | 'successful talk' ⟶ 'cheering up by means of compliment')

Thus, we obtain a labelled transition system, which consists of an infinite tower of states $\mathcal{L} = \langle R_0^n, R_1^n, \ldots R_i^n \ldots \rangle_{i \to \infty} \subset {}^*D^n$ and a collection $\mathcal{T} = \{P^n\}$ of labels (transitions, actions) over them. The set $\mathcal{L}$

consists of infinite streams from $^*D^n$, the set $\mathcal{T}$ consists of performative propositions.

Our dialogue is formalized as an infinite sequence (stream) of subsets from D:
$$R_0^n \alpha_0 R_1^n R_2^n \alpha_0 R_3^n R_4^n \alpha_0 \ldots,$$
where α_0 is a performative proposition '*you are looking good*' as a label, $R_0^n, R_2^n, R_4^n, \ldots \in \{$'*successful talk*', '*unsuccessful talk*'$\}$, $R_1^n, R_3^n, R_5^n, \ldots \in \{$'*cheering up by means of compliment*', '*not cheering up by means of compliment*'$\}$.

Let us notice that '*successful talk*' may be presented by a set of facts from D, e.g. by the following: '*the girl is happy*', '*she is in good spirits*'. Hence, each level R_i^n from the infinite tower $\mathcal{L} = \langle R_0^n, R_1^n, \ldots R_i^n \ldots \rangle_{i \to \infty}$ is presented as a subset of D that is a partial relation to fix just some facts for the successful talk with uttering P^n and some facts for the unsuccessful talk wth uttering P^n. As a result, the infinite tower $\mathcal{L}$ is egual to the type of situation $^*R_j^n \cup \neg^* R_j^n$ for the performative atomic proposition P^n.

Definition 2. *Let P_i^n be a performative atomic proposition. Then its emotional truth is expressed as follows:*

P_i^n *is emotionally true in* $^*\mathfrak{D}$ *(symbolically* $^*\mathfrak{D} \models P_i^n$) *iff there exists a relation* $^*R_j^n$ *in* $^*\mathfrak{D}$ *such that the evaluation function V does not give the meaning* $\{^*\bot\}$ *for P_i^n on* $^*R_j^n$.

From this definition we entail that the Liar paradox is emotionally true in any model $^*\mathfrak{D}$.

4 Conclusion

In the paper, I have proposed a strong extension of the idea of pragmatic (or quasi-) truths and obtained some models to explicate the idea of emotional (performative) truths. Within these models we can interpret performative atomic propositions expressing different atomic illocutionary acts as they are given in illocutionary logic [13].

The main idea of this paper was inspired by the investigations in truth theory performed by the Brazilian logicians [2], [3], [4], [5], [6], [7], [9].

References

[1] Aldrich, V.C. Analytic A Posteriori Propositions, *Analysis*, Vol. 28, No. 6, 1968.

[2] Béziau, J-I. Truth as a Mathematical Object, *Principia*, 14 (1), 2010, 31-46.

[3] Bueno, O., de Souza, E.G. The concept of quasi-truth, *Logique & Analyse*, 153-154, 1996, 183-199.

[4] Costa, N. C. A. da, Bueno, O., French, S. A coherence theory of truth, *Manuscrito*, 30 (2), 2007, 539-568.

[5] De Sousa, R., Morton, A. Emotional Truth, *Proceedings of the Aristotelian Society*, Supplementary Volumes, 76, 2002, 247-263; 265-275.

[6] De Sousa, R. The Good and the True, *Mind*, 83, 1974, 534–51.

[7] De Sousa, R. Truth, Authenticity, and Rationality, *Dialectica*, 61, 2007, 323-345.

[8] Kripke, Saul. *Naming and Necessity*. Oxford UP, 1980.

[9] Mikenberg, I., Da Costa, N.C.A., and Chuaqui, R. Pragmatic Truth and Approximation to Truth, *Journal of Symbolic Logic*, 51, 1986, 201-221.

[10] Robinson, A. *Non-Standard Analysis. Studies in Logic and the Foundations of Mathematics*. North-Holland, 1966.

[11] Searle, J.R. *Speech acts; an essay in the philosophy of language*. Cambridge: Cambridge University Press, 1969.

[12] Searle, J.R. *Expression and meaning: studies in the theory of speech acts*. Cambridge: Cambridge University Press, 1979.

[13] Searle, J.R. and Vanderveken D. *Foundations of Illocutionary Logic*. Cambridge: Cambridge University Press, 1984.

[14] Tarski, A. The Semantic Conception of Truth and the Foundations of Semantics, *Philosophy and Phenomenological Research*, 4, 1944, 341-376.

Paraconsistentization through antimonotonicity: towards a logic of supplement

Hilan Bensusan
Gregory Carneiro

Department of Philosophy, University of Brasília, Brazil

Abstract

As part of an investigation into non-Taskian logics, we introduce a method of paraconsistentization by generating the minimal, fully non-monotonic counterpart of a logical system. Antimonotonic systems are such that no addition in the premises can preserve conclusions. Further, antimonotonic systems are not only paraconsistent but also minimal in an important sense. Antimonotonicity is also a crucial feature of what Derrida intended by his idea of a *logic of supplement*.

1 Introduction

Any system, formal or not, is revised or abandoned when a refutation or a contradiction is found. Refutation is a negation of an axiom, often through a negation of one or more of its consequences. Contradiction is the commitment, in the system, of both something and its negation. Negation is behind any episode of conclusions loss. In fact, conclusions are otherwise preserved by assumptions of monotonicity, that is, that adding new premises or new information would not harm conclusions. Monotonicity is a common assumption —it amounts to postulating that we are hiking to reach something like the top of the Mount Fuji, every step up is in the right direction. Likewise, negation is a step downward and if it doesn't occur, nothing can harm the attained conclusions.

Indeed, given a formula φ and a sets of formulas Γ and Δ and relation of consequence $\vdash$, usually logical systems satisfy the following condition:

$\vdash$-**Monotonicity**: If $\Gamma \vdash \varphi$ and $\Gamma \subseteq \Delta$ then $\Delta \vdash \varphi$

Further, negation itself has embedded assumptions of monotonicity. Classical systems assume that negation prefixes harms conclusion but only in odd numbers. In contrast, some non-classical systems posit that the negation prefix is never innocuous —and therefore double negation cannot fully restore the original conclusions. Those non-classical systems can be viewed as dropping the assumption that whenever negation promotes neither refutation nor contradiction it is monotonic. The discussion introduces the crucial issue when monotonicity is at stake: are additions always innocuous? It is indeed possible that addition is effectively as harmful to attained conclusions as negation —or rather, that negation is harmful because it is an addition. If the landscape is not like Mount Fuji, a step up will distance one from the top while a step down is not necessarily a step backwards. What would happen if we drop the assumption that adding premises is at least non-regressive? This paper sketches an approach to formal systems that assume no monotonicity. In other words, our concern is with systems where no conclusion can be safe when something new is added (even if nothing is negated).

A formal system where no argument is monotonic proves to be paraconsistent. Edelcio de Souza, with Alexandre Costa-Leite and Diogo Dias (2016), have studied strategies to paraconsistentize a logical system —to make a non-paraconsistent logic into one. We show here that when all monotonicity is dropped from a system, it is paraconsistentized.

2 Supplement as a novel (yet old) logical notion

Jacques Derrida introduced the idea of a non-innocuous addition under the name of supplement. While a complement *completes* something, a supplement only *makes clear that what is supplemented is incomplete* and admitted a supplement. The basic intuition can be explained through simple examples. An ordinary car is only shown to have been incomplete when it is supplemented with automated devices that make much of the driving easier —or redundant. An ordinary plate of fries could be shown to be incomplete when it is supplemented by some drops of vinegar. Neither the automated parts of the car nor the vinegar are complements to what was already there in any sense, they are supplements in the sense that they promote a change when the supplemented result replaces the unsupplemented original. Derrida first mentions 'supplement' in the *Of Grammatology* while discussing Rousseau's[1] use of the word while describing what writing does to the spoken word. Derrida then writes: "Either writing was never a simple 'supplement', or it is urgently necessary to construct a new logic of the 'supplement'"'(*Of Grammatology*, 7). The simple "supplement" would be an innocuous addition and this is not what writing is for it changes our relation to the spoken word —ignorance and forgetfulness become acceptable, as Plato remarks in the *Phaedrus*(274c-275b). Further, the spoken word – like the car without the automated supplements or the fries without vinegar —is rendered incomplete by the very non-necessary addition of the supplement of writing. Derrida pictures supplement as forcefully two-fold: a supplement is an exterior, unnecessary and unpredictable addition and it scarcely leaves nothing as they were.

Derrida understands the logic of supplement as a departure from "the logic of identity"(*Of Grammatology*, 61). If we define + as the addition of a formula to a set of formulas, this can be understood as follows:

(i) $\varphi \vdash \varphi$
(ii) $\varphi + \psi \nvdash \varphi$

If (1) is the case, so is (2). Hence, nothing can be established once and for all and supplementarity is therefore incompatible with monotonicity. The principle of identity is not indifferent to supplement. Indeed, a logic of supplement is such that none of its arguments are monotonic.

Often, the attempts to represent the supplement logically focused on metalogical discussions. They do offer intuition about the nature of supplement and, in particular, with its relation to contradictions. In fact, the logic of supplement is suppose to be in contrast with any logic of identity, as we have just seen. Accordingly, Graham Priest, in *Beyond The Limits of thought*, understands the supplement as an instance of a broader *inclosure schema* defined for a totality T. Let id be the identity function that takes from x to itself, and let $diag(f_A)$ be the diagonalization[2], defined for a one-place function f on one set A,

---

[1]"Languages are made to be spoken, writing serves only as a supplement to speech... Speech represents thought by conventional signs, and writing represents the same with regard to speech. Thus the art of writing is nothing but a mediated representation of thought". Rousseau, apud Derrida (*Of Grammatology*, 144).

[2]The diagonalization is a very common technique used to expose many different paradoxes and to prove important results, as the well know Cantor's theorem and Gödel's incompleteness theorems. Observing the pattern in the application of this

as the set of x in A that are such that x is not a member of the object delivered when f_A is applied to x —what is denoted by $f_A(x)$:

(1) T exists
(2) if $x \subseteq T$ either
    (2a) $diag(id_x) \notin x$
    (2b) $diag(id_x) \in T$

Transcendence (2a) and Closure (2b), Priest claims, are the basis for any contradiction that arise in the attempt to go beyond (*Beyond*, 3). In Priest's words, for any contradiction

> "the limit of what can be expressed; the limit of what can be described or conceived; the limit of what can be known; the limit of iteration of some operation or other, the infinite in its mathematical sense. [...] There is a totality (of all things expressible, describable, etc.) and an appropriate operation that generates an object that is both within and without the totality. I will call these situations Closure and Transcendence, respectively". (*Beyond*, 4.)

The formal presentation of the coincidences of the *inclosure schema* and the *supplement* follows the pattern of one trying to express something not expressible. Because of the supplement, no all-encompassing totality is possible. Further, the supplement is rendered unintelligible by the very logic of identity that grounds intelligibility by assuming no addition can alter what something is.

Paul Livingston, in *Derrida and formal logic: formalising the undecidable*, summarizes this issue as a mixture of three central elements. First, it takes in account the *syntax*, the language involving a given logic, for example. Second, the *totality*, that in the context of Gödel's results can be considered as totality of a decision procedure or simply proofs. Finally, the *in-closure*, that means that any linguistic system can be

> "closed only at the price of the inherent paradox of tracing its limits, and open just insofar as this paradoxical closure also operates as the diagonalization that generates a contradictory point that is both inside and outside".

---

method, Priest proceeds to demonstrate that the diagonalisation is the common denominator the expressibility paradoxes, which should be seen as "inherent in the object of discourse" (*Beyond*, 140).

So, diagonalization would not just be central in Gödel's argument for the incompleteness theorems, as well in many other formal results, but also the procedure that expose the transcendence —what can not be expressed. This proximity, as it is, seems to capture the behavior of what we described as the supplement.

A logic of supplement, however, should be more than a remark about enclosure and diagnonalization. In the next section, we elaborate on what we take to be the crucial feature of a logic of supplement: a radical departure from monotonicity.

3 Antimonotonicity and paraconsistentization

Monotonicity is a salient feature of classical reasoning —it is accordingly captured by classical logic. As we have seen, it ensures that any addition is innocuous to identity assertions like $\varphi \vdash \varphi$. In a monotonic system, (ii) above is not true for any ψ. The so called Taskian logics defined by a ordered pair (F, $\vdash$), where F is set and $\vdash$ is a relation between the powerset of F and F, are those where the consequence relation satisfies reflexivity, transitivity and monotonicity. Those are often also called well-behaved logics. To have an initial idea of what we could find beyond the scope of Taskian logics, consider an antilogic $\bar{L}$ defined as

$$\Gamma \vdash_{\bar{L}} \varphi \text{ iff } \Gamma \nvdash_L \varphi$$

The antilogic (see Bensusan et al. 2015) can be defined for any logic L, Tarskian or not. We can then define, however, an anticlassical system as the antilogic of classical (propositional) logic. Such an anticlassical system is not Tarskian for what makes it non-monotonic makes it also unable to satisfy the other two requisites (see Béziau et al. 2015). Indeed, if the concept is considered to the fullest, non-monotonic systems can behave in interesting and surprising ways. Still, it is easy to see that in the anticlassical system some of its arguments are monotonic, but not all.

In contrast, we take the logic of supplement to require all of its argument to be non-monotonic. That is, no addition is innocuous. If we accordingly define a logic L as a ordered pair (F, $\vdash$), we define anti-monotonicity as follows:

$\vdash$-**Antimonotonocity**: If $\Gamma \vdash \varphi$, then $\Gamma' \nvdash \varphi$, for all Γ' such that $\Gamma \subset \Gamma'$.

From now on we will omit the scope of the property whenever is clear that it is applies to the consequence relation ($\vdash$). We can also define:

Non-antimonotonocity: If $\Gamma \vdash \varphi$, then $\Gamma' \vdash \varphi$, for at least one Γ' such that $\Gamma \subset \Gamma'$.

That is to say, the negation of antimonotonicity is equivalent to a weaker form of monotonicity. More generaly, given any logic L, its antimonotonic counterpart $\widehat{L}$ is such that

Antimonotonization: If $\Gamma \nvdash_{\widehat{L}} \varphi$, then $\Gamma \nvdash_L \varphi$ or $\Gamma \vdash_L \varphi$, in which case there is a $\Gamma^* \subset \Gamma$ such that $\Gamma^* \vdash_L \varphi$.

Once no addition can be made to the premises that will not spoil the argument, an antimonotonic logic is akin to minimal logical systems where only the smallest proof (or the smallest entailment) is valid. In fact, it is easy to see that antimonotonic systems are in that sense **minimal**. Antinomonotonization is, in this sense, minimization. For example, the antimonotonic counterpart of the classical propositional logic is such that $p \vdash p$ but $p \ \& \ q \nvdash p$.

Minimal systems are paraconsistent for whatever a given logic has as falsity constant (usually denoted as $\bot$), the principle of explosion will fail in all cases that the premises are not solely $\bot$. For instance, we can affirm that

$$\bot \vdash_{\widehat{L}} \varphi,$$

for any φ, but if we take any α and add it to the premises, this makes the argument invalid:

$$\bot \cup \alpha \nvdash_{\widehat{L}} \varphi.$$

In other words, minimal systems are such that contradiction fails to prove anything in any occasion —they are paraconsistent (see Da costa, 1963). There is a deep and interesting connection between antimonotonicity, minimality and paraconsistency in the logic of supplement. If it is an antimonotonic system, it is also minimal and paraconsistent. Because it is a logic where no addition is innocuous, no addition can collapse the system; an addition that brings a contradiction to the premises makes the previous conclusions, as any addition, false. In fact, if $\Gamma \vdash \varphi$ then $\Gamma + \psi \nvdash \varphi$ and, in particular, if $\{\psi\} \subset \Gamma^3$.

In order to show how the concept of antimonotonocity is connected with paraconsistency, it is interesting to note that the antimonotonization can be seen as instance of paraconsitentization. The latter was introduced to study an unified method for producing paraconsistent logic out of any given logical system by Edelcio De Souza, Alexandre Costa-Leite and Diogo Dias (2016). Through category theory and universal logic, they were able to create a *paraconsistentization functor* ($\mathbb{P}$). In the original formulation, this functor is the result of the construction of a *endofunctor* on CON, the category of consequence structures, constituted by consequence structures as objects and homomorphisms as CON-morphisms. For a precise characterization, the authors take (X, Cn) to be the usual consequence structure, and a new operation $Cn_\mathbb{P} : \wp(X) \longrightarrow \wp(X)$ such that for all $A \subseteq X$:

$$Cn_\mathbb{P}(A) := \bigcup \{Cn(A') : A' \subseteq A, \text{Cn-consistent}\}$$

This operation unable to take $x \in Cn_\mathbb{P}(A)$ if and only if there is $A' \subseteq A$ Cn-consistent such that $x \in Cn(A')$. In other words, an conclusion is present in $Cn_\mathbb{P}$ just in case it can be obtained from a smaller set of consistent premises. For all consequence structures (X, Cn) that satisfies *ex falso quodlibet*[4], joint consistency[5] and the conjunctive property[6], holds as a theorem that (X, $Cn_\mathbb{P}$) is paraconsistent[7].

If there is certain inconsistent set in X for a normal structure (X, Cn), the paraconsistentized version of it shall not have such inconsistent set. In similar way, the antimonotonic version of a explosive logic becomes not-explosive once a smaller set of premises can be designated in a given argument. Minimization —and antimonotonization —is a form of paraconsistentization. For instance, in the antimonotonic version of

---

[3]Work on paraconsistency and antimonotonicity was carried out by the authors with Agnes Caiado and Emanuel Paiva; some results were compiled in Bensusan, Carneiro et al. (2019).

[4]Using the structure provided in the original paper, that means a (X, Cn) that for all $A \subseteq X$, if $x \in X$ such that $x, \neg x \in Cn(A)$, then $Cn(A) = X$.

[5]A system (X, Cn) satisfies joint consistency iff there is $x \in X$ such at $\{x\}, \{\neg x\}$ are consistent but $\{x, \neg x\}$ is Cn-inconsistent.

[6]A (X,Cn) satisfies the conjunctive property iff for all x, y $\in$ X, there is $z \in X$ such that $Cn(\{x, y\}) = Cn(\{z\})$.

[7]The authors present the propositional paraclassical logic as a particular application of the procedure described.

propositional logic, the inconsistent set o premises {p, ¬p} can not derive q, for p and q atomics, because the argument would not minimal; in the paracosistentized version, the set $u \in X$, such that $\{u\}$ is Cn-inconsistent, in a structure (X, Cn), such u would not be present in the structure (X, Cn$_\mathbb{P}$), because Cn$_\mathbb{P}$ contains no inconsistent sets[8]. In the case of paraconsistentization through antimonotonization we get rid of more arguments than would be necessary to preserve consistency in a given consequence structure. As the minimal counterpart of a logic is paraconsistent, to antimonotonicize a logic is to paraconsistentize it. The force of minimality seems to be strong enough to avoid explosion and to stop contradiction from deriving everything.

4 The supplement and its oppositions

The property of monotonicity and related ones can be applied not only to derivations (the relation of consequence ⊢) but also to its failure (to $\Gamma \nvdash \varphi$). If a failure or absence of derivation is monotonic, every addition to it is innocuous:

⊬-Monotonicity: If $\Gamma \nvdash \varphi$ and $\Gamma \subseteq \Delta$ then $\Delta \nvdash \varphi$

A logical system which satisfies this property is such that nothing can be added to ensure a conclusion. It is a system where nothing could be supplemented (in the sense elaborated above) to a failure to derive. Clearly, systems that hold such feature would hold that $\Gamma \vdash \varphi$ only if Γ is empty (It is a system of axioms only).

Computer scientists have explored issues related to monotonicity, mostly connected to the management of large datasets. In these contexts, it could be the case that if a item does not satisfies the minimum requirement to pass a test, all its extensions will also going to fail the same test. In other words, if the item fails a test, every addition is innocuous and will not improve its performance. This is a case where it is failure itself which is monotonic. Interestingly, Han, Pei & Kamber coined the term *antimonotonicity* to describe the monotonicity of failures (see Han, Pei and Kamber, 2011). Notice that ⊬-monotonic systems are in fact different from ⊢-antimonotonic ones for in the latter if $\emptyset \nvdash \varphi$, there could still be an α such that $\alpha \vdash \varphi$. Conversely, in a ⊬-monotonic

---

[8] Proposition 3.6 in the original paper.

system, if $\emptyset \vdash \varphi$ there could be an α such that $\alpha \vdash \varphi$. A minimal system, which is ⊢-antimonotonic, still admits that $\Gamma \nvdash \varphi$ but, if $\Gamma \subseteq \Delta$, $\Delta \vdash \varphi$ and therefore is not ⊬-monotonic. Conversely, a ⊬-monotonic system is not minimal for whenever $\Gamma \vdash \varphi$ - and this is the case only when Γ is empty - $\Gamma + \alpha \vdash \varphi$ for any α. We can see that the properties of ⊢-antimonotonic and ⊬-monotonic are therefore disjunct. Any one of two could be true of a given logic while the other is false —and they both could be true or false of one logic.

Further, we can also define ⊬-antimonotonicity, for systems where additions are not innocuous with respect to failure:

⊬-Antimonotonicity: If $\Gamma \nvdash \varphi$ and $\Gamma \subseteq \Delta$ then $\Delta \vdash \varphi$

Here failure can always be supplemented for additions bring in conclusions. Here the supplement leads to triviality for in such systems, everything can be derived by adding new premises. From anything including from contradictions, *quodlibet sequitur*. Well-behaved logics are not ⊬-antimonotonic for such systems enable anything to be derived. Classical logic and possibly all Tarskian systems, which are ⊢-monotonic, are neither ⊬-monotonic nor ⊬-antimonotonic; rather they are simply ⊬-nonmonotonic. A ⊢-nonmonotonic system could also be ⊬-nonmonotonic, and the anticlassical logic mentioned above is both.

It is enough for a system to be ⊬-antimonotonic to enable anything to be derived —to be a trivial logic. It is worth mentioning that the notions of ⊢-antimonotonicity and ⊬-antimonotonicity combined would make any logic circular (and trivial, of course): if one adds premises the derivation is suppressed, but adding premises again would restore the derivation —in this sense, the two notions are incompatible.

In contrast, if a logic is both ⊢-antimonotonic and ⊬-monotonic, it is empty. For, if a logic has both these properties, for all Δ such that $\Gamma \subset \Delta$, if $\Gamma \vdash \varphi$, then $\Delta \nvdash \varphi$, and if $\Gamma \nvdash \varphi$, then $\Delta \nvdash \varphi$. In any case, $\Delta \nvdash \varphi$; nothing is ever derived.

We can begin to study the relation between these properties in terms of a geometry of oppositions (Blanché, 1966; Moretti, 2009). ⊢-Monotonicity and ⊢-antimonotonicity cannot both be true but both can be false of a given logic, they are in an opposition by contrariety. Additionally, ⊢-nonmotonicity and ⊢-non-antimonotonicity cannot be both false but can be both true of a given logic, they are subcontraries. These last two properties can be seen as subalterns with respect to

⊢-monotonicity and ⊢-antimonotonicity respectively. We can sketch the following figures:

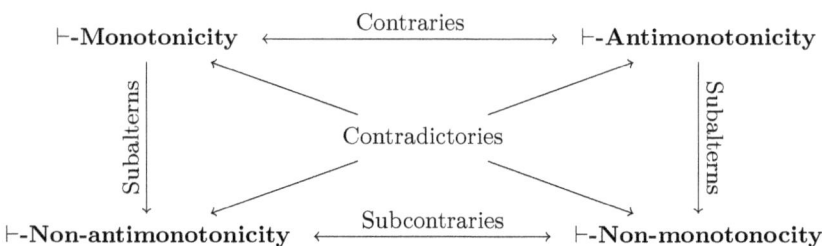

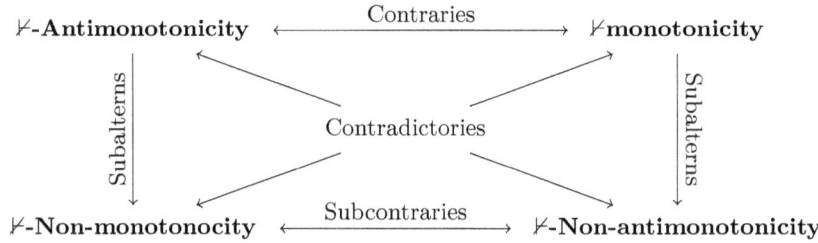

Additionally, we have pointed at some relations between the properties in the two squares. We have seen that some of these properties combined result in empty or trivial logics. We conjecture that further developments could expand the relations around the concepts involving antimonotonicity towards a geometric figure that could integrate the two squares above. For now, it is enough to observe that these two squares give us a wide image of the relations between premises and conclusion —a image that would be undoubtedly incomplete without the idea of the logic of supplement.

5 Conclusion

Supplement, as an addition that collapses inferences and challenge identities, have been proven to provide a strategy of paraconsistentization. If a logic is antimonotonicized —converted to its minimal version where nothing can be added to its arguments —it becomes paraconsistent. The logic of supplement is ⊢-antimonotonic. It is clearly a non-Tarskian logic. We have shown some interesting features of the ⊢-antimonotonic property in comparison to other related notions such as ⊢-nonnomotonicity and ⊬-monotonicity. Work on non-Tarskian logics as much as on the logic of supplement promise to reveal further frontiers of the capacity to rigorously study controlled inferences. The logic of supplement, in particular, with its complete rejection of innocuous addition opens the possibility of understanding how inferences can find their homes far away from the safe harbour of fixed and unchallenged identities.

References

[1] Bensusan, Hilan; Costa-Leite, Alexandre; de Souza, Edelcio G. *Logics and their galaxies*. In: *The Road to Universal Logic*, Birkhäuser, Cham, pp. 243-252, 2015.

[2] Bensusan, Hilan; Carneiro, Gregory; Caiado, Agnes; Paiva, Emanuel. *Antimonotonicity, minimality and paraconsistency*. Unpublished, 2019.

[3] Beziau, J. Y.; Buchsbaum, A. *Let us be antilogical: Anti-classical logic as a logic*. In: *Let us be logical*, edited by Amirouche Moktefi, Alessio Moretti and Fabien Schang. London: College Publications, 2013.

[4] Blanché, Robert. *Structures intellectuelles: Essai sur l'organisation systématique des concepts*. Paris: Librairie Philosophique J. Vrin, 1966.

[5] da Costa, Newton. *Calculs propositionnels pour les systèmes formels inconsistants*. Compte Rendu Acad. des Sciences, v. 257, pp. 3790-3793, 1963.

[6] Derrida, Jacques. *Of Grammatology*. Baltimore, Maryland: Johns Hopkins University Press, 1976.

[7] de Souza, Edelcio G.; Costa-Leite, Alexandre; Dias, Diogo HB. *On a paraconsistentization functor in the category of consequence structures*. Journal of Applied Non-Classical Logics, v. 26, n. 3, pp. 240-250, 2016.

[8] Grooters, Diana; Prakken, Henry. *Two aspects of relevance in structured argumentation: Minimality and paraconsistency*. Journal of Artificial Intelligence Research, v. 56, pp. 197-245, 2016.

[9] Han, Jiawei; Pei, Jian; Kamber, Micheline. *Data mining: concepts and techniques*. Elsevier, 2011.

[10] Łukaszewicz, Witold. *Non-monotonic reasoning: formalization of commonsense reasoning*. UK: Ellis Horwood Limited, 1990.

[11] Livingston, Paul. *Derrida and formal logic: formalising the undecidable*. Derrida Today, 3(2), pp. 221-239, 2010.

[12] Moretti, Alessio. *The geometry of logical opposition*. PhD Thesis, Université de Neuchâtel, 2009.

[13] Plato. *Phaedrus*. Oxford: Oxford University Press, 2002.

[14] Priest, Graham. *Beyond the limits of thought*. Cambridge: Cambridge University Press, 1995.

Part 4

Philosophy and history of logic

Hans Hahn and the Foundations of Mathematics

Diogo H.B. Dias

State University of the North of Paraná, Brazil

Abstract

Hans Hahn can be considered as the real founder of the Vienna Circle. Nonetheless, there are few studies on his philosophical work. This paper investigates Hahn's Philosophy of Mathematics, with special emphasis on his proposal concerning the foundations of mathematics. This analysis can cast some light on the influence Hahn had on the Circle's debate on these matters, thus contributing for a broader understanding of the philosophical landscape of the Vienna Circle.

1 Introduction

This paper aims at highlighting some essential aspects regarding Hans Hahn's Philosophy of Mathematics, as well as his relation with the Vienna Circle.

Hahn is a member of the so called First Vienna Circle, together with Phillip Frank, Neurath and von Mises[1]. Not only that, but he is also considered by Frank as the real founder of the Circle[2]. Moreover, according to Menger[3], Hahn is responsible for introducing into the Circle the discussion regarding logic and the foundations of mathematics[4]. Therefore, for a comprehensive understanding of these discussions inside the Circle it is essential to return to Hahn's thought. Thus, this investigation is a first step of this broader analysis.

---

*Dedicated to Edelcio Gonçalves de Souza, my first logic teacher, on his 60th birthday.

[1] For an account of the activities of the First Circle, cf. Haller (1991).
[2] Cf. Stadler (1997), p. 7.
[3] Cf. Menger (1980).
[4] He is also responsible for Schlick's nomination for the position previously occupied by Mach and Boltzman; for including the reading of Wittgenstein's *Tractatus* in the Circle's meetings and for advising Gödel's PhD.

2 Foundations of mathematics

In this period, as it is known, three main philosophical schools were competing for the role as the foundation of mathematics, namely: logicism, formalism and intuitionism.

Hahn endorses logicism, which consists, in general terms, in an attempt to reduce mathematics to logic. But only saying that is to forget that this discussion is part of a broader picture inside the Vienna Circle, which is the formulation of a scientific conception of the world with an empiricist character. In the *Manifesto* written by the Circle, and signed by Carnap, Hahn and Neurath (1929), the basic tenant of this conception is presented, that is: there are only two forms of knowledge; one through experience, and another one through logical transformation.

In this sense, the choice of a foundation for mathematics must suit these philosophical considerations. Hence, the fundamental question formulated by Hahn regarding mathematics is: "How is the empiricist position compatible with the applicability of logic and mathematics to reality?" (Hahn (1931), p. 32). This is the main requirement that such foundation should fulfill. For, in developing an empiricist's theory, according to Hahn, we have to deal with the fact that logic and mathematics seem to provide knowledge about the world and, besides that, they are absolute and universal.

Thus, his main foundational concern is not to find a formal proof of the non-existence of contradictions in mathematics - as the formalist would pursue -, nor to carry out a reduction of mathematics to a primordial intuition, but an explanation of the "compatibility of mathematics with an empiricist position" (Sigmund (1995b), pp. 236-7). Of course this is not a new philosophical problem. The novelty rests in Hahn's proposal.

The traditional attempt to develop an empiricist account of mathematics have always failed given that, in the final analysis, they understood logic as the study of the most general properties of objects. To overcome this difficulty, Hahn claims that

> "logic does not in any way deal with all objects, and it does not deal with any objects at all: *it only deals with the way we talk about objects* (...). And the certainty and universal validity of a proposition, or better, its irrefutability, flows precisely from this, that it says nothing about any objects". ((Hahn , 1933a), p. 29, emphasis on the original)

In the same sense, on another paper, he says that

> "logic first comes into being when - using a symbolism - people *talk about the world*, and in particular, when they use a symbolism whose signs do *not* (as might at first be supposed) stand in an isomorphic one-one relation to what is signified." (Hahn (1929), p. 40, emphasis on the original)

Put in other words, logic is "a set of directions for making certain transformations within the symbolism we employ" (Hahn (1930a), p. 24). And this is what he calls the "tautological character of logic" (Hahn (1930a), p. 23).

3 Hahn's novelty

These quotes contain the essential features of Hahn's philosophy of mathematics. First, it is clear that his logicism is different from Russell's. For, according to the English philosopher's views on logic at that time, logic deals with individuals, properties of individuals, properties of properties of individuals and so on.

Nonetheless Hahn believes that the formal apparatus developed by Russell, even requiring reformulations, can be adjusted for a foundation of mathematics. The fundamental change would be in its philosophical interpretation. Thus, logic would no longer speak about the world, but only about the way in which we talk about the world.

Second, there is a strong influence of Wittgenstein. Hahn's general conception of logic is openly taken from the *Tractatus* (1922). Even though Hahn had opposed to some specific topics of this work - such as regarding the possibility of a metalanguage -, he takes Wittgenstein's position that logic deals only with tautologies.

But here we can see another novelty introduced by Hahn. Although he accepts the tautological character of logic, he does not take the notion of tautology in the same strict sense of Wittgenstein, that is, as being true solely due to its logical form. Hahn, in his turn, uses this concept, as he himself asserts, in a broader way. But what does this enlargement of the concept of tautology consists of and why is it necessary?

In Wittgenstein's notion of logic, it was possible to apprehend formal properties of the world through an analysis of language. According to him, "Logic is not a body of doctrine, but a mirror-image of the world. Logic is transcendental" (Wittgenstein (1922), 6.13). Therefore, the logical form is not merely an abstraction, but has also a transcendental function. It is precisely this transcendental character that must be abandoned, according to Hahn, for it consists on what he and Neurath call the metaphysical traps of the Tractatus. Hahn says that "it is a big mistake to infer the structure of the world from the structure of language" (Hahn (1930b), p. 8). Here it is clear why the symbolism of logic cannot be in an isomorphic relation to what is meant. If this was the case, it would be possible to pass from a logical analysis of language to a metaphysical analysis of reality.

Thus even if Wittgenstein's notion of tautology refused Russell's view that logic deals with objects' most abstract properties, it still kept a transcendental function. And this is the restricted view of tautology that Hahn needs to enlarge in case he wants to free logic from any relation with the world. Also, Hahn extend the tautological character of logic to mathematics, which Wittgenstein never accepted, due to the non-logicality of some mathematical axioms, such as that the axioms of choice and infinity.

4 Hahn's logicism

These considerations consist in the core of Hahn's answer for the relation between empiricism and logicism: given that it is only possible to obtain knowledge from experience, or from logical transformation; that knowledge of reality is only possible from experience, and that logic does not have any empirical character, but consists only in tautological transformation, then mathematics must also consist in these transformations. And, therefore, if we prove that mathematics is part of logic, a radical empiricism becomes compatible with a logical foundation of mathematics.

Hence, the solution for the problem concerning the foundation of mathematics may be expressed in the following passage:

> "I assume, like Russell, that for describing the world (or better: a section of the world) we have at our disposal a system of predicative functions, of predicative functions of predicative functions, etc. - though, unlike Russell, I do not believe that the predicative functions are something absolutely given, something we can point out in the world. Now the

> description of the world will turn out differently according to the richness of this system of predicative functions; we therefore make certain *assumptions* about its richness (...). Now the whole of mathematics arises out of the tautological transformation of the requirements we make about the richness of our system of predicative functions". (Hahn (1931), pp. 35-36, emphasis on the original)

This way, the core of the solution is in the word *assumptions*. Then, although Hahn agrees with Wittgenstein on the non-logical character of some axioms, such as the axiom of choice, Hahn considered that accepting it or not was a pragmatic matter to be taken when constructing a language. This position also differs essentially from Frege's and Russell's. The debate on the foundations of mathematics turns into a purely pragmatic question and, to Hahn, the main pragmatic question was the compatibility between logic and mathematics with empiricism.

For this reason Hahn chooses logicism, since intuitionism, in the final analysis, rests on a primitive intuition with no connection with the empirical world and, which, in Hahn's words, consists in a "mystical" concept[5]; and formalism, by already pressuposing finite arithmetic, "cannot be regarded as a theory of the foundations of mathematics" (Hahn (1931), p. 32).

It is important to note that claiming that logic and mathematics are tautologies does not mean that they are trivial. It could be strange to accept that all mathematical and logical demonstrations are nothing more than tautological transformations, that is, different ways of saying the same thing. Nonetheless, according to Hahn, this strangeness comes from forgetting that we are not omniscient beings. "An omniscient subject", says Hahn, "needs no logic, and contrary to Plato we can say: God never does mathematics" (Hahn (1930a), p. 23).

Although Hahn is a great mathematician, he never attempts to demonstrate the tautological character of the mathematical statements. Nonetheless, he believes that it is easy to see the tautological character of finite arithmetic. For instance, the expression "$5 + 7 = 12$"[6] is tautological in the sense that it is obtained merely through manipulations and transformation of sentences. In this particular case the result can

---

[5]Cf. Hahn (1931), p. 32.

[6]The choice of the example is not random; it is precisely the same example used by Kant to explain that arithmetic is synthetic a priori.

be derived from the definition of numbers and operations between them, and from Peanos' axioms. However, this reduction is not clear in other fields of mathematics.

Another important aspect of his solution is that, in principle, it is not affected by Gödel's Incompleteness Theorems. Even though Hahn never writes about this subject, he surely knows Gödel's results. Sigmund (1995a) conjectures that this absence is due to the fact that Hahn considers Gödel's results as mathematical theorems and not philosophical insights. Hahn (1934) seems to reply the possible objection of incompleteness towards logicism when he asserts that an absolute proof of consistency is impossible; this kind of proof is always relative to another system. And he claims that this is not a problem for logicism, since the demand of an absolute knowledge, of an absolute certainty, is absurd and intangible in any sphere of knowledge. And this is a thesis that, although in an incipient way, it is already present in 1919, in a review that Hahn made of a book on Number Theory by Aldred Pringshem[7]. One of the main reasons for choosing this Book was Pringshem's intention to drive intuition from any mathematical demonstration. Let us then analyze briefly Hahn's position on mathematical intuition.

In Hahn (1933b), Hahn tries to show that intuition is not a source of geometrical knowledge. This field is particularly chosen because it is considered as a typical example of knowledge grounded on intuition. Hahn's strategy is to present several examples of geometrical objects and counterintuitive results, making it clear that intuition has been gradually abandoned even in geometry, which seemed to be its original domain in mathematics. The examples presented by Hahn were already known among the mathematicians of his time, and were used, for instance, by Poincaré[8], and were called "pathological examples". However, they were used to defend a limitation of intuition, and not to banish it. They were also referred as monsters because, at first sight, they seem impossible or paradoxical.

Hahn's final result of this analysis is that

> "Every theorem of geometry appears as the (tautological) implication $P \to Q$, where the antecedent P is the logical product of the axioms and the consequent Q the theorem in question. In this way the axioms

---

[7]Cf. Hahn (1919).
[8]Cf. Poincaré (1889).

no longer appear as self-evident though unprovable truths, but as assumptions from which deductions are made; and the basic concepts no longer appear as objects incapable of being dissected further by definition though capable of being grasped immediately by intuition, but merely as logical variables". (Hahn (1930a), pp. 26-7)

In even more abstract terms, we can see geometry as a branch of logic. Since the axioms are relations between variables that, on their turn, represent basic concepts, geometry is nothing but a specific kind of relational system. Hence, it belongs to the theory of relations, which is part of logic.

Curiously, while attempting to explain the limits of intuition, Hahn manages to create visual representations of these monsters. Not only that, but also to deal with one of them, he ends up creating what later came to be called fractal geometry. Some commentators, such as Erhard Oeser, claims that Hahn's reasoning, in the final analysis, has as result "the possibility of an extension of intuition beyond just the visual intuition, rather than proving the failure of intuition." (Oeser (1995), p. 251). Besides that, Hahn recognizes that the non-intuitive character of a mathematical construction becomes intuitive when it is clear how to apply it in the empirical sciences. In the end of this paper, Hahn reiterates that intuition is not capable of providing a priori knowledge, but consists merely in "force of habit rooted in psychological inertia" (Hahn (1933b), p. 101).

5 Final remarks

To conclude, it is important to state some future lines of investigation. At the beginning, we saw that Hahn argues that logic only provides guidelines for the manipulation of symbols, and that this thesis was partly an objection to the transcendental character of logic defended by Wittgenstein. It is necessary to investigate more deeply this point, and to verify, as Uebel suggests[9], whether this thesis consists in an anticipation of - and, therefore, influences - Carnap's Principle of Tolerance. For, in the final analysis, this conventionalism proposed by Hahn does not favor some set of rules over others. In being so, we can interpret this thesis as an important step towards tolerance and logical pluralism.

---
[9] Cf. Uebel (2005).

Note that Hahn wrote this in 1929, and that Carnap's Principle of Tolerance was only explicitly stated in his *Logical Syntax of Language*, fist published in 1934, when Hahn had already passed away.

Moreover, it is necessary to analyze some technical aspects of the logicism proposed by Hahn. In the Second Conference on Theory of Knowledge in the Exact Sciences, in Konigsber, in 1930, having Carnap, Heyting and von Neumann defending, respectively, logicism, intuitionism and formalism, Hahn made it clear that his position differs in essential points from those of Frege, Russell and Wittgenstein. There was a huge discussion regarding some mathematical axioms, such as the axiom of choice and infinity. Many argued that they consisted in extra-logical axioms and, since they were used on the reduction of mathematics to logic, logicism had failed. Here, Hahn's position, as well as his originality, is evident:

> "Whether a certain proposition is or is not valid (e.g., the proposition about the cardinal number of the set of powers or the proposition about well-ordered sets) depends on the requirements we have made about the richness of the underlying system of predicative functions, or if you want to call them that, on the *axioms*; the question about the *absolute* validity of such propositions is completely senseless". (Hahn (1931), p. 36, emphasis on the original)

Therefore, the axioms of choice and infinity will be valid according with the expressive power of the chosen systems. Given that they are not necessary for the notion of deduction, they can be inserted as mere hypothesis. Hence, they are not treated as independent axioms, but used as antecedent in conditional mathematical statements. In this way, the reduction of mathematics to logic is no longer absolute, but relative to certain hypothesis, which is in agreement with a certain relativism presented before. Note that, once again, we have an allusion to a logical pluralism, given that, according to the choices regarding the richness of the language, we can formulate different logics.

References

[1] Carnap, R. (1937). *The Logical Syntax of Language.* London: Kegan Paul.

[2] Carnap, R.; Hahn, H.; Neurath, O. (1929). *Wissenschaftliche Weltauffassung. Der Wiener Kreis. Vienna*: Gerold. English translation: Neurath, O. The Scientific World Conception. The Vienna Circle. In: Neurath, M.; Cohen, R. S. (eds.) (1973) *Empiricism and Sociology.* Dordrecht: Reidel, pp. 299-318.

[3] Hahn, H. (1919). Besprechung von Alfred Pringsheim, *Vorlesungen über Zahlen und Funktionenlehre*, Göttingsche gelehrte Anzeigen no. 9-10, pp. 321-338. English translation: Review of Alfred Pringsheim, *Vorlesungen über Zahlen und Funktionenlehre.* Hahn, H. *Empiricism, Logic and Mathematics.* In: McGuiness, B. (ed.) Dordrecht: Reidel, pp. 51-72.

[4] Hahn, H. (1929). Empirizismus, Mathematik, Logik. *Forschungen und Fortschritte* 5. English translation: Empiricism, Mathematics and Logic. Hahn, H. *Empiricism, Logic and Mathematics.* In: McGuiness, B. (ed.) Dordrecht: Reidel, pp. 39-42.

[5] Hahn, H. (1930a). Die Bedeutung der wissenschaftlichen Weltauffassung insbesondere für Mathematik und Physik. *Erkenntnis* 1, pp. 96-105. English translation: The Significance of the Scientific World View, Especially for Mathematics and Physics. Hahn, H. *Empiricism, Logic and Mathematics.* In: McGuiness, B. (ed.) Dordrecht: Reidel, pp. 20-30.

[6] Hahn, H. (1930b). *Überflüssige Wesenheiten (Occam's Rasiermesse)*, Wolf, Vienna. English translation: Superfluous Entities, or Occam's Razor. Hahn, H. *Empiricism, Logic and Mathematics.* In McGuiness, B. (ed.) Dordrecht: Reidel, pp. 1-19.

[7] Hahn, H. (1931). [Beitrag zur] Diskussion zu Grundlagenfragen der Mathematik, *Erkenntnis* 2, pp. 135-141. English translation: Discussion about the Foundations of Mathematics. Hahn, H. *Empiricism, Logic and Mathematics.* In: McGuiness, B. (ed.) Dordrecht: Reidel, pp. 31-38.

[8] Hahn, H. (1933a). *Logik, Mathematik, Naturerkennen*, Gerold, Vienna. English translation: Logic, Mathematics and Knowledge of Nature. In McGuiness, B. (ed.) *Unified Science*, Dordrecht: Reidel, 1987, pp. 24-45.

[9] Hahn, H. (1933b). Die Krise der Anschauung. In: *Krise und Neuaufbau in den exakten Wissenschaften. Fünf Wiener Vorträge*, Vienna: Deuticke, pp. 41-64. English translation: The Crisis in Intuition. Hahn, H. *Empiricism, Logic and Mathematics*. In: McGuiness, B. (ed.) Dordrecht: Reidel, pp. 73-102.

[10] Hahn, H. (1934). Gibt es Unendliches? In: *Alte Probleme—Neue Lösungen in den exakten Wissenschaften. Fünf Wiener Vorträge*, Deuticke, Vienna. English translation: Does the Infinite Exist? Hahn, H. *Empiricism, Logic and Mathematics*. In: McGuiness, B. (ed.) Dordrecht: Reidel, pp. 103-131.

[11] Hahn, H. (1980). *Empiricism, Logic and Mathematics*. McGuiness, B. (ed.) Dordrecht: Reidel.

[12] Haller, R. (1991). The First Vienna Circle. In: Uebel, T. *Rediscovering the Forgotten Vienna Circle*. Dordrecht. Ed.: Kluwer.

[13] Menger, K. .(1980). Introduction. In: Hahn, H. *Empiricism, Logic and Mathematics*. In: McGuiness, B. (ed.) Dordrecht: Reidel, pp. ix-xviii.

[14] Oeser, E. .(1995). Crisis and Return of Intuition in Hans Hahn´s Philosophy of Mathematics. In: Depauli-Schimanovich, W.; Köhler, E.; Stadler, F. (eds.) *The Foundational Debate. Vienna Circle Institute Yearbook* [1995] (Institut 'Wiener Kreis' Society for the Advancement of the Scientific World Conception), vol 3. Springer, Dordrecht, pp. 247-258.

[15] Poincaré, H. (1889). La logique et l'intuition dans la science mathematique et dans I 'enseignement. In: *L'enseignement mathematique* 1, pp. 157-162.

[16] Sigmund, K. (1995a). A Philosopher's Mathematician: Hans Hahn and the Vienna Circle, *The Mathematical Intelligencer* 17, 16-29.

[17] Sigmund, K. (1995b). Hans Hahn and the Foundational Debate. In: Depauli-Schimanovich, W.; Köhler, E.; Stadler, F. (eds.) *The Foundational Debate. Vienna Circle Institute Yearbook* [1995] (Institut 'Wiener Kreis' Society for the Advancement of the Scientific World Conception), vol 3. Springer, Dordrecht, pp. 235-245.

[18] Stadler, F. (1997). *Studien zum Wiener Kreis*, Suhrkamp, Frankfurt a.M. English translation: *The Vienna Circle. Studies in the Origins, Development and Influence of Logical Empiricism*, Vienna: Springer, 2001.

[19] Uebel, T. (2005). Learning Logical Tolerance: Hans Hahn on the Foundations of Mathematics. *History and Philosophy of Logic*, 26, 2005, pp. 175-209.

[20] Wittgenstein, L. (1922). *Tractatus Logico-Philosophicus*. English translation: Ogden, C. K. London: Routledge & Kegan Paul, 1983.

A FIRST SURVEY OF CHARLES S. PEIRCE'S CONTRIBUTIONS TO LOGIC: FROM RELATIVES TO QUANTIFICATION

Cassiano Terra Rodrigues

Aeronautics Institute of Technology (ITA), Brazil

Abstract

In this paper, I will shortly present C. S. Peirce's development of his algebra of logic. Starting from his logic of relatives, since the very beginning Peirce was concerned with quantification. So, around 1870 already Peirce had devised ways of dealing with quantification in an algebraic manner, even though he had not yet devised a specific notation for the quantifiers themselves. Peirce's concerns were focused both upon Boole's conflation of logic and mathematics and upon De Morgan's rigid restraint of logic to relations. Striving to distinguish the specific difference between logic and mathematics, Peirce came then to a multiple quantification system from which a very original conception of logic as semiotic arises.

To my friend Edelcio, parmerense, porém gente boa.

Any account of Charles S. Peirce's logic is not an easy thing to accomplish[1]. Peirce is nowadays considered as one of the greatest logicians

---

[1] For Peirce's works, I mostly follow the usual international convention among Peirce's scholars for abbreviations and quotations, as follows: 1) CP, followed by volume and paragraph numbers: *Collected Papers of Charles Sanders Peirce*. Ed. by: C. Hartshorne and P. Weiss (vols. 1-6); A. Burks (vols. 7-8). Cambridge, MA: Harvard University Press, 1931-1958. 8 vv.; 2) EP, followed by volume and page numbers: *The Essential Peirce: Selected Philosophical Writings*. Ed. by: N. Houser and C. Kloesel (vol. 1: 1867-1893); "Peirce Edition Project" (vol. 2: 1893-1913). Bloomington; Indianapolis: Indiana University Press, 1992-1998. 2 vv.; 3) LF, followed by volume and page numbers: *Logic of the Future: Writings on existential graphs*. Edited by Ahti-Veikko Pietarinen. Berlin; Boston: Walter de Gruyter GmbH, 2020.; 4) W, followed by volume and page numbers: *Writings of Charles S. Peirce: A Chronological Edition*. Ed. by "Peirce Edition Project". Bloomington; Indianapolis: Indiana University Press, 1982-2000. 7 vv.

of the 19th century, one of the greatest logicians of all times indeed, together with Aristotle, Ockham, Leibniz, Boole, Frege and anyone else one might think of. His work is deemed as a source of original ideas for all who study logic. But it was not like this some forty years ago, when the mainstream narrative of the development of logic would include Peirce as a relatively important but nonetheless minor figure, at best an obscure forerunner of ideas better worked out by other logicians (Frege, Russell, Tarski, for instance). Peirce would then deserve a minimal recognition, and his contributions to logic could be packed up into some 5 or 6 pages, or no recognition and no pages at all[2].

Now everything seems to have changed. Peirce's importance seems clear to everyone interested in logic, its philosophy and its history[3]. But his ideas are yet not as known as Frege's or Booles, for instance, and his works only recently began to be published in more reliable editions[4]. This is crucial for a proper evaluation of his contributions. Some introductory remarks are needed to give the reader a bit more of context.

Peirce began his work on logic very early. He himself tells the tale he fell in love with logic when he was 12 years old (1851), and grabbed his older brother Jem's copy of Richard Whately's (1787-1863) *Elements of logic*; and ever since Peirce defined himself as a logician[5]. His main works in the field began to be developed in the late 1860's and early 1870's, when upon George Boole's algebra of logic and Augustus De Morgan's logic of relations Peirce developed his own logic of relatives. In the 1880's, about four years after and independently from Frege, Peirce in-

---

[2]For instance, see W. Kneale and M. Kneale 1962. In J. van Heijenoort 1967, Peirce is quoted 18 times, but not a single one of his major papers is included. The idea one gets is he is not totally absent from the history of logic, but is just a noble forerunner. See Anellis 2012 for a consistent approach to van Heijenoort's omission of Peirce.

[3]Several works have contributed for a more just evaluation of Peirce's importance in the history of logic. The bibliography has recently grown up to a point it is impossible to keep up with everything. Some indications might serve as didactic introductions from which the reader may find deeper approaches by her/him-self: H. Putnam 1982; N. Houser 1994; R. Dipert 1995; G. Brady 2000; N. Houser, D. D. Roberts and J. van Evra (eds.) 1997; I. H. Annelis 2012; C. Terra Rodrigues 2017.

[4]For more on the difficulties of editing Peirce's works, see N. Houser 1992 and A. De Tienne 2014. The most recent account is by M. Keeler 2020. The newest edition of Peirce's work on logic up to date is the *Peirceana* collection to be published in 3 volumes by De Gruyter, edited by Francesco Bellucci and Ahti-Veikko Pietarinen, the first volume of which collection is referenced in note 1 above).

[5]See W 1: xviii.

troduced quantifiers in an algebraic symbolic system as variable binding operators, having invented the very term "quantifier", according to J. Lukasiewicz[6]. If only for this, he can be considered together with Frege as the founder of contemporary symbolic logic. The contemporary standard logical notation owes indeed much more to Peirce than to Frege, since not only Russell's and Whitehead's *Principia Mathematica* relies heavily upon Peirce's and Schroeder's works; Tarski's works also take a lot from Peirce's[7]. Peirce was also the first to apply those connective to electric circuits, having defined the complete table of sixteen forms of the binary connectives for the first time in 1880[8]. Around the same time, Peirce had discovered and proved the functional completeness of the joint denial and its dual, the alternative denial, from which he also developed a proto-truth-table-device, decades before the more known work of L. Wittgenstein and H. M. Sheffer. He also worked on modal logic and began developing a three-valued system of his own. Dissatisfied with his linear notation, Peirce sought to develop a diagrammatic system of logical graphs, as well as he continued to deepen his approach to higher order and modal logic in a rich and yet not fully examined correspondence with E. Schroeder[9]. In his later years, in his bulky correspondence with Lady Victoria Welby and William James, Peirce developed a broader conception of logic not restricted to deductive symbolic logic, but extended to a *quasi*-formal and general theory of signs, his semiotics, or *semeiotic*, as he himself preferred to call just for etymological reasons. Peircean semeiotic is divided into: *speculative grammar*, the analytical and classificatory study of signs; *critical logic*, the study of the validity and justification of the forms of reasoning; and *methodeutic* or *speculative rhetoric*, the theory of logical methods and their applications to all areas of knowledge. This includes perhaps Peirce's most famous achievements, such as the distinction between icons, indexes and symbols as kinds of signs relative to the objects they signify, and the three basic forms of reasoning, namely, deduction, induction, and a distinct third kind of reasoning he called abduction or retroduction, which he claimed characterizes the logic of scientific discovery within the con-

---

[6] J. Lukasiewicz 1934.
[7] see Tarski 1975; Anellis 1997.
[8] See Clark 1997.
[9] See Houser 1991; Brady 2000; Anellis 2004 and 2012; Salatiel 2011 and 2017; Bellucci and Pietarinen 2016; Hintikka 1997; Legris 2016; Terra Rodrigues 2017.

text of his pragmatism[10]. A great part of this production was published in the form of independent yet related articles during Peirce's lifetime, since he was never able to publish a whole book of his own.

Much more could be enlisted as Peircean contributions to logic. But of course I cannot pretend to present all of them in this single article. Having to choose, I decided then to concentrate upon Peirce's logic of relatives, for it contains what I judge are the basics for understanding Peirce's approach to *exact logic*, as he later in his life began to say, following Ernst Schroeder [11]. As I hope, this is the first of a series of articles about Peirce's contributions to the study of logic. I have dealt before in other places and in a scattered manner with some other topics, and my treatment of the subjects profited from Edelcio's knowledge. So, I also hope he finds this one as interesting and intriguing as I found his helpful remarks were for me.

From relatives to quantification

The first thing we have to know is that Peirce was the first logician to develop a consistent and workable expressive notation for Boole's calculus, by combining it with the logic of relations De Morgan and he himself developed.[12] So, Peirce's logic is not simply an algebra. In point of fact, the main objection Peirce makes to Boole is the latter's equation of logical propositions to algebraic equations. For Peirce, logic is not and cannot be something like a *lingua characterica* devised in a mathematical fashion. His conception of language as a calculus[13] is an instrumental one, for one question is to reason, the other is how to understand reasoning regardless of the code expressing it. So, the main question becomes how to clearly distinguish logical from mathematical operations in a system of signs, and Boole's first attempts were very insightful but far from adequate. First of all, if only because Boole's calculus is a helpful instrument for making deductions. Besides, Boolian algebra is restricted to quantitative operations between classes, not comprising the whole of reasoning processes logical implication involves.[14] And this leads to a

---

[10]See Savan 1988; Santaella 1995; Terra Rodrigues 2011.
[11]See CP 3.616-618, 1902.
[12]I have given a deeper treatment of the subject in Terra Rodrigues 2017.
[13]This point was more notoriously put forward by J. Hintikka 1997. But this is not the whole picture. See my Terra Rodrigues 2017 p. 462 ff. for some qualification on this point.
[14]See W2: 12 seq., 1867.

second, but not at all less important reason why Peirce's algebra of logic is original: he sees no reason why to get entirely rid of the idea of a proposition as a subject-predicate structure. For Peirce, it is all a matter of predication, but understood in a novel way, as the following will make it clear.

So, instead of the sign for mathematical identity = used by Boole, Peirce uses a sign he himself devised for the primary and fundamental relation of illation, expressed by *ergo*: $\prec$ [15]. Peirce uses the sign also to mean inclusion, so that $\prec$ functions as a sort of subsumption operator, at once subsuming implication ($\rightarrow$) in propositional logic, subset inclusion ($\subseteq$) in class logic, and logical consequence ($\vDash$) in a metatheoretical level. In this manner, all logical relations can be defined solely upon the formal characters of illation.[16] Take, for instance, the following quotation:

> "Logic supposes inferences not only to be drawn, but also to be subjected to criticism; and therefore we not only require the form P∴C to express an argument, but also a form, $P_i \prec C_i$, to express the truth of its leading principle. Here P_i denotes any one of the class of premises, and C_i the corresponding conclusion. The symbol $\prec$ is the copula, and signifies primarily that every state of things in which a proposition of the class P_i is true is a state of things in which the corresponding propositions of the class C_i are true"[17].

By identifying this illative relation as the most basic logic operation, Peirce understands propositions themselves as basic forms of inference. For instance, the proposition "A is B" can be interpreted as a rudimentary sort of reasoning itself, namely, "any A is B and any B is A" [18]. Even terms could be interpreted in this way. Boole's system in fact is defective for expressing particular and hypothetical propositions.[19] This means that the logic of relations was a natural outcome of an attempt (Peirce's) to overcome the expressive difficulties of Boole's algebraic calculus, and not the converse, as it is usual to assume when one restricts the study of the history of modern logic to the artificial Peano-Frege-Russell narrative framework.

---

[15] See W 4: 170, 1880.
[16] For more detail, see Terra Rodrigues 2017.
[17] W 4: 166, 1880.
[18] W 2: 60, 1867.
[19] W2: 421, 1870.

Now, given his interest in dealing with the composition of classes and relations, Peirce gives a step De Morgan was unable to give, making quantifications quite explicit. De Morgan was working with relational compositions, such as "X is a lover of a servant of Y", or as in a most famous example, "Every man is an animal, therefore every head of some man is the head of some animal".[20] His concern was with inferences that could not be represented syllogistically, so the problem was how to quantify the predicate and still maintain the logical nature of the inference.

Peirce, in his turn, preferred to work with a somewhat different kind of propositions, such as "lover of a woman" or "lover of a servant of a woman", so the relative places of terms could be more easily identified. In his 1870 article on a proper notation for Boole's system, Peirce introduced ways of dealing with quantified expressions, but not specific signs for the quantifiers themselves. He accomplished this by introducing relational expressions, which he called "logical terms", with gaps for the insertion of variables, from one to three gaps, or more. These terms are named according to the number of gaps in the following way:

1. "Absolute terms" are one-gaped terms, like "a_____". Absolute terms are like single functional terms, possible expressions of individuals.

2. "Simple relative terms" are also one-gaped terms, but they do not refer to a single object, rather expressing dual relations, like "father of_____" or "lover of_____".

3. "Conjugative terms" are at least two-gaped terms, and their distinctive character is that "they regard an object as medi-um or third between two others, that is, as conjugative". For instance, "giver of_____ to_____", or "buyer of_____ to_____ from_____"[21].

Later on, Peirce identified simple relative terms and conjugative terms as "blank forms of propositions" [22], calling the latter predicates as well.

---

[20]The example is not exactly what De Morgan writes. De Morgan is of course aware of the problem, but only in his second article on the syllogism (De Morgan 1850) he deals with such quantified expressions even though without quantifiers; so, not yet in his 1847 *Formal Logic*.
[21]W 2: 365, 1870.
[22]EP 2: 299, 1907.

These blank forms are truly schematic patterns for various possible formulas; the blanks are places to introduce individual signs as variables, so it is possible to refer to different individuals in different possible logical situations. This makes Peirce's system capable of expressing both Boole's class operations and De Morgan's relational operations. Besides, already in 1870 it is also almost fully functionally complete.

Completeness comes with the adoption of specific signs for the nowadays widely known universal and existential quantifiers, a term Łukasiewicz claims was invented by Peirce[23] himself. The very first introduction of those signs is documentedly accomplished by Peirce in 1882, in a letter to his student O. H. Mitchell.[24] In 1883, Peirce used specific signs for quantifiers in print for the first time, in one of his contributions to the only whole volume he organized as a professor while at Johns Hopkins.

The first basic feature of Peirce's linear notation is the use of juxtaposed subscript letters to the side of the operative sign. These subscripts are called indexes, that is, deictic signs pointing to specific items within the universe of discourse, like "a pointing finger"[25]. Subscripts also serve to indicate which terms are linked by a certain relation and in which specific order. For instance, if l denotes the relation of loving, then l_{ij} signifies "i loves j", while l_{ji} signifies "j loves i", and l_{ii} signifies "i loves him-/her-/it-self" ("i" and "j" are indexes for whatever individuals are in this love relation). The notation very easily allows for expressing reflexivity, as in the last example, and convertibility, as in $l_{ij} = l_{ji}$: "j is loved by i". It also allows for expressing binary or higher level relations or predicates, as for instance, Peirce's favourite example of a relation:

$$g_{ijk} : \text{"i gives j to k"} .$$

Now, Peirce adds to this system two specific signs for quantifying relations, namely, Greek letters $\prod$ – for *product* – and $\sum$ – for *sum*, respectively denoting the universal quantifier and the existential quantifier[26]. So, for instance, if x denotes whatever property:

1. $\prod_i x_i = x_s x_p x_c x_m x_y \ldots$, etc. $\prod_i x_i$ means x is a property of *all* the individuals denoted by i. For instance, "All men are mortals"

---

[23] J. Łukasiewicz 1930, p. 147.
[24] See C. Terra Rodrigues 2017, p. 454 ff.
[25] W 5: 163, 1885.
[26] W 5: 178–180, 1885.

means that Socrates, *and* Pelé *and* Chaplin, *and* me, *and* you ... are all mortals.

2. $\sum_i x_i = x_s + x_g + x_c + x_y \ldots$, etc. $\sum_i x_i$ means x is a property of *at least one* of the individuals denoted by i. For instance, "Some men are philosophers" means that Socrates, *or* Gramsci, *or* Chaplin, *or* you ... are philosophers [27].

The propositions "Everybody loves Chaplin" and "Everyone loves someone" can then be respectively expressed as follows: $\prod_i l_{iC}$ (C as an index for the individual Chaplin) and $\prod_i \sum_j l_{ij}$ (j for *jemand*, German for *someone*). So, the coefficient and the signs for quantifiers are enough to represent all propositions of traditional logic. In fact, much more.[28]

Two final remarks are noteworthy. First, in this famous paper from 1885, "On the algebra of logic: A contribution to the philosophy of notation", Peirce remarks the symbols $\prod$ and $\sum$ were chosen to make the notation as *iconic* as possible, that is, as representative of the form of the operations as possible. Indeed, the quantifiers are defined in terms of potentially infinite conjunctions and disjunctions, with the care not to straightforwardly identify them with sum and product, since "the individuals of the universe may be innumerable" [29], but not necessarily are, as the examples show. Second, although G. Frege has chronological precedence over Peirce for the invention of the quantifiers, Peirce's system was an independent achievement, being just as Frege's functionally complete for first-order predicate calculus. Notwithstanding, Peirce wanted to avoid any understanding of his logic as a universal language, "like that of Peano"[30], preferring instead to adopt a meta-theoretical approach aiming at second-order logic since the very beginning[31]. This makes Peirce's approach akin to a contemporary model-theoretic one, as Hintikka has suggested,[32] but it also puts Peirce in a long-termed tradition of symbolic thought, according to which logic is not restricted the study of deductive forms of human reasoning, but extends itself to inquiring about how thought embodies itself in semiotic processes in the world.[33] These subjects are to be reserved for a further occasion.

---

[27] W 5: 180, 1885.
[28] The power of Peirce's symbolism can be seen for instance in W 5: 181.
[29] W 5: 180.
[30] CP 4.424, c. 1903.
[31] W 2: 56, 1867; W 5: 185, 1885; see Dipert 1997.
[32] J. Hintikka 1997.
[33] See J. Legris 2016; C. Terra Rodrigues 2017, p. 450.

References

[1] Anellis, Irving H. 1997. Tarski's development of Peirce's logic of relations. *In*: Houser, Roberts, and van Evra, *op. infra cit.*, pp. 271-303.

[2] Anellis, Irving H. 2004. The genesis of the truth-table device. *In*: *Russell: The Journal of Bertrand Russell Studies*, n.s. 24 (Summer 2004), pp. 55-70.

[3] Anellis, Irving H. 2012. How Peircean was the "Fregean revolution" in logic? *In*: *Logicheskie Issledovaniya / Logical Investigations*, vol.18 (2), pp. 239-272.

[4] Bellucci, Francesco and Ahti-Veikko Pietarinen. 2016. From Mitchell to Carus: Fourteen Years of Logical Graphs in the Making. *In*: *Transactions of the Charles S. Peirce Society*, vol. 52 (4), pp. 539-575. DOI: 10.2979/trancharpeirsoc.52.4.02.

[5] Brady, Geraldine. 2000. *From Peirce to Skolem: a neglected chapter in the history of mathematical logic.* Amsterdam: Elsevier Science B.V.

[6] Clark, William G. 1997. New light on Peirce?s iconic notation for the sixteen binary connectives. *In*: Houser, Roberts, and Van Evra, *op. infra cit.*, pp. 304-333.

[7] De Morgan, Augustus. 1847. *Formal Logic: Or, the calculus of inference, necessary and probable.* London: Taylor and Walton.

[8] De Morgan, Augustus. 1850. On the symbols of logic, the theory of the syllogism, and in particular of the copula, and the application of the theory of probabilities to some questions of evidence. *In*: *Transactions of the Cambridge Philosophical Society*, vol. 9 (1856), pp. 79-127.

[9] De Tienne, André. 2014. 1914-2014: One hundred years of editing and publishing Peirce (Commens Working Papers no. 4). Retrieved from *Commens: Digital Companion to C. S. Peirce*, http://www.commens.org/papers. Access date: March 2nd, 2020.

[10] Dipert, Randall. 1995. Peirce's underestimated place in the history of logic: a response to Quine. *In*: Ketner, Kenneth L. (ed.). *Peirce and Contemporary Thought*. New York: Fordham University Press, pp. 32-58.

[11] Van Heijenoort, Jean. 1967 (ed.). *From Frege to Goedel: a source book in mathematical logic, 1879-1931*. Cambridge, MA: Harvard University Press.

[12] Hintikka, Jaakko. 1997. The Place of C.S. Peirce in the History of Logical Theory. *In*: *The Rule of Reason: The Philosophy of C.S. Peirce*. Edited by Jacqueline Brunning and Paul Forster. Toronto, CA: University of Toronto Press, pp. 13-33.

[13] Houser, Nathan. 1991. The Schroeder-Peirce correspondence. *In*: *The Review of Modern Logic*, vol. 1 (2-3), pp. 206-236.

[14] Houser, Nathan. 1992. The fortunes and misfortunes of the Peirce papers. *In*: *Signs of Humanity*, vol. 3. Michel Balat and Janice Deledalle-Rhodes (eds.). General editor: Gérard Deledalle. Berlin: Mouton de Gruyter, pp. 1259-1268.

[15] Houser, Nathan; Roberts, Don D.; Van Evra, James (eds.). 1997. *Studies in the Logic of Charles Sanders Peirce*. Bloomington; Indianapolis: Indiana University Press.

[16] Keeler, Mary. 2020. The Hidden Treasure of C. S. Peirce's Manuscripts. *In*: *Chinese Semiotic Studies*, 2020, 16 (1), pp. 155-166.

[17] Kneale, William C. and Martha Kneale. 1962. *The Development of Logic*. Oxford: Clarendon Press, 1985 (reprint).

[18] Legris, Javier. 2016. Los gráficos existenciales en la historia de la lógica simbólica. *In*: *Charles S. Peirce: Ciencia, filosofía y verdad*. Catalina Hynes y Jaime Nubiola (eds.). San Miguel de Tucumán, AR: La Monteagudo, pp. 59-68.

[19] Lukasiewicz, Jan. 1930. Investigations into the sentential calculus. Translated from the German version by S. McCall. *In*: *Selected Works*. Edited by L. Borkowski. Amsterdam; London: North Holland Publishing Company; Warszawa: PWN – Polish Scientific Publishers, 1970, pp. 131-152.

[20] Lukasiewicz, Jan. 1934. On the history of the logic of propositions. Translated from the German version by S. McCall. *In*: *Selected Works*. Edited by L. Borkowski. Amsterdam; London: North Holland Publishing Company; Warszawa: PWN – Polish Scientific Publishers, 1970, pp. 197-217.

[21] Putnam, Hilary. 1982. Peirce the logician. *In*: *Historia Mathematica: Journal of the International Commission on the History of Mathematics*, vol. 9 (3), pp. 290-301.

[22] Salatiel, José Renato. Aspectos filosóficos da lógica trivalente de Peirce. *In*: *Kínesis*, vol. III (5), pp. 31-42.

[23] Salatiel, José Renato. 2017. Tempo, modalidade e lógica trivalente em Peirce e Łukasiewicz. *In*: *Kínesis*, vol. IX (20), pp. 151-173.

[24] Santaella, Lucia. 1995. *Teoria geral dos signos: semiose e autogeração*. São Paulo: Ática.

[25] Savan, David. 1988. *An Introduction to C. S. Peirce's Full System of Semiotics*. 2nd ed. Toronto, CA: Victoria University Press. Toronto Semiotic Circle Monographs, Working Papers and Prepublications vol. I.

[26] Tarski, Alfred. 1975. *Conferências na UNICAMP em 1975 / Lectures at UNICAMP in 1975*. Transcrição e organização por Leandro Suguitani, Jorge Petrucio Viana, Ítala M. Loffredo D'Ottaviano. Campinas, SP: Editora da UNICAMP, 2016.

[27] Terra Rodrigues, Cassiano. 2011. The method of scientific discovery in Peirce's philosophy: deduction, induction, and abduction. *In*: *Logica Universalis*, vol. 5 (1), pp. 127-164.

[28] Terra Rodrigues, Cassiano. 2017. Squaring the unknown: the generalization of logic according to G. Boole, A. De Morgan, and C. S. Peirce. *In*: *South American Journal of Logic*, vol. 3 (2), pp. 415-481.

On the very idea of choosing a logic: the role of the background logic

Jonas R. B. Arenhart[∧]
Sanderson Molick[∨]

[∧]Federal University of Santa Catarina, Brazil
[∨]Federal University of Rio Grande do Norte, Brazil
[∨]Ruhr-University of Bochum - Germany

Abstract

Logical anti-exceptionalism is the view that logic is not special among the sciences. In particular, anti-exceptionalists claim that logical theory choice is effected on the same bases as any other theory choice, i.e., by abduction, by weighting pros and cons of rival views, and by judging which theory scores best on a given set of parameters. In this paper, we first present the anti-exceptionalists favourite method for logical theory choice. After spotting on important features of the method, we discuss how they lead to trouble when the subject matter of choice is logic itself. The major difficulty we find concerns the role of the logic employed to evaluate theory choice, or, more specifically, the role of the metalanguage employed to run the abductive method. When rival logical theories are being evaluated and compared, we argue, it is difficult not to beg some important questions; the metalanguage introduces biases difficult to avoid. These difficulties seem to be inherent to the method described. We suggest that they put some constraints on the scope of application of the method of abductive theory choice in logic and on the kind of disputes the anti-exceptionalist may plausibly expect to solve with it. We end the paper with some suggestions for how the anti-exceptionalist may address these issues on this front.

1 Introduction

Logic is typically conceived as being *a priori*, necessary, and analytic. In this traditional view, at least *prima facie*, there is no sense attached to the idea of choosing a logic, or of revising logic, in the face of any kind of (conflicting) evidence. Now, despite its venerable credentials, this traditional view has been attacked, among others, by Quine, and most recently, by the so-called *anti-exceptionalists*. Hjortland characterizes logical anti-exceptionalism thus:

> Logic isn't special. Its theories are continuous with science; its method continuous with scientific method. Logic isn't a priori, nor are its truths analytic truths. Logical theories are revisable, and if they are revised, they are revised on the same grounds as scientific theories. [Hjortland, 2017, p. 632]

The anti-exceptionalist plan for logical theory revision is that whatever it is that counts as our current logical system, it may be replaced by a more suitable system after all relevant matters are considered, just like Newtonian physics was replaced by the Special Theory of Relativity, so to say. This possibility has captured the attention of many philosophers who are fond of the idea of having *a method* for logical revision and logical theory choice that works just like theory choice in other sciences[1]. As Routley has argued,

> Choice of a logical theory is a special case of the choice of a theory or a system, and choice of these does not differ in principle from choice of such diverse items as a new house, a winner (e.g. of a gymnastics or equestrian contest), or of a recording of a symphony. [Routley, 1980, p. 81]

In this sense, the plan for logical theory choice sounds rather simple: choose some features that count as important virtues a system of logic ought to have (explanatory power, capacity of systematization and simplicity, for instance), evaluate how well the competing logical systems fare according to those virtues, and choose the one that scores best. The idea seems simple, and employs a method we seem to be familiar with when choosing a new car or a new umbrella. Far from being a non-sense, logical revision — from this perspective — is just part of the scientific enterprise of finding the theory that best squares with the evidence we currently have; in the case of logic, the concern is with inferences, but there is nothing special about it, the process is similar to any other process of theory choice.

---

[1] Logical revision and logical theory choice are used almost as synonymous in the anti- exceptionalist literature. Here, despite our reservations concerning it, we follow common practice.

However, in spite of its attractivenes, to assume this analogy between logical and scientific theories is not exempt from problems. In a nutshell, in the following paper we shall explore the following concern: given that the process of theory choice requires inferences to be made, and that these require that a system of logic is already settled to guide the inferential steps, the process of *logical theory choice* seems to presuppose the use of a logic, and this fact, we shall argue, leads us to beg the question against those in disagreement over what concerns the most appropriate logic to be used. As we shall explain, this type of choice procedure is vulnerable to some kinds of circularity, thus leaving room for non-rational features[2].

The paper proceeds as follows. In section 2 we briefly revise the anti-exceptionalist method for logical theory choice. In section 3 we advance two major arguments against this method. As we have already mentioned, the arguments concern the relation between the logic we use to evaluate logical choice and the evidence in favour or against distinct systems. We conclude in section 5 by suggesting that these difficulties may be overcome if the anti-exceptionalist could better specify the sort of logical disputes to which the theory choice method being discussed is applicable. We also indicate lines in which this suggestion may be carried out.

2 The anti-exceptionalist basic tenets

In this section, we shall provide for a clear assessment of the main features of anti-exceptionalism view on logical theory choice. There are certainly further aspects of anti-exceptionalist views of logic, such as its modal status and analyticity issues, but we shall not discuss them here. We shall concentrate on logical theory choice and bring to light two special features of the process recommended for such.

The first aspect of the versions of anti-exceptionalism that are being taken into account here, and that must be further specified, is that it is widely assumed that we use a logic for reasoning in natural language.[3] This involves the so-called *canonical application of logic*, the use of logic for studying the validity of inferences in natural language, as opposed

---

[2]Some of these problems are already known by authors such as Hjortland [2017] and Woods [2017]. Our purpose in the following paper is to explore how these elements play a role in the process of decision by abductive means.

[3]See Priest [2006].

to a purely mathematical study of logic on the one hand, as well as to the applications of logic in technology; for instance, in the study of electric circuits. In other words, it is assumed that natural language does embody a logic (the so-called *logica utens*, in medieval terms), and when one considers logical revision, or choice of a logical system, one is talking about this logic. As [Woods, 2017, p. 02] puts it, the target cases of logical revision that concern the typical anti-exceptionalists deal with "our most general canons of implication", our *"background logic"*.

That means that whenever we make inferences about any subject, in particular about the most appropriate system of logic, we are already using logic, where the logic in use is the logic of natural language. Although that seems reasonable enough, as we shall see, this fact engenders difficulties for the anti-exceptionalist. It is not as if anti-exceptionalists try to pretend that no logic is needed; rather, they try to minimize the effects of the background logic in the process of logical theory choice by the rational evaluation of the theoretical virtues of disputant logical theories. Hence a natural problem is to know whether (or how) this is possible. As an example of an anti-exceptionalist that clearly deals with this issue, [Priest, 2016, p. 51] comments on that topic, claiming that there seems to be no urgent problem in that:

> But some logic (and arithmetic) is necessary. Which? The logic (and arithmetic) we have. If we were trying to establish logical knowledge from first principles, then any use of logic would generate a vicious regress. But we are not: our epistemic situation is intrinsically situated. We are not *tabulae rasae*. In a choice situation, we already have a logic/arithmetic, and we use it to determine the best theory — even when the theory under choice is logic (or arithmetic) itself.

[Routley, 1980, p. 94] makes a similar case by arguing that at some point one will have to rely on natural language and the informal reasoning conduced in this language; he also claims that this informal reasoning must be reproducible in the system one claims to be the best candidate for correct system. We shall take this to be enough evidence for the claim that logic is involved in the choice of a logical theory, and that we "adopt" the logic we have in order to discuss logical theory revision, provided that *this claim* makes sense.

The second aspect of anti-exceptionalism we wish to spot on concerns the methodology of logical theory choice. According to the anti-exceptionalist tenets, recall, theory choice proceeds just as in any case of theory choice for any scientific theory. For this one must first choose some relevant factors on the basis of which the systems will be evaluated, and according to a measure attributing to each system how well it fares according to each factor. A weighted sum of the values is calculated and determines which system scores best in the end. The factors to be taken into account in the evaluation include simplicity, capacity of systematization, fruitfulness, economy (Ockham's razor), but are not limited to these.

Let us briefly present some of the features most praised in a logical system, according to some anti-exceptionalists.[4] They seem to be uncontroversial, but we shall discuss whether this is really the case later:

1 **extensive scope**: logic is the science with the most extensive scope; it applies overall. Systems that do satisfy this requirement score better than those that do not apply in some specific situations (e.g. not dealing with intensional contexts).

2 **conformity to the facts**: there may well be logical facts, some claims that no one can deny that an appropriate logic should account for (for instance, that a conditional is false when its antecedent is true and its consequent is false). A system of logic not accounting for the logical facts is ruled out as inadequate.

3 **accountability of the data**: our linguistic practices may provide important data that a logical system may have to account for. The data are somehow 'soft', theory laden, and one may sometimes reject the data if a theory has many other relevant virtues.

4 **explanatory power**: it is not enough to catalog the valid inferences. A logical theory must explain why such inferences are valid or invalid, i.e., give an account of validity that illuminates the valid and invalid consequences.

Now, suppose we have agreed on a list of factors that must be taken into account in logical theory choice, among which the above factors may be included. We provide a list of such factors:

---

[4]Here we follow the list presented in Routley [1980], but see also Priest [2016].

$$c_1, c_2, c_3, \ldots c_n$$

Distinct factors may even be evaluated differently. For instance, simplicity may be less important than conformity to the facts and/or explanatory power. Consistency may be also less important for some (e.g. paraconsistent logicians), and not even counted as a relevant factor that a system of logic must possess. This difference in the importance of each factor is reflected in the anti-exceptionalist model by assigning each criterion c_i a weight w_i, which is taken into account in the evaluation process. In the end, once every criterion receives a weight and a value according to a measure m, we have what Priest [2016] calls a *rationality index* for theories, a weighted sum of each of the criteria:

$$\rho(T) = m(c_1)w_1 + m(c_2)w_2 + \ldots + m(c_n)w_n \qquad (1)$$

Although it is clear that while the method operates on a given list of relevant factors, it is not clear how to motivate the selection of some factors as having priority over others. For instance, Routley draws a distinction between *heavyweight* and *lightweight* factors, in which the former includes theoretical factors like scope, conformity to the data and explanatory power, and the latter includes aesthetic factors like simplicity and elegance. In a different perspective, Williamson's anti-exceptionalist defense of classical logic is based on prioritizing factors like scope, elegance and simplicity.[5]

Furthermore, even where authors coincide in choosing some factors as of greatest importance, there may be disagreement over how to properly understand them. For instance, consider adequacy to the data and conformity to the facts. These may or may not be distinct factors, depending on how one further specifies the terms 'data' and 'facts'. Routley [1980] distinguishes between data and facts, while Priest [2016] does not. For the sake of argument, in this paper we shall not distinguish between data and facts. What is relevant for us is that even if there is agreement that a logical theory must be faithful to the data and/or facts, it is not clear which facts and/or data are relevant. Routley [1980], in particular, presents the following list of Facts that must be accounted for by a system of logic:

- Fact 1) Much of our discourse is intensional (while classical logic is extensional).

---

[5]See Williamson [2016].

- Fact 2) Much of philosophical discourse is about the non-existent.

- Fact 3) There are inconsistent non-trivial theories and inconsistent non-trivial situations (while classical logic is explosive in the face of inconsistency).

Notice: consistency is not welcome here! If we take this list at face value, classical logic fails to meet the facts. And as we have already mentioned, the choice of factors to be taken into account in logical choice is not without problems. As the reader may foresee, the discussion over which are supposed to be the relevant facts may also bring in a great deal of trouble, for the very choice of relevant facts may be detrimental to the rationality of the choice procedure.

In the following, we explore two kinds of problems related to the anti-exceptionalist choice method: 1) the role of the logic we have as the base logic for logical theory choice, and 2) the role of the background logic in the metatheory and the selection of the relevant logical facts.

3 No neutral metalanguage

The anti-exceptionalist recommends that logical theory choice must be carried through by the logic we use in a given language, i.e., we should employ *the logic we have* for running the choice procedure. In this section, we shall start by arguing that the logic we have may play a major role in the process of logical choice.

We start with the Kripkean objection that logic is not revisable (see Berger [2011]). Kripke argued that the very idea of adopting a logic does not make sense, in light of the fact that adoption of a logic already presupposes that a logic is given. We shall leave this more skeptical ring aside, dealing with a challenge for the claim that one can coherently change logic when a logic is already given. The argument indicates that the metalanguage we do employ impacts on the possibility of evaluating evidence against our current system. This makes the role of the logic we have much more relevant to the evaluation of a dispute than the anti-exceptionalists are willing to concede. The logic we use in the evaluation of distinct candidates to revise it impinges on the very result of the evaluation Berger [2011].

For the Kripkean argument, the desired conclusion is reached by a kind of thought experiment. We shall call it the perverse inference

(PI) argument, and it runs as follows. Suppose someone believes that from 'every x is B' it follows logically that 'x is **not** B' (this is the perverse inference). We may also assume that the user of PI does not accept universal instantiation (UI), given that this would make for an inconsistent set of rules (not impossible, of course, but let us not take it into account for the moment). Consider an opponent attempting to call the user of PI to her senses by arguing that this inference is fallacious and the logic containing it should be dropped. It seems plausible to suppose that the contender would have to claim something along the following lines: 'look, every instance of PI is fallacious, so that this inference you made is fallacious' (this is an instance of universal instantiation). The friend of PI may agree on the relevant data (every instance of PI is fallacious), but disagree on what results from it and on the need of revision. *Nota bene*: there may even be agreement between the two contenders over the truth of the premise, without that implying that the user of PI could agree that she needs to change logic; she may simply not get to the claim that some particular inference of hers is fallacious when she applies her accepted forms of reasoning. The user of the rule PI could claim that, by using the rules of inference she accepts, even if the contender is correct in claiming that every instance of PI is fallacious, the conclusion the contender wishes her to accept does not follow. In fact, by using PI we have: 'Every instance of the rule PI is fallacious, therefore, this instance is not fallacious'. As a result, the evidence available for both, friend and foe of PI, may be the same, but the *logic the friend of PI* has as her background logic may not allow her to see that the PI rule must be revised. The patterns of inference we already *use* won't allow us to change our inference rules in these cases [Berger, 2011, p. 185]. Basically, once a set of inference rules is assumed, we can't see the problem with them, because we are always operating with them to judge the data available. In other words: the claim that some set of inference rules is fallacious can't be justified when one employs that same set of inference rules. The trouble with those inferences must be seen 'from the outside', as it were, given that someone using that set of inference rules will not think she is inferring illegitimately.[6]

---

[6]One might object that Kripke's example is too borderline, since, in many logical disputes, the disputants agree on some (or perhaps even most) inference rules. However, even if they disagree over one inference rule relative to a single logical constant, it is not clear that one disputant will then be able to "adopt" the point of view of the

This last remark leads us to our second point, which generalizes the first one. The Kripkean argument shows that the logic adopted in the metatheory determines which inferences are accepted and therefore brings trouble to any process of theory choice. We shall argue further that this kind of consideration may be expanded to other features of the choice choice procedure. In particular, the metalanguage and metalogic we have (or think we have) infiltrates in the process of theory choice not only by the inferences accepted, but also by interfering on how we judge simple issues such as *the choice of relevant factors* for theory evaluation. Philosophical agendas infiltrate, consciously or unconsciously, in these discussions. Consider, for instance, the logical facts which a system must accommodate in order to be appropriate. What is taken to be a logical fact, or the data, is already logic-laden, as it were, and the facts that must be taken into account already reveal the preferences of those in the dispute. Problems of this kind are already known in the philosophy of science, where the available data needs to be described within the language of the old theory and therefore are susceptible to all the biases inflicted by the old theory.

Hence, our claim is that choice of the relevant factors on logical evaluation is very much purpose driven, and the purposes one has in mind as the most relevant ones determine the factors that weight more. Our focus will be on the broad features of a logical system. Given that logic is involved with many important concepts, it is also open to bias infiltration in any consideration of theory choice. As [Priest, 2016, p. 39] puts it:

> The central notion of logic is validity, and its behaviour is the main concern of logical theories. Giving an account of validity requires giving accounts of other notions, such as negation and conditionals. Moreover, a decent logical theory is no mere laundry list of which inferences are valid/invalid, but also provides an explanation of these facts. An explanation is liable to bring in other concepts, such as truth and

---

other, so that Kripke's worries could be overcome. There are many very interesting cases of logical disputes of this kind, in which the minor difference in the assumption/rejection of the inference rule in dispute carries with it many consequences that imply the change in a number of philosophical assumptions by each party. Such is the case of the dispute between paraconsistent logics and classical logic, with its far-reaching consequences for our theories of truth.

meaning. A fully-fledged logical theory is therefore an ambitious project.

That is, logical theorizing is already involved in basic matters such as the meaning of the connectives and truth, not only logical consequence. In fact, logical consequence and the logical vocabulary are often intertwined, so that it is not clear how to changing one without altering the other. When discussing logical theory choice, these features are also involved. Furthermore, when one assumes, as anti-exceptionalists typically do, that a logic must be available for us to actually use it in the process of logical theory choice, these items (connectives and their meanings, a theory of truth or, at least, a view on how truth behaves) are also assumed as settled in the logic we use. As a result, the logic one uses impacts on theory choice not only with its notion of logical consequence, but also with its accompanying meaning for the connectives and (importantly) its available notion of truth.

In order to illustrate how the argument of the impact of the features of the metalanguage would run in this broader scenario, let us focus on the informal semantic characterization of logical consequence:

Def[**Logical consequence**] A follows from B iff in every case in which formulas in B are true, A is also true.

One obtains a specific notion of logical consequence provided that the very concept of 'cases' is made more precise. What is the range of the quantifier in the definition of logical consequence? The cases that one needs to have available are the cases that make the premises and conclusions of inferences holding or not.[7] One evaluates inferences on the set of cases available.

This issue hinges on the data that must be accounted for by any candidate system of logic, and on the facts that logic must convey. Recall Routley [1980] enumerating the 'facts' that must be accounted for: Fact 1) Much of our discourse is intensional; Fact 2) Much of philosophical discourse is about the non-existent; Fact 3) There are inconsistent non-trivial theories and inconsistent non-trivial situations.

The facts to be accounted for already reveal some features of an intended underlying logic. Let us focus on fact 3. The claim that there

---

[7]We use the neutral 'holding' instead of true or false to allow cases where there may be more than just the two truth values.

are inconsistent non-trivial theories makes it analytic that the underlying logic must be paraconsistent. In fact, that encompasses the very definition of paraconsistency, so that it results analytically that a paraconsistent logic must be adopted if we are to take those facts into account (see also Michael [2016].). In other words: one cannot even state 'the facts' appropriately unless a paraconsistent negation is assumed at work in the metalanguage. Indeed: consider a classical logician using her classical connectives and concepts sincerely stating that 'there are inconsistent non-trivial theories and inconsistent non-trivial situations'. That would be self-refuting! On the other hand, a paraconsistent logician saying that is merely a reflection of the definition of paraconsistency. So, the logic one uses in the metalanguage affects the very account of the data and of the facts.

This general kind of difficulty infiltrates from the mere appraisal of the data available to the proper assessment of the most appropriate set of rules of inference to deal with those data. That is, in order to evaluate the available inferences, one must, in this case, already accept that some of the cases available comprise inconsistent non-trivial theories or situations (or worlds). That is precisely what the classical logician will deny. In this case, there is a disagreement over what counts as a legitimate case, or a legitimate fact that a system of logic must take into account. This makes for both contenders, paraconsistent and classical logicians, using incompatible evidence, as seen from their own point of view.

Other features of the data or the cases that must be taken into account are similarly logic-laden. The idea that inconsistent cases must be taken care of in the scope of the quantifier 'for every case', allowing for instance that some propositions are both true and false in some cases (instantiating thus a truth value glut), or rather other way around, that every case is consistent (no gluts available), depends on the logic employed to legislate over the cases. That is, one cannot legitimately claim that some cases are available to the evaluation of propositions where contradictions obtain, for instance, without beforehand having settled that propositions are allowed to be evaluated in such situations as legitimate cases. Logic has priority over the cases by constraining the behavior of the truth values. It is precisely in this sense that the evidence available depends on the logic we assume beforehand. As a further example, not involving the notion of logical consequence, think of paraconsistent set theories based on naive principles of set formation that lead to sets such as Russell's set; the data available for these theories are simply denied by the classical logician for their very threat of inconsistency.

When these difficulties are plugged in with the typical claim by the anti-exceptionalist, things get even more obscure. Consider the claim that the logic in the metalanguage (the one in which talk about the object system is performed) and the logic chosen as the correct one (the one that scores best) should be the same. [Routley, 1980, p. 94] is clear on this subject:

> The choices of system and metasystem — more generally, system and extrasystematic adjuncts — are by no means entirely independent. It is not satisfactory for example, to reject classical logic systemically, e.g. as involving mistakes or illegitimate assumptions (such as the law of excluded middle), and to use it metasystemically without further ado or qualification; for to do so would be to proceed by what are confessedly mistaken paths.

[Priest, 2006, p. 98] puts the same point about the meaning of the logical operators (which are related to logical consequence, to be sure):

> Any intuitionist or dialetheist takes themself to be giving an account of the correct behaviour of certain logical particles. Is it to be supposed that their account of this behaviour is to be given in a way that they take to be incorrect? Clearly not. The same logic must be used in both "object theory" and "metatheory".

However, given that a metatheory is required in order to evaluate the logical choice, and once it is assumed that it must be the same logic that is available both in metatheory as in the object language, troubles arise. If we follow the advice of Routley and Priest, and choose to use in the metalanguage the account we think is correct, the evidence available will be relative to the choice of metasystem. For instance, once one has chosen a paraconsistent negation, it will be available to her that some facts may be contradictory without triviality. Those facts will not be available for a classical logician, though. Classical and paraconsistent logicians, in this setting, are talking past each other. Even if one chooses the metatheory of a paraconsistent logic able to recover the rationale of classical logic in consistent situations, the defender of classical logic may argue that the full power of classical logic is not present, and that many of the advantages of classical logic were sacrificed for little gain.

The anti-exceptionalist may avoid this "incommensurability" between theories by requiring that the disputants must at least share the set of logical facts. This seems to result as a minimal desiderata for the kind of dispute able to be settled by anti-exceptionalist means. This move, of course, significantly shrinks the range of logical disputes treated by the anti-exceptionalist, and goes on a different direction than that pursued by authors such as Williamson [2016] and Routley [1980] in order to settle the debate between defenders of paraconsistent logics and defenders of classical logic.

The further relevant questions to be raised are: what other desiderata are required to hold for sensible application of the anti-exceptionalist method for logical theory choice? How to characterize the set of logical disputes open for treatment by current anti-exceptionalist means? When rival theories are in dispute for the description of a set of facts, the elements of the theory are present not only in the object language, but in the metalanguage as well.

A clear example of this type of problem may be seen in Priest's ([Priest, 2006, chap.4]) discussion of Boolean negation. Given that Priest does not agree with Boolean negation, he feels free to use De Morgan negation in the metalanguage to characterize Boolean negation (in the object language). This has as a result that one cannot prove, in the object language, that Boolean negation is explosive (the inferences required for that are not available in the metalanguage). However, if a friend of Boolean negation could do the same, and characterize De Morgan negation in her own terms, then, it seems, De Morgan negation could also lead to results such as ex falso. It seems there is no easy way out of this kind of question begging scenario, when the supposition of being using 'the right logic' in the metalanguage is in force. Therefore, a natural problem is to know how to perform a non-biased choice procedure in these scenarios (assuming the parties in dispute are, indeed, comparable) [8].

---

[8] See the discussion in Arenhart and Melo [2017]. In Anderson et al. [1992] another example of this kind is introduced by the authors. Given the presence of De Morgan and Boolean negation in the object-language, if De Morgan negation is adopted in the metalanguage, then both negations collapse. However, if Boolean negation is assumed, then the relevantist can claim not to understand the classical reasoner. This is considered by the authors as illustrative of an incompatibility between the

Even when a new theory emerges against the accepted theory so that the adherents start to accept the peaceful coexistence of both, the metalanguage of the old theory is still present and spreads across all the disputants in question. This is the case because rival theories are born out of the background of the old theories. Examples of this kind are found not only in logic, but also in mathematics or in physics.[9]

4 Theory-choice loops

Our next argument against the feasibility of the method presented earlier comes from Woods [2017]. According to Woods, using a specific metalanguage in the evaluation process also engenders loops in the choice of the most appropriate logic. In a nutshell, the argument runs by creating loops in the choice of a logical system. Once a system is chosen and adopted due to its best results in the theory selection method, when the anti-exceptionalist choice procedure is performed again, now using the newly adopted system as background logic, it leads one to choose the rival (old) system back again. Hence when one changes back to the "old" logic, one sees that the rival system scores better again. And so on. The logic we use determines the evaluation of the evidence, and in some particular cases, the logic we use seems always to imply that we would be better off changing the logic. Woods's loops illustrate how the choice of relevant factors seems not to provide enough grounds for theory choice. Some kind of choice underdetermination still arise in face of the relevant factors conjoined with the adoption of a background logic.

The most prominent example of such loops concerns a discussion between classical logic and the relevant logic T (for Tennant). The existence of a loop in logical choice is clearly illustrated here. Assuming classical logic in the metalanguage, one is able to show that T recaptures classical logic in the object language level. This opens up the possibility of obtaining all of classical mathematics that depends on the use of classical logic. Also, given that proofs in T are more informative, T seems to be preferable. That is, T scores better than classical logic, because its proofs are more informative, and one loses nothing of classical mathematics. So, a choice of T is advisable. However, once T is assumed

---

worldviews of the disputants.

[9]For instance, quantum mechanics needs classical physics to account for the results of its experiments. Non-classical logic is sometimes said to need classical set theory as a background.

as the logic we have and use, it is part of the language in which we evaluate the evidence for logical choice. So, let us run the method of logical choice again. When the metalanguage is T, there is no way to recapture classical logic in T, and T cannot reproduce classical mathematics. Although T is more informative on its proofs, the recapture of classical logic in order to have classical mathematics is much more important, so that it is preferable to have classical mathematics than a more informative deduction system. In this sense, from the point of view of T, classical logic is preferable. And then, the loop is created.

Similar loops seem to arise in cases of relevant logics in general, or in non-contractive logics, non-transitive logics [Woods, 2017, p. 16]. These systems recapture classical logic only when classical logic is already available in the metalanguage. So, from the point of view of classical logic, these systems should be adopted, given that they have clear advantages over classical logic when reasoning with so-called versions of naive theory of truth are concerned. However, when those systems are adopted and become the system we use, they cannot be used to recapture the full power of classical logic, and then, they fall short of providing for classical mathematics. Again, the result is that it is preferable to have classical mathematics than these treatments to the paradoxes of self-reference. Thus, from the point of view of such sub-structural logics, classical logic should be adopted as preferable. The loop reappears.

A possible solution is found by following a suggestion of Bueno [2010]. Bueno argues that disagreements about which logic to choose must proceed by employing a logic, but that this logic need not be the same logic that is under evaluation. The logic in the metalanguage does not need to be the same as the logic in the object language. That is, we may disagree on which logic to choose for a given purpose (inferences in natural language, say), but may agree on which logic to use when we conduce disputes about that. For instance, it is possible that we could agree that we may use classical logic to debate over which logic to use when dealing with inferences in natural language.

Bueno's strategy goes in a direction already pointed out in Dummett [1991]. For Dummett, in order to solve logical disputes the disputants must agree on a metalanguage that is completely insensible to the object-language. As described by the author:

> What is needed, if the two participants to the discussion are to achieve an understanding of each other, is a semantic

> theory as insensitive as possible to the logic of the metalanguage. Some forms of inference must be agreed to hold in the metalanguage, or no form of inference can be shown to be valid or to be invalid in the object-language; but they had better be ones that both disputants recognise as valid. Furthermore, the admission or rejection in the metalanguage of the laws in dispute between them ought, if possible, to make no difference to which laws come out valid and which invalid in the object-language.(...) If both disputants propose semantic theories of this kind, there will be some hope that each can come to understand each other; there is even possibility that they may find a common basis on which to conduct a discussion of which of them is right. [Dummett, 1991, p. 55]

Dummett's concern with an agreement relative to the metalanguage comes from the fact that he takes it to be a pernicious principle to require the coherence between the metalanguage and the object-language of the disputant theories. The reason for this is that when this coherence is achieved, the defender of a non-classical logic can always resist arguments in favour of a classical law rejected by the non-classical adopter, namely, by claiming that the argument assumes the validity of the law in the metalanguage. However, this same counter-attack is often presented by defenders of classical logic against attempts of showing how to recover the classical derivations within non-classical theories.

Could then Dummett-Bueno's strategy work for the purposes of the anti-exceptionalist? It seems it couldn't, for many distinct reasons. First, assuming that the logic we use to discuss adoption of logics may be distinct from the systems that are under discussion, we beg the question against the logical monist, who accepts that only one logic must be true. As a second point, the logical monist may claim that discussions as to the most reasonable system of logic involve cases of inferences in natural language, so that if we agree on which logic to use in *this discussions*, then we have already settled the issue. Third: by claiming that we must agree on a metalogic, one could ask: how is this agreement achieved? By the method of the anti-exceptionalist? But then, the problems we have just examined reappear, and the suggestion amounts to no real progress at all. On the other hand, if the metalogic is not chosen by these standards, then, there are other means by which to choose a logic, and the relevance of the anti-exceptionalist method for logical choice is lost. Both horns of the dilemma seem to lead to trouble. On the one hand, to require the coherence between object-language and metalanguage may not allow us to characterize the dispute as a *genuine* logical dispute; on the other hand, dispensing such coherence may lead to the irrelevance of the anti-exceptionalist method.

These arguments have shown, again, that the metalogic one has in the background, in other words, the logic we use in conducting logical theory evaluation, plays a pivotal role much more detrimental to the choice of a logical system than the anti-exceptionalists seem willing to concede. In order to settle these issues, the anti-exceptionalist must adequately characterize — probably by restricting — the set of logical disputes their method is supposed to apply for.

5 Conclusion: possible routes

In the present contribution we have exhibited a set of problems related to the anti-exceptionalist strategy of selecting logical theories through abductive means. We argued that all these obstacles arise from the idea that logical theory choice has to be performed from a background logic. For this, we presented different types of 'intrusions' that the background logic may employ during the process of theory choice.

On one hand, if the principle of uniformity between theory and metatheory is to be demanded as a desiderata for logical theory choice, then it is not clear how to avoid the biases inflicted by the background logic. On the other hand, if uniformity is not demanded, then any choice procedure also seem to result problematic.

All difficulties raised in the previous section point to a limitation of the anti-exceptionalist method due to the absence of an adequate characterization of the set of logical disputes intended to be accounted. Even if the anti-exceptionalist drops the assumption that we employ our background logic in the choice procedure, she still ought to establish what kind of logical dispute she takes to be genuine and susceptible to be settled by the proposed abductive means. Based on what has been discussed, we introduce bellow a set of desiderata for a possible logical theory comparison, which an anti-exceptionalist will have to take into account in order to avoid the difficulties raised here:

- **Set of logical facts**: As discussed in Section 3.1, a minimal desiderata for the existence of a genuine logical dispute seems to be that the disputants share the same set of logical facts. One may suggest that a difference in the set of logical facts can be handled by the existence of a translation between the vocabularies of the logics in dispute. However, even when charitable interpretations of the principle in dispute are available for one of the disputants, problems like the ones mentioned in Section 3 appear again, namely the choice of metatheory may intrude the description of the relevant factors from the point of view of the disputant.

- **Coherence between object-language and metalanguage**: The horns of this dilemma were discussed in Section 3.2. However, it seems that for the pluralist this is no practical requirement. The pluralist might accept very well that the metalanguage of an old theory is kept within the disputant logic. The relevant question then is why would the pluralist want to choose among logics, to begin with? A local pluralist in the sense of Da Costa and Arenhart [2018] may just want to find the (provisionally) best tool for a specific job. The anti-exceptionalist ought to seriously take the issue of how pluralist or monist commitments might infiltrate into the choice procedure. Anti-exceptionalists like Hjortland [2017] have defended that the anti-exceptionalist ought to promote a form of ecumenism. However, it is not clear how ecumenical one can be when choice of metatheory is in play.[10]

- **Agreement on the set of heavyweight epistemic virtues**: Another important desiderata for a genuine logical dispute in the sense intended by the anti-exceptionalist is that the parties in dispute agree at least on the set of heavyweight epistemic virtues, i.e. the set of epistemic virtues they take to be most important. Many anti-exceptionalist arguments talk past each other because different epistemic virtues are prioritized.

It might very well be the case that the fulfillment of these conditions will significantly reduce the range of application of the anti-exceptionalist method. It shall remain as a future work the development of a precise characterization of the desiderata above. However, to establish them would clarify the usefulness and the very possibility of coherently choosing logics by abductive methods. The anti-exceptionalist

---
[10]See Read [2006] on this matter.

might prefer to work and produce logics that she is sure to be comparable through abductive means.

References

[1] Alan Ross Anderson, Nuel D Belnap Jr, and J Michael Dunn. *Entailment, Vol. II: The Logic of Relevance and Necessity*. Princeton University Press, 1992.

[2] Jonas R. Becker Arenhart and Ederson Safra Melo. Classical negation strikes back: Why priest's attack on classical negation can't succeed. *Logica Universalis*, 11(4):465–487, 2017.

[3] Alan Berger. Kripke on the incoherency of adopting a logic. *Saul Kripke*, pages 177–207, 2011.

[4] Otávio Bueno. Is logic a priori? *The Harvard Review of Philosophy*, 17(1):105–117, 2010.

[5] Newton Da Costa and Jonas R Becker Arenhart. Full-blooded anti-exceptionalism about logic. *The Australasian Journal of Logic*, 15(2):362–380, 2018.

[6] Michael Dummett. *The logical basis of metaphysics*, volume 5. Harvard university press, 1991.

[7] Ole Thomassen Hjortland. Anti-exceptionalism about logic. *Philosophical Studies*, 174(3):631–658, 2017.

[8] Michaelis Michael. On a "most telling" argument for paraconsistent logic. *Synthese*, 193(10):3347–3362, 2016.

[9] Graham Priest. *Doubt Truth to be a Liar*. Oxford University Press, 2006.

[10] Graham Priest. Logical disputes and the a priori. *Princípios: Revista de Filosofia (UFRN)*, 23(40):29–57, 2016.

[11] Stephen Read. Monism: The one true logic. In *A logical approach to philosophy*, pages 193–209. Springer, 2006.

[12] Richard Routley. The choice of logical foundations: Non-classical choices and the ultralogical choice. *Studia Logica*, 39(1):77–98, 1980.

[13] Timothy Williamson. Abductive philosophy. In *The Philosophical Forum*, volume 47, pages 263–280. Wiley Online Library, 2016.

[14] Jack Woods. Logical partisanhood. *Philosophical Studies*, pages 1–22, 2017.

Some remarks on logical realism and logical pluralism[1]

Lorenzzo Frade
Abilio Rodrigues

Department of Philosophy, Federal University of Minas Gerais, Brazil

Abstract

Logical realism is the view according to which the truths of logic describe a reality that is independent of our beliefs, conceptual schemes, and linguistic practices. Logical pluralism is the view according to which there is more than one correct account of logical consequence. At first sight, these two views seem to be incompatible. In this paper, we present an approach to logical pluralism that makes it possible to reconcile different accounts of logical consequence with a realist view of classical logic.

1 Introduction

Realism about some field or object is usually characterized as the view according to which the truths about that field or object do not depend on our beliefs, linguistic practices, and mental processes. So, for example, if one says she is a realist about mathematics, she is endorsing a view according to which mathematical objects exist and are as they are independently of our conceptual schemes, mental processes, and linguistic practices.

In the 20th century logic as a field of study was divided into different disciplines, but it is fair to say that the main concept of logic is the concept of logical consequence, and the central question of logic is *what follows from what*. So, when one says she is a logical realist, she is endorsing a view according to which the laws of logic, like the laws of nature, are matters of fact, and whether or not a conclusion follows from

---

[1]The second author acknowledges support from *CNPq* (Conselho Nacional de Desenvolvimento Científico e Tecnológico) research grant 311911/2018-8. Parts of sections 4.1 and 4.2 of this text have been already published in the text *On epistemic and ontological interpretations of intuitionistic and paraconsistent paradigms* [13].

a set of premises does not depend on our conceptual schemes, linguistic practices, and mental processes.

Frege conceived the laws of logic as laws of being true – which must be understood rather as laws of preservation of truth – and thought of logic as an investigation of reality analogous, in some sense, to empirical sciences.

> [Logic] has much the same relation to truth as physics has to weight or heat. To discover truths is the task of all sciences; it falls to logic to discern the laws of truth [21, p. 289].

As expected, since he was a realist with respect to mathematics, and his logicist project was to prove that arithmetic could be grounded on logic, Frege endorsed a realist view of logic.

> If being true is thus independent of being acknowledged by somebody or other, then the laws of truth are not psychological laws: they are boundary stones set in an eternal foundation, which our thought can overflow, but never displace [20, p. 13].

Logical realism, indeed, provides a traditional and powerful way of understanding the nature of logic that can be traced back to Aristotle. Although in his *Metaphysics*, when defending the principle of non-contradiction (PNC), to him the most certain of all principles, Aristotle makes clear that it has to do with reality, thought, and language[2], there is a consensus that in its main formulation PNC is a claim about reality: it cannot be the case that the same *property* belongs and does not belong to the same *object*, at the same time and in the same respect (1005b15). Thus, Aristotle conceived PNC primarily with an ontological character, like a general law of nature.

More recent characterizations of logical realism can be found in Rush [40] and McSweeney [28]. For Rush, logical realism is the claim that logic, like mathematics, is concerned with the rules of a world that exists independently of us:

> both logic and mathematics might be understood as applicable to a world (either the physical world or abstract world) independent of our human thought processes [40, p. 13].

---

[2]See [2], respectively, 1005b15, 1005b25, 1011b15.

McSweeney offers a similar but more concise characterization of logical realism as

> the view that there is mind-and-language independent logical structure in the world [28, p. 1].

So, once it is accepted that it is reality, the world, that makes a true proposition true, the truths of logic do not depend on us, but rather depend on the 'logical structure of the world'. This is how we understand logical realism.

2 Logic and logical realism

Being a realist about logic does not necessarily require the endorsement of a particular logic. But it is clear that the principle of excluded middle,

(PEM) $\vdash A \vee \neg A$

which is valid in classical logic, fits well with the realist claim that given a proposition A, either A is true or $\neg A$ is true, independently of us and what we know, or could possibly know, about A.[3] It is precisely the rejection of a realist view of mathematics that was the motivation of Brouwer's intuitionism and the rejection of excluded middle as a universally valid principle [see, e.g. 8].

The principle of explosion, according to which anything follows from a pair of contradictory propositions,

(EXP) $A, \neg A \vdash B$,

is also classically valid. Note that EXP is not an inference that is really applied in reasoning. Nobody concludes an arbitrary proposition B from a pair of contradictory propositions A and $\neg A$ precisely because contradictory propositions, in both classical and intuitionistic logic, cannot be

---

[3]Excluded middle is not the same as bivalence (BIV). The latter is a metatheoretical principle that says that a proposition A is either true or false, or more precisely, that the semantics of a given formal system has two semantic values, which in the case of classical logic are the true and the false. Since both hold in classical logic, saying that either A is true or A is false and saying that either A is true or $\neg A$ is true, end up being tantamount in the classical framework. However, PEM and BIV are not equivalent: it may be that PEM holds in a many-valued semantics (e.g. the logic of paradox [34]), or that PEM does not hold in a two-valued semantics (e.g. non-deterministic valuation semantics for intuitionistic logic [25]).

simultaneously true. EXP, in a certain sense, compared to PNC, is just a more effective way of saying that contradictions cannot be simultaneously true – as Aristotle (correctly) claimed some time ago. Let us call *classical* a view of truth that endorses excluded middle and rejects true contradictions, and so is in line with classical logic.

Dialetheism is the view according to which some contradictions are true. A dialetheia is defined as a proposition A such that both A and its negation $\neg A$ are true, and dialetheism claims that there are dialetheias [e.g. 36, 37]. Together with a realist conception of logic, dialetheism requires that reality is in some sense contradictory, or that there are contradictory things in the world. A good reason for accepting EXP rests on the view according to which reality is not contradictory in any sense whatsoever – a view that is a basic methodological tenet for scientific investigations, philosophy, and the ordinary use of language. But once one accepts that there are, or there might be, real contradictions, as dialetheists do, of course one must reject EXP.[4]

In a two-valued semantics, the principles of explosion and excluded middle characterize classical negation. The latter is defined by two conditions: a pair of propositions A and $\neg A$ cannot be simultaneously true, nor simultaneously false. So, one and at most one between A and $\neg A$ holds. Logics in which excluded middle is not valid, like intuitionistic logic, are called paracomplete, and logics that reject the validity of explosion are called paraconsistent.

A realist view of classical logic is suitable in justifying both PEM and EXP, and rejecting them led to the development of some non-classical logics. But one should not draw the conclusion that a non-classical logic is obliged to reject a realist view on logic. A dialetheist can stick to the claim that logic is grounded on the logical structure of the world, that in this case allows some contradictions. In the case of intuitionistic logic, the point is a more subtle one. Brouwer rejects a realist view of mathematics and claims that logic is just a description, in the language of mathematics, of patterns of mental constructions [see 8, Sect. 1], which arguably makes Brouwer's intuitionism a form of idealism. But

---

[4]The expression 'real contradictions' of course has to be understood as a *façon de parler*, since strictly speaking only sentences, propositions, formulas etc. can be contradictory. But our point is that if a proposition $A \wedge \neg A$ is true, there must be something in reality that makes it true: a contradictory object, fact, truthmaker, property, whatever. This is indeed how dialetheism is usually understood [see e.g. 13, 17, 18, 27, 30].

once the thesis that mathematical objects are constructed by the human mind is accepted, mathematics turns out to be a sector of objects that has to be described by a logic that cannot be classical logic.

3 Logical pluralism

Logical pluralism is the view according to which there is more than one correct account of logical consequence, or more than one correct logic. Logical monism, on the other hand, is the claim that there is only one correct logic. The fact that there are several different accounts of logical consequence as formal systems with a syntax and/or a semantics mathematically defined is not enough to qualify as logical pluralism. In the philosophically interesting sense, logical pluralism requires that different logics are, indeed, *applied* to analyze and describe real-life contexts of reasoning.

At first sight, logical pluralism and logical realism may seem two irreconcilable positions. Once it is accepted that logic is a theory about reality, how can there be different but correct theories about reality? The answer is simply that there cannot be. Indeed, if logical consequence is defined as preservation of truth, if truth depends on reality in the sense that reality determines unequivocally the truth or falsity of a given proposition, and given that there is only one reality, it is hard to see how a logical realist could be a pluralist without saying that there are different notions of truth. In our view, it is not an available alternative to endorse a form of alethic relativism and to claim that different accounts of logical consequence are truth-preserving. We think that the only sense in which a truth-predicate could be regarded as a two-place predicate, like 'x is true according to y', would be 'x is true according to reality', which is in fact a one-place predicate.[5]

Indeed, as already mentioned above, for a full-blooded intuitionist, classical logic and constructive mathematics are incompatible because mathematical objects cannot be described by classical logic. So, such an intuitionist would be a monist about logic. Priest, as expected, defends a monist view according to which the paraconsistent logic LP (the logic of paradox) is the unique correct logic [see, e.g. 34, 35]. Another view that rejects PEM but is not restricted to the domain of mathematics is Dummett's anti-realism [see 1, Sect. 6]. Nevertheless, it should be

---

[5]On alethic relativism, see [3, Sect. 4.3].

clear that when a dialetheist like Priest endorses a paraconsistent logic, or an anti-realist defends some paracomplete logic, this is not pluralism about logic. Rather, the anti-realist and the dialetheist, as well as a staunch defender of classical logic like Quine, are logical rivals, each one defending a particular view about 'the right logic'.

3.1 Two views on pluralism in logic

3.1.1 Carnap's pluralism

One of the precursors of logical pluralism is Carnap. In an often-quoted passage, we read:

> *In logic, there are no morals.* Everyone is at liberty to build his own logic, i.e. his own form of language, as he wishes. All that is required of him is that, if he wishes to discuss it, he must state his methods clearly, and give syntactical rules instead of philosophical arguments. [10, p. 52]

Carnap's pluralism results from language variance: different linguistic frameworks give rise to different logics. It is not our purpose here to analyze in detail Carnap's pluralism. But in our view, the idea of pluralism based only on language variance without further considerations about why and how the languages at stake differ from each other yields an uninteresting form of pluralism. Note, moreover, that pluralism based only on language variance is not capable of replying to a criticism like the one presented by Quine [38, Chap. 6] according to which the defenders of two different logics whose connectives have different meanings are not, in fact, in disagreement because they are talking about different things.[6]

As has been mentioned above, there is a distinction between logics that can be, and logics that cannot be applied to formalize real-life reasoning and explain meanings. Following Copeland [14], let us call them, respectively, applied logics and pure logics. Clearly, different logics that are not applied logics provide trivial examples of pluralism. Carnap's passage quoted above seems to be in line with the fact that nowadays we find in the literature plenty of pure logics, but even these logics do not come without conceptual motivation. However, as interesting as

---

[6]On Quine's views on non-classical logics, see Sect. 5 footnote 10.

the study of these pure logics may be, it is clear that they do not provide philosophically relevant examples of logical pluralism. In fact, the derogatory tone in which Carnap refers to philosophical arguments is closely connected to the idea of the 'rejection of metaphysics' that pervaded philosophy in the first half of the 20th century. But what justifies the rejection of explosion in Priest's dialetheism and in Belnap-Dunn four-valued logic (cf. Sect. 4.2.2 below), as well as the invalidity of excluded middle in intuitionistic logic, are philosophical arguments.

It is also worth noting that, as a result of Carnap's pluralism, logic loses at least two of its main features, namely, universality and normativity. It may not be a problem if one is mainly interested, as Carnap was, in eliminating any 'metaphysical ingredient' from philosophical discussion, which in the end may be tantamount to eliminating philosophy itself. However, we do think that a logic has to be normative and universal to be legitimately called logic. Besides, but no less important, that 'metaphysical ingredient', that should be rather called a conceptual ingredient, is not only unavoidable in philosophy, logic, and sciences, but is a necessary condition for understanding as best as possible what is going on in philosophy, logic, and sciences.

With respect to the requirement of normativity, we cannot see what the point is in building a language with a bunch of arbitrary syntactical rules without a conceptual motivation. That is, without an intended interpretation, a context to be formalized, or some formal and deductive properties to be investigated. The logician is just not free to do whatever she wants. Moreover, if the proposal is to express the deductive behavior of some context of reasoning, obviously there will be constraints, given by the very context of reasoning that is the object of analysis. Again, the anti-realist, as well as the dialetheist, cannot propose the logic she wants, and neither the classical logician, who is constrained by a classical notion of truth. These constraints have a normative character, for in each case it is established which inferences are sound. And if the point is not preservation of truth, such as the approaches to paraconsistency in terms of preservation of evidence and information [7, 12, 30, 39] that will be seen in Sect. 4.2 below, the invalidity of disjunctive syllogism, for example, has a normative character because it is not sound from the viewpoint of preservation of evidence/information.

3.1.2 Beal and Restall's pluralism

A more substantial and sustained approach to pluralism is proposed by Beall and Restall [4, 5, 6] (hereafter, B&R). Unlike Carnap's pluralism, B&R's pluralism does not depend on language but relies on a sort of vagueness of the standard model-theoretic notion of logical consequence defined in terms of cases: a conclusion A follows from a set of premises Γ if and only if any case in which each premise in Γ is true is also a case in which A is true. B&R argues that a case can be specified in different ways, and these different ways of specifying what a case is yield different logics. They propose the *generalized Tarski thesis*:

> (GTT) An argument is valid$_x$ if and only if, in every case$_x$ in which the premises are true, so is the conclusion [6, p. 29].

Pluralism results from different specifications of x. If x are Tarskian models, we get standard classical validity. By means of other specifications of x, we get intuitionistic logic when cases are stages of intuitionistic Kripke models [6, p. 62], and a paraconsistent and paracomplete relevant logic when cases are Barwise and Perry's situations [6, p. 49]. Note, however, that although different kinds of cases are allowed, logical consequence is still defined in terms of truth-preservation. In our view, this is the main problem with B&R's approach because it is hard to see how they can stick to different kinds of cases preserving truth without falling into alethic relativism.[7]

4 Two epistemic approaches

In this section, we will see two epistemic approaches to paraconsistency and paracompleteness that may be reconciled with the ontological, and realist, character of classical logic.

4.1 Intuitionistic logic

From the viewpoint of a convinced intuitionist, classical and intuitionistic logic cannot be reconciled because the former is just wrong. But a compromised position that is perfectly compatible with a realist conception of mathematical objects is already found in Heyting [22], where we read that

---

[7]Beall and Restall pluralism will be discussed in more detail in [9].

> Here, then, is an important result of the intuitionistic critique: *The idea of an existence of mathematical entities outside our minds must not enter into the proofs.* I believe that even the realists, while continuing to believe in the transcendent existence of mathematical entities, must recognize the importance of the question of knowing how mathematics can be built up without the use of this idea [22, p. 306].

Heyting remarks open the possibility of a peaceful coexistence between classical and constructive mathematics, and so between classical and intuitionistic logic – something that was not feasible in the end of 1930s, given the animosity between Brouwer and Hilbert.[8]

The central point that allows the reconciliation of classical and intuitionistic logic is to understand that the latter is not expressing preservation of truth, but rather preservation of availability of a constructive proof. Thus, a realist view of mathematical objects does not exclude an interest in intuitionistic logic as a study of such objects from the perspective of maybe more informative constructive proofs. So, the claim that in a given circumstance both A and $\neg A$ do not hold means only that there is no constructive proof of them, independently of the question whether any of them has been proved true by non-constructive (classical) means.

A result of this reading of classical and intuitionistic logic as 'talking about different things' is that the meanings of classical and intuitionistic connectives are not the same. Indeed, excluded middle holds in classical logic because it is always the case that at least one between A and $\neg A$ is true, and does not hold in intuitionistic logic because it may be that there is no constructive proof of A nor of $\neg A$, no matter if A or $\neg A$ has been proved true.

This idea of a peaceful coexistence between classical and intuitionistic logic that depends on the meanings of the connectives has been proposed and worked out by Prawitz [33]:

> the classical as well as the intuitionistic codification of deductive practice is fully justified on the basis of different meanings attached to the involved expressions (...)
>
> When the classical and intuitionistic codifications attach dif-

---

[8]On the *Grundlagenstreit*, the conflict between Brouwer and Hilbert that culminated in the exclusion of Brouwer from the board of *Mathematische Annalen* in 1928, see [13, footnote 5] and [41].

> ferent meanings to a constant, we need to use different symbols (...)
>
> The classical and intuitionistic constants can then have a peaceful coexistence in a language that contains both. [33, p. 28]

Prawitz' ecumenical system that combines classical and intuitionistic connectives has been further investigated in [32] and [31].

The idea that intuitionistic logic does not talk about truth can be traced back to Kolmogorov [24], who proposed a reading of intuitionistic logic in terms of solutions of problems instead of truth, and Heyting [22, p. 307], where we read that an assertion of A means in classical logic that A is true, while in intuitionistic logic it means that it is known how to prove A. Accordingly, a formula $A \vee \neg A$ means different things in classical and intuitionistic logic, and this is not exactly because the connectives $\neg$ and $\vee$ have different meanings; on the contrary, classical and intuitionistic connectives have different meanings because classical and intuitionistic assertions of A mean different things. So, classically, PEM means that either A or $\neg A$ is true, while intuitionistically it means that either it is known how to prove A or how to prove $\neg A$, or as we prefer to put it, either a proof of A or a proof of $\neg A$ is (in some sense) available. And of course when we are talking about availability of proof instead of truth, PEM is not universally valid. So, there is nothing wrong in accepting the universal validity of PEM in the classical reading on the one hand, and still endorse the "intuitionistic critique" mentioned by Heyting (page 329 above) on the other hand. Indeed, if intuitionistic logic is understood as not being concerned with preservation of truth but rather with preservation of availability of constructive proofs, there is no conflict with classical logic, as is the point of Prawitz' ecumenical system [32, 33].

4.2 On epistemic contradictions

In his seminal paper on paraconsistency, Jaśkowski [23] presented three conditions that a paraconsistent logic should satisfy: (i) it must be non-explosive in the presence of a contradiction; (ii) it should be rich enough to enable practical inference; (iii) it should have an intuitive justification. Condition (i) is just what defines a paraconsistent logic. Condition (ii) is required if a logic is designed to give an account of real-life contexts of reasoning, and condition (iii), in our view, is a consequence of (ii) because a plausible and intuitive justification is a necessary condition for (ii). In fact, conditions (ii) and (iii) are closely connected with a central question for the philosophy of paraconsistency: how to explain, in a plausible and intuitive manner, the nature of contradictions accepted by paraconsistent logics? We do not think that dialetheism provides a convincing answer to this question.

In order to accept contradictions without endorsing dialetheism we need a non-explosive negation not committed to the truth of a pair of propositions A and $\neg A$. Such a negation can only occur in a context where what is at stake is a property weaker than truth, in the sense that a proposition may enjoy that property without being true. So understood, paraconsistency can be combined with a realist view of truth that endorses excluded middle and rejects true contradictions.

4.2.1 Inconsistency in sciences

The fact that contradictions are unavoidable in empirical sciences is pointed out by Nickles [29], who sees empirical sciences as "non-monotonic enterprises in which well justified results are routinely overturned or seriously qualified by later results", and 'non-monotonic' implies 'temporally inconsistent' [29, p. 2]. These provisional contradictions are not dialetheias. They may be a result of limitations of our cognitive apparatus, failure of measuring instruments and/or interactions of these instruments with phenomena, provisional stages in the development of theories, or even simple mistakes that in principle could be corrected later on. In all these cases, contradictions do not belong to reality but, rather, are related to thought and to the process of the acquisition of knowledge. We call them epistemic contradictions.

The notion of non-conclusive evidence provides an account of contradictions without commitment to their truth that explains contradictions in empirical sciences. The notion of evidence for the truth (resp. falsity) of a proposition A is explained in [12, Sect. 2] as reasons for believing in or accepting A as true (resp. false). Such reasons may be non-conclusive, and in their turn can be explained as *justifications that might be wrong*.[9]

In [12] two paraconsistent logics have been proposed and studied. BLE (the basic logic of evidence), which is Nelson's logic $N4$ interpreted in terms of preservation of evidence, and LET_J (the logic of evidence and

---

[9]We return to the notion of evidence in the end of Sect. 4.2.2 below.

truth), a logic of formal inconsistency and undeterminedness [cf. 11, 26] that extends $BLE/N4$. LET_J is a paracomplete and paraconsistent logic equipped with a classicality operator ∘ that recovers classical negation by means of the following inference rules:

$$\frac{\circ A \quad A \quad \neg A}{B} \; EXP^\circ \qquad \frac{\circ A}{A \vee \neg A} \; PEM^\circ$$

The unary operator ∘ divides the propositions of LET_J into two groups: one subjected to BLE, another subjected to classical logic [12, Sect. 4]. Analogously to Prawitz' ecumenical system that combines classical and intuitionistic logics (cf. Sect. 4.1 above), LET_J combines classical logic and the paraconsistent logic BLE in the same formal system. But while in Prawitz' system different symbols are used for classical and intuitionistic connectives, in LET_J the connective ∘ works like a context switch that changes the meanings of the connectives according to the context.

The logic BLE preserves evidence instead of truth. The underlying idea is that a proposition A follows from a set Γ just in case there cannot be a circumstance in which there is evidence available for all the propositions in Γ but no evidence available for A. So, not only explosion, but several classically valid inferences fail. The following inferences are not valid in BLE because they do not preserve evidence:

1. $B \vdash A \vee \neg A$,
2. $A, \neg A \vdash B$,
3. $\neg A, A \vee B \vdash B$,
4. $A \to B, A \to \neg B \vdash \neg A$,
5. $A \to B, \neg B \vdash \neg A$.

The counterexamples are straightforward. For 1, a circumstance without any evidence for both A and $\neg A$; for 2 and 3, conflicting evidence for both A and $\neg A$ but no evidence for B. The implication of BLE, like classical and intuitionistic implication, is not relevant: it validates the principle known as *ex quodlibet verum*,

(EQV) $\vdash A \to (B \to A)$,

tantamount in natural deduction systems to the vacuous discharge of an assumption and in sequent calculus to the structural rule of weakening.

So, given B, for any A, $A \to B$ follows. A circumstance in which there is conflicting evidence for B and $\neg B$ but no evidence for $\neg A$ shows that 4 and 5 are not valid in BLE. Note, however, that the following inferences are all valid in LET_J:

6. $\circ A, B \vdash A \vee \neg A$,
7. $\circ A, A, \neg A \vdash B$,
8. $\circ A, \neg A, A \vee B \vdash B$,
9. $\circ A, \circ B, A \to B, A \to \neg B \vdash \neg A$,
10. $\circ A, \circ B, A \to B, \neg B \vdash \neg A$.

Disjunctive syllogism and explosion, for example, are valid from the viewpoint of preservation of truth (cf. 7 and 8), but invalid from the viewpoint of preservation of evidence (cf. 2 and 3), and so on.

4.2.2 Paraconsistency and information

Databases, in particular non-structured ones like the web, are prone to having contradictions. Inconsistent databases were the motivation of Belnap [7] when he proposed a four-valued semantics for the logic of first-degree entailment (FDE), a paracomplete and paraconsistent propositional logic that is the implication-free fragment of $BLE/N4$.

Belnap thinks of FDE as a local logic, that is, a logic to be used by a computer that receives information from different sources which may not be a hundred percent reliable. A logic such as this does not substitute classical logic because they are going to be used in different contexts. The information received by the computer can be incomplete or contradictory, and four scenarios are possible, represented by the semantic values *True*, *False*, *Neither*, and *Both*. These values are explained with the notion of a computer 'being told', so they mean, respectively, 'just told true', 'just told false', 'told neither true nor false', and 'told both true and false'. Since a computer may 'be told' that a proposition A is true even if A is not true, and vice-versa, the propositions that receive one of these four values are true or false independently of these values. Belnap's proposal is clearly pluralist, since different logics are used in different contexts, its underlying idea is that classical logic and FDE are talking about different things – namely, truth and information – and it explains paraconsistency by means of a property weaker than truth, which may be characterized as information that is not a hundred percent reliable.

About 30 years after Belnap's paper, Dunn explained a basic notion of information as what remains when we subtract "justification, truth, belief, and any other ingredients such as reliability that relate to justification" from Plato's analysis of knowledge as justified true belief [19, p. 589]. Accordingly, information is something expressible by means of a proposition, is objective, does not imply belief, and does not need to be true. In [39, Sect. 2.2.1], the notion of non-conclusive evidence is explained as information in the sense of Dunn [19] to which a non-conclusive justification has been added.

5 Logical realism and logical pluralism

If one sticks to the claim that logic is a matter of preservation of truth and keeps the generally uncontroversial idea that a true proposition is made true by reality, it is hard to see how one could be a pluralist about logic. What makes logical realism and logical pluralism two incompatible views is the idea that any account of logical consequence has to be defined in terms of preservation of truth.

The interpretation of paraconsistent logic and intuitionistic logic in terms of availability, respectively, of evidence/information and constructive proof, sketched in the sections 4.1 and 4.2 above, shows that these logics can be justified with no need of rejecting a classical view of truth, and so they can be combined with classical consequence in terms of preservation of truth. And the central idea that makes it possible is to understand paraconsistent, intuitionistic and classical logic as talking about different things.[10] The more effective argument in defense of this approach to pluralism is the fact that the readings of intuitionistic and paraconsistent logics sketched in the Sect. 4.1 and 4.2 above are successful in describing correct reasoning in the respective contexts.

Logics of formal inconsistency (*LFI*s) have a character intrinsically pluralist because, from the very beginning, since the seminal writings of da Costa [15, 16], the idea was to establish a distinction between what

---

[10]That non-classical logics and classical logic talk about different things is the point of Quine's slogan 'change of logic, change of subject' [38]. But the moral Quine wants to draw is that a non-classical logic is simply not logic. We have of course a completely different view. What was a problem for Quine for us is what makes possible a pluralist view of logic.

da Costa called well-behaved and not well-behaved propositions.[11] The logic LET_J is an LFI, and the claim that how a property weaker than truth can be transmitted from premises to conclusion is worth studying and deserves to be called logic is the point of the logics presented and studied in [12, 39]. The invalidity of EXP in LET_J is explained by means of the notion of evidence, which can be non-conclusive and even wrong. So, there may be contradictory evidence for a proposition A. Thus, the motivation for paraconsistency is not some contradictory character of reality. Rather, paraconsistency has to do with how to reason correctly in contexts where contradictory information or conflicting evidence occur.

In [13] it is argued that Brouwer's idealism provides an ontological interpretation of intuitionistic logic, while dialetheism is an ontological interpretation of paraconsistency. We cannot see how these positions could be combined with a pluralist view of logic. On the other hand, the interpretations of intuitionistic logic in terms of preservation of constructive proof and of paraconsistent logic in terms of preservation of evidence are called epistemic in [13] because both have an epistemic ingredient, a conclusive justification in the case of a constructive proof, and a (maybe) non-conclusive justification, in the case of evidence. The epistemic interpretation makes a pluralist view possible in which intuitionistic, paraconsistent, and classical logic do not conflict with each other. Note, besides, that these readings of intuitionistic, classical and paraconsistent logic keep the universal character of logic because, in each case, they can be understood as universal accounts of constructibility, truth and evidence. They also keep the normative character of logic because, again, in each case, they say which inferences are and are not allowed. In this way, intuitionistic and paraconsistent logics achieve a 'peaceful coexistence' with a classical notion of truth, and therefore with classical logic. This is how logical pluralism and logical realism can be combined without conflict.

References

[1] M. Alexander. Realism. In E. N. Zalta, editor, *The Stanford Encyclopedia of Philosophy (Winter 2019 Edition)*. 2019.

[2] Aristotle. Translation of Aristotle's Metaphysics. In R. McKeon,

---

[11]On how LFIs further developed da Costa's seminal ideas, see [11].

editor, *The Basic Works of Aristotle*. Random House, New York, 1941.

[3] M. Baghramian and J. A. Carter. Relativism. In E. N. Zalta, editor, *The Stanford Encyclopedia of Philosophy (Winter 2019 Edition)*. 2019.

[4] J. C. Beall and G. Restall. Logical pluralism. *Australasian Journal of Philosophy*, 78(4):475–493, 2000.

[5] J. C. Beall and G. Restall. Defending logical pluralism. In *Logical Consequence: Rival Approaches Proceedings of the 1999 Conference of the Society of Exact Philosophy*. Stanmore: Hermes, 2001.

[6] J. C. Beall and G. Restall. *Logical Pluralism*. Oxford University Press, Oxford, 2006.

[7] N. D. Belnap. How a computer should think. In G. Ryle, editor, *Contemporary Aspects of Philosophy*. Oriel Press, 1977.

[8] E. Campos and A. Rodrigues. Some remarks on the validity of the principle of explosion in intuitionistic logic. In E. Almeida, A. Costa-Leite, and R. Freire, editors, *Seminário Lógica no Avião*. Brasília: Lógica no Avião, 2019.

[9] G. Cardoso, W. Carnielli, and A. Rodrigues. On Beall and Restall's logical pluralism and paraconsistency. Submitted, 2020.

[10] R. Carnap. *The Logical Syntax of Language*. Routledge (2001), 1937.

[11] W. Carnielli, M. Coniglio, and A. Rodrigues. Recovery operators, paraconsistency and duality. *Logic Journal of the IGPL*, 2019.

[12] W. Carnielli and A. Rodrigues. An epistemic approach to paraconsistency: a logic of evidence and truth. *Synthese*, 196:3789–3813, 2017.

[13] W. Carnielli and A. Rodrigues. On epistemic and ontological interpretations of intuitionistic and paraconsistent paradigms. *Logic Journal of the IGPL*, 2019.

[14] B. J. Copeland. Pure semantics and applied semantics. *Topoi*, 2:197–204, 1983.

[15] N. da Costa. *Sistemas Formais Inconsistentes (Inconsistent Formal Systems, in Portuguese)*. Habilitation thesis, Universidade Federal do Paraná, Curitiba. Republished by Editora UFPR, Curitiba (1993), 1963.

[16] N. da Costa. On the theory of inconsistent formal systems. *Notre Dame Journal of Formal Logic*, 4:497–510, 1974.

[17] N. C. da Costa and S. French. Inconsistency in science: A partial perspective. In J. Meheus, editor, *Inconsistency in Science*. Dordrecht: Springer, 2002.

[18] N. C. da Costa and D. Krause. Physics, inconsistency, and quasi-truth. *Synthese*, 191(13):3041–3055, 2014.

[19] J. M. Dunn. Information in computer science. In P. Adriaans and J. van Benthem, editors, *Philosophy of Information. Volume 8 of Handbook of the Philosophy of Science*, pages 581–608. Elsevier, 2008.

[20] G. Frege. *The Basic Laws of Arithmetic*. University of California Press, 1893.

[21] G. Frege. The Thought: A Logical Inquiry. *Mind* (1956), 65:289–311, 1918.

[22] A. Heyting. On intuitionistic logic. In P. Mancosu, editor, *From Brouwer To Hilbert: The Debate on the Foundations of Mathematics in the 1920s*. Oxford University Press (1998), 1930.

[23] S. Jáskowski. Propositional calculus for contraditory deductive systems. *Studia Logica* (1969), 24:143–157, 1949.

[24] A.N. Kolmogorov. On the interpretation of intuitionistic logic. In P. Mancosu, editor, *From Brouwer to Hilbert: The Debate on the Foundations of Mathematics in the 1920s*. Oxford University Press (1998), 1932.

[25] A. Loparic. Valuation semantics for intuitionistic propositional calculus and some of its subcalculi. *Principia*, 14(1):125–133, 2010.

[26] J. Marcos. Nearly every normal modal logic is paranormal. *Logique et Analyse*, 48:279–300, 2005.

[27] E. Mares. Semantic dialetheism. In Priest, Beall, and Armour-Garb, editors, *The Law of Non-Contradiction: new philosophical essays*. Oxford University Press, 2004.

[28] M. M. McSweeney. Following logical realism where it leads. *Philosophical Studies*, 176(1):117–139, November 2017.

[29] T. Nickles. From Copernicus to Ptolemy: inconsistency and method. In J. Meheus, editor, *Inconsistency in Science*. Dordrecht: Springer, 2002.

[30] S. Odintsov and H. Wansing. On the methodology of paraconsistent logic. In H. Andreas and P. Verdée, editors, *Logical Studies of Paraconsistent Reasoning in Science and Mathematics*. Springer, 2016.

[31] L. C. Pereira. On Prawitz' ecumenical system. In T. Piecha and P. Schroeder-Heister, editors, *Proof-theoretic Semantics – Assessment and Future Perspectives – Proceedings of the Third Tubingen Conference on Proof-Theoretic Semantics, Tubingen, 27-30 March 2019*, 2019.

[32] L. C. Pereira and R. O. Rodriguez. Normalization, soundness and completeness for the propositional fragment of Prawitz' ecumenical system. *Revista Portuguesa de Filosofia*, 73:1153–68, 2017.

[33] D. Prawitz. Classical versus intuitionistic logic. In E. Haeusler, W. Sanz, and B. Lopes, editors, *Why is this a Proof? Festschrift for Luiz Carlos Pereira*. College Publications, 2015.

[34] G. Priest. The logic of paradox. *Journal of Philosophical Logic*, 8(1):219–241, 1979.

[35] G. Priest. Logical pluralism. In *Doubt Truth to be a Liar*. Oxford University Press, 2006.

[36] G. Priest. Paraconsistency and dialetheism. In D. M. Gabbay and J. Woods, editors, *Handbook of the History of Logic (Vol. 8)*. Elsevier, 2007.

[37] G. Priest, F. Berto, and Z. Weber. Dialetheism. *The Stanford Encyclopedia of Philosophy (Fall 2018 Edition)*, 2018.

[38] W. V. O. Quine. *Philosophy of Logic*. Harvard University Press (1986), 1970.

[39] A. Rodrigues, J. Bueno-Soler, and W. Carnielli. Measuring evidence: a probabilistic approach to an extension of Belnap-Dunn logic. *Synthese*, in print.

[40] P. Rush. *Logical Realism*. Cambridge University Press, Cambridge, 2014.

[41] D. van Dalen. The war of the frogs and the mice, or the crisis of the *Mathematische Annalen*. *The Mathematical Intelligencer*, 12(4):17–31, 1990.

Modal rationalism, logical pluralism, and the metaphysical foundation of logic

Duško Prelević

Faculty of Philosophy, University of Belgrade, Serbia

Abstract

It is argued in this paper that modal rationalism can be reconciled with logical pluralism together with the metaphysical foundation of logic. Namely, ideal conceivability is extensionally more akin to metaphysical modality than to logical modality, given that the space of conceivable worlds (scenarios) is (like the space of metaphysically possible worlds) narrower than the space of logically possible worlds, which opens room for the claim that the space of conceivable scenarios and the space of metaphysically possible worlds are coextensive. These considerations corroborate the claim that even if it turns out that modal monism (a view that logical modality and metaphysical modality are extensionally equivalent) is false, modal rationalism might still be true. At the end of this paper, I address pluralism in geometry (according to which there is more than one correct system of geometry) to show that modal rationalism is compatible with that view either.

Introduction

In contemporary modal epistemology, a lot of attention is paid to the question of what makes a good justification of our beliefs about metaphysical modality. According to modal rationalism, which will be assessed in this paper, conceivability (properly understood) is a good evidence that some states of affairs are metaphysically possible. In David Chalmers's settings (see Chalmers 1996; 1999; 2002, 2010 for more details), ideal positive primary conceivability entails primary (or counter-*actual*) possibility, which, together with established nonmodal facts, entails metaphysical (or counterfactual) possibility.[1] Here, "ideal positive

---

[1] The distinction between counteractual (or primary) and counterfactual (or secondary) possibility is typically drawn in the two-dimensional semantic framework

primary conceivability" refers to envisaging a pertinent scenario (or a situation) that verifies a proposition one is conceiving of,[2] and which is undefeatable by better reasoning (Chalmers 2002:§§1–3).[3]

In that respect, Kripkean cases of necessary *a posteriori* statements (which might, at first sight, seem to be counterexamples to modal rationalism),[4] such as "Water is H_2O", are understood by modal rationalists as statements in which nonmodal component (in our purported example, that liquid with properties that we typically ascribe to water is H_2O) is knowable *a posteriori*, while their modal component (that they are necessary) is knowable a priori (through conceptual analysis and deduction).[5]

Many potential counterexamples to modal rationalism have been proposed by now, and there is an ongoing debate with regard to their plausibility (see, for example, Chalmers 2002; Vaidya 2015; Prelević 2013). Relatedly, it is usually taken for granted that modal rationalism relies upon modal monism, a view that logical modality and metaphys-

---

(various interpretations of such a framework are presented in Garcia-Carpintero and Macià 2006). The latter evaluates possibilities relative to the truths of the actual world, while the former evaluates possibilities independently of that. Chalmers (2002:§3) draws a parallel distinction between primary and secondary conceivability: the latter is *a posteriori*, because it relies upon our knowledge of the truths about the actual world), while the former is *a priori* since it does not depend on it.

[2]Chalmers also holds that ideal negative primary conceivability entails primary possibility (this is what he calls "strong modal rationalism"), where negative conceivability refers to conceiving of a scenario (or a situation) which itself does not determine whether a conceived proposition is true or not (like in the cases of vague predicates, and the like). Strong modal rationalism presupposes that all truths are scrutable from a compact class of truths and therefore it is incompatible with the existence of inscrutable truths (Chalmers addresses this issue in detail in Chalmers 2012).

[3]Undefeatability by better reasoning is, according to Chalmers (2002:§1), a hallmark of ideal conceivability, which separates it from *prima facie* and *secunda facie* conceivability.

[4]Actually, Kripke himself has famously used strategies that are characteristic of modal rationalist approach to criticize materialism in philosophy of mind (Kripke 1980). Nonetheless, many philosophers hold that his views about essence and modality are opposite to the modal rationalist view.

[5]Namely, conceptual analysis shows in the purported example that water is a colorless, odorless, tasteless (*etc.*) liquid (that is water's primary intension). From this information and *a posteriori* information that such a liquid is H_2O, we can infer that water's molecular structure is H_2O (that is water's secondary intension; see Chalmers 2002:161–165 for more details).

ical modality are extensionally equivalent (see, for example, Chalmers 1999:§3.5.; Vaidya 2008:192–193). While Chalmers (1999) thinks that modal monism should be preferred over modal dualism (a view that both logical modality and metaphysical modality are primitive) for the sake of simplicity and other methodological reasons, many philosophers disagree mainly because they believe that the source of logical modality and metaphysical modality is not the same, that is, because the former is, unlike the latter, rooted in the makeup of the actual world (which, according to essentialists, include essential properties as well; see, Mallozzi 2018 for more details).

In what follows, I argue that logical pluralism, a view that there is more than one correct system of logic (see, Russell 2019 for more details) together with a metaphysical foundation of logic (which will be presented in due course) and a proper understanding of conceivability, allow us to endorse modal rationalism independently of whether modal monism (as defined above) is true or not. Namely, even if it turns out that modal monism is false, modal rationalism still might be true because conceivability, as it will be shown in due course, can be understood so that it is extensionally more akin to metaphysical modality than to logical modality. At the end of this paper, I address pluralism in geometry (which consists in the existence of more than one correct system of geometry) to show that modal rationalism is compatible with that view either.

1 Grounding logic in metaphysics

Although many philosophers believe that logical necessity is absolute (see, for example, Hale 2013:Ch.4 for more details), many alternatives to classical logic have been proposed by now.[6] For example, intuitionists restrict the universal validity of the law of excluded middle (see, for example, Heyting 1971); the proponents of three (or many) valued logic (Lukasiewicz 1968) are willing to restrict the universal validity of

---

[6] An overview of general ideas about various systems of logic and the metaphysical foundation of logic, which are presented here and in the next passages, are adapted from Prelević 2017a:§2.3.2, in which these issues are put into the context of the debates in modal metaphysics. As it will be shown in due course, I go a step further in this paper and argue that modal rationalism is compatible with logical pluralism.

principle of bivalence; philosophers willing to adopt paraconsistent logic (see, for example, da Costa, Beziau and Bueno 1995 for more details) are ready to restrict the validity of the law of noncontradiction (adopting the *ex falso quodlibet* principle at the same time); possibilists hold that all sentences are possible (Mortensen 1989). trivialists hold that all sentences are true (see, for example, Estrada-Gonzáles 2012), while nihilists hold that there are no laws of logic (see, for example, Russell 2018). It is also worth mentioning that even *modus ponens* have been challenged by purported counterexamples (McGee 1985). Thus, classical logic is not the only option available, and logicians are by no means unanimous as to which logical system ought to be accepted.

These considerations give rise to the claim that instead of searching for a universally accepted logical system, it is more appropriate to hold that there are classes of possible worlds (or galaxies) compatible with corresponding logical systems. That is, every logic has its own class of (logically) possible worlds (see Bensusan, Costa-Leite and de Souza 2015 for more details). Logical pluralism, if true, shows that logical necessity is not absolute, yet it does not entail that there is no room for absolute necessities, given that modal space is not limited to the space of logical possibilities and necessities. The friends of absolute necessities might still hold that, for example, metaphysical necessity is absolute (in the ontological sense), which means that although it is rooted in the makeup of the actual world (see, for example Mallozzi 2018), it holds in all metaphysically possible worlds no matter what else was the case.[7] This idea goes hand in hand with a metaphysical grounding of logic since it enables us to make a choice between competing systems of logic in view of a previously adopted metaphysical theory.

In order to get a grip on such an approach, let us recall Vaidya's (2006) view that there are three possible attempts of establishing which logic ought to be accepted: one can try to make the choice within logic itself; one can ground logic on physical facts; and finally, one can ground logic on metaphysics. The first approach cannot be conducted in a noncircular way, since if we, for example, try to refute paraconsistent logic

---

[7]Relatedly, Bob Hale (2013) thought that both logical and metaphysical necessities are absolute. On the other hand, some philosophers (most notably Nathan Salmon; see Salmon 1981 for more details) think that metaphysical necessity is not absolute, and that a system of modal logic weaker than S4 is suitable to it.

by using a *reductio ad absurdum* proof, we would do that on pain of begging the question against the law of noncontradiction and paraconsistent logic thereof. On the other hand, if we try to ground logic on physical facts, it is likely that such an enterprise would end up in committing a naturalistic fallacy, given that logic is normative, and therefore it does not seem to be reducible to descriptive facts. Bearing this in mind, Vaidya maintains that the metaphysical foundation of logic is the only remaining option for those who try to make a choice between various logical systems.

Relatedly, Tuomas Tahko (2009) argues that the law of noncontradiction is in fact a metaphysical law. Here, Tahko appeals to Aristotle, who claimed that "the same attribute cannot at the same time belong and not belong to the same subject in the same respect" (Aristotle 1984:1005b19-20). Tahko argues that recently proposed examples purported to show that the law of noncontradiction fails in the actual world are not convincing.

In addition, it is noticeable that different logical systems are built up for different purposes, as well as that not all of them are purported to provide us with principles that are likely true in the actual world. Actually, some logical laws are likely false in the actual world. For example, if essentialism is true, then trivialism and possibilism are false. The same goes for logical nihilism and paraconsistent logic if the law of noncontradiction (considered to be a metaphysical principle) holds in the actual world, and the like.

Along these lines, it is possible to recall that very often the main motivation of building new systems of logic is to provide a formal analysis of information in one way or another (see, for example, Carnielli 1999:§1). Here, Makinson's paradox of the preface (see Makinson 1965) might serve as an illustration. Namely, when the author of a book says in the preface that she is responsible for all errors in the book, the situation is paradoxical because, on the one hand, the author would correct those errors had she noticed them before publishing, while, on the other hand, she herself is well aware of the fact that even if she did her best in preparing the penultimate draft of the book, errors could still remain. The upshot is that in this particular situation the author is rational albeit she knows that she holds logically incompatible beliefs, which might suggest that the law of noncontradiction need not be accepted as axiom.

Still, Makinson's envisaged scenario (and the whole logic purported to handle the paradox of the preface) depicts the author's epistemic situation only, and it by no means shows that in reality her book both do contain and do not contain any errors (see, for example, Prelević 2017a:110; 113). Likewise, intuitionists are not willing to accept the law of excluded middle as axiom primarily because they think that a disjunction should be considered true only if we already know which of its disjuncts is true (otherwise we should refrain from making a judgment). In the case of the principle of bivalence, Łukasiewicz (1968) famously claimed that our choice between accepting this principle and restricting it depends on whether we are determinists (who believe that future is closed) or indeterminists (who believe that future is open). This, again, shows that the choice between competing logical systems might depend on previously accepted metaphysical theories.

2 Ideal conceivability and logical possibility

Now, if considerations from the previous section are correct, then logical pluralism and the metaphysical foundation of logic license a view that logical modality and metaphysical are not coextensive, given that, for instance, restricting the universal validity of the law of noncontradiction is logically possible whilst being (according to some philosophers) metaphysically impossible. Does this mean that modal rationalism would be false in that case, given that it relies upon the assumption that logical modality and metaphysical modality are coextensive? I think this conclusion does not follow, because the space of conceivable worlds (scenarios) is narrower than the space of logically possible worlds. In that respect, conceivability is more akin to metaphysical modality than to logical modality. This opens room for the claim that the space of conceivable scenarios and the space of metaphysically possible worlds are coextensive.

Yet, it should be noticed that some philosophers, such as Graham Priest (see Priest 2016 for more details), think that impossible worlds are conceivable, as well as that a proper semantics of conceivability requires impossible worlds. Priest also holds that Chalmers's notion of ideal positive primary conceivability is "not a very useful notion for mere mortals" (Priest 2016:*fn*.37) since it appeals to infinite and infallible *a priori* reasoners.

As for Priest's last point, it should be stressed that ideal positive primary conceivability can be understood in at least two ways, either in the non-idealized sense, or in the idealized sense. The later consists in envisaging a pertinent counter-actual *scenario* that verifies a proposition one is conceiving of, while the former consists in envisaging corresponding *situation* (a part of a scenario) that verifies the very proposition (see, for example, Prelević 2011; 2013:§1.5.4.; 2015a:§2 for more details; Cf. Chalmers 1996:67).

Now, it is likely that the idealized sense of ideal positive primary conceivability requires massive idealization that consists in specifying all (infinitely many) details about an envisaged world (see, for example, Hale 2013:§10.3. for more details), and that such an idealization seems to be of no use for mere mortals, as Priest has pointed out. However, the non-idealized sense of ideal positive primary conceivability is, contrary to what Priest has claimed, available to mere mortals, while at the same time it is reasonable to suppose that, when it comes to ideal conceivability, there is a corresponding scenario into which the envisaged situation is embedded.

Bearing this in mind, Priest's thesis that impossible worlds are conceivable turns out to be contentious: ideal conceivability is simply incompatible with the existence of impossible worlds and restricting the law of noncontradiction. Of course, some other intentional notions, such as understanding, desire, and the like, are more suitable to allow contradictions.[8] Be that as it may, conceivability (properly understood) does not belong to this category.[9]

---

[8] It is also interesting to notice here that some philosophers hold that intentional notions, such as hope, exclude contradictions (see, for example, Andrew Chignell's interpretation of Kant's views of rational hope in Chignell 2014).

[9] That conceivability should not be conflated with logical possibility is stated, for example, by Douglas Rasmussen, who says that "no-one can conceive or imagine a contradiction" (Rasmussen 1983:522). However, Rasmussen does not address the problem of logical pluralism, but simply identifies contradiction with logical impossibility.

Relatedly, it should be stressed that ideal positive primary conceivability should be separated from imaginability, intuitions, understanding, supposing, and the like.[10] Unlike imaginability (as usually understood), conceivability does not require visualization in any relevant sense (see, for example, Costa-Leite 2010). On the other hand, ideal conceivability is epistemically more demanding than a mere supposing of a hypothetical situation (see, for example, Horvath 2015) since it presupposes in addition *checking* the coherence of an envisaged situation or scenario. Conceivability also differs from understanding in a sense in which we can understand contradiction (for example, when addressing a *reductio ad absurdum* proof; see, for example Vaidya 2010:817) that is by no means ideally conceivable. Ideal positive primary conceivability differs from intellectual seeming either, which is at the heart of George Bealer's *moderate rationalism*, according to which rational intuition is fallible and nonetheless reliable guide to metaphysical modality (see, for, example, Bealer 2002 for more details). Unlike intellectual seeming above, ideal conceivability cannot be defeated by better reasoning. For example, the Naive set theory is, on the one hand, intuitive but, on the other hand, not ideally conceivable due to Russell's paradox (Chalmers 2002:155).

Of note, it is also a bit surprising that Priest (2016:2658) uses Goldbach's conjecture as an example purported to show that conceivability does not entail possibility. Such an example seems to be irrelevant here, since we already know (on *a priori* grounds, due to Feferman's completeness theorem; see Feferman 1962) that Peano arithmetic is consistent, which implies that Goldbach's conjecture has determinate truth-value (albeit we do not know at the moment whether it is true or not). So we know that either Goldbach's conjecture or its negation is (necessarily) true, yet, in the absence of a corresponding mathematical proof, we should hold Goldbach's conjecture (or its negation) *prima facie* negatively conceivable rather than ideally conceivable (see Chalmers 2002:160). However, Chalmers (2012:262) argues that an idealized reasoner would be in position to determine the truth-value of

---

[10] Ideas presented in this passage are elaborated in detail in Prelević 2013:§1.5.3. Relatedly, in Prelević 2014, the distinction between ideal positive primary conceivability and (various senses of) intuition are used for establishing the difference between conceivability arguments and thought experiments and avoiding the objection (which can be found, for example, in Dennett 1991) that conceivability arguments in philosophy of mind are intuition pumps.

every statement of arithmetic (for example, by applying the omega rule or something along these lines).[11]

All in all, ideal conceivability, to which modal rationalists appeal, presupposes the validity of the law of noncontradiction and therefore cannot be spelled out in systems of logic that restrict its universal validity. If so, then the space of conceivable worlds is (like the space of metaphysically possible worlds) narrower than the space of logically possible worlds.

The same holds even if we restrict our attention to classical logic[12] since here logical possibility can be counted as a prerequisite for ideal conceivability without being sufficient for it. This becomes more apparent if we recall a common view that there are analytic truths that are not logically true. For example, sentence "All bachelors are unmarried" is true in virtue of meaning of terms "bachelor" and "unmarried", yet it is by no means a tautology (see, for example, Zalta 1988:57; cf. Prelević 2013:9; 2017a:23). It also seems plausible to say that married bachelors are both inconceivable and metaphysically impossible.

However, conceivability (likewise metaphysical modality) need not be considered coextensive with analyticity either,[13] given that it seems that there are analytic truths that are conceivably (and actually) false. Pluralism in geometry seems to be a good illustration here (examples of contingent logical and analytic truths can be found in Zalta 1988). So, let us turn to this case in order to get a better grip on the relation between ideal conceivability and logical possibility.

As is well known, there are competing axiomatic systems in geometry, which are, on the one hand, considered consistent, while, on the

---

[11]The cases of higher-set theory, such as the Continuum hypothesis or its negation, are more complicated (see Chalmers 2012:263–264), but anyway they are related to the question as to whether ideal negative primary conceivability entails primary possibility (which depends on whether there are inscrutable truths or not; see Chalmers 2002:§8 for more details), rather than whether ideal positive primary conceivability entails primary possibility, in which I am mainly interested in this paper.

[12]This is probably how philosophers typically understand modal monism (see Vaidya 2008 for more details).

[13]This is evident in the case of metaphysical modality, since Kripkean cases, such as the statement "Water is H_2O", are typically counted as metaphysically necessary and synthetic *a posteriori*.

other hand, they are incompatible to each other. For example, in Euclidean geometry sum of angles of a triangle is equal to the straight angle, which is not the case in hyperbolic and (Riemannian) spherical geometries (in the former, the sum of angles of a triangle is less than than straight angle, while in the later the sum of angles of a triangle is greater than than straight angle). Relatedly, it is well known that Einstein was aware of the fact that he could choose whether to accept a simpler physics and a more complicated (non-Euclidean) geometry or a simpler geometry and a more complicated physical theory (by introducing "universal forces" that have the power to deflect light rays, and so on; see, for example, Reichenbach 2006; BonJour 1998:221; Howard 2005:38; cf. Tahko 2008). As a physicist, he chose the first option, but the second option was available either.

Now, it might be interesting to see how modal rationalists could address this issue: on the one hand, we typically hold that geometrical truths are necessary and *a priori*; on the other hand, we either live in Euclidean or in a non-Euclidean space, so if, for example, a non-Euclidean geometry is true, then it is necessarily true and therefore Euclidean geometry is necessarily false. Given that all these systems of geometry are conceivable,[14] it seems that conceivability does not entail metaphysical possibility, contrary to what the proponents of modal rationalism typically hold.

Here, modal rationalists could respond in the following way: they might say that competing systems of geometry are *a priori* and (geometrically) necessary in their domains, yet just one of them is metaphysically (secondarily) necessary. Those systems are *a priori* because they are deductive systems with corresponding axiomatizations that establish boundaries of conceivability within them. For example, in Euclidean geometry it is inconceivable that sum of angles of a triangle is not equal to the straight angle, while in hyperbolic and spherical geometries it is inconceivable that sum of angles of a triangle is equal to the straight angle.

This can be represented in S5 system for modal logic by using the distinction between accessibility relation understood as equivalence, that

---

[14]This is what makes this case different from the case of logical pluralism addressed in the previous section.

is, as a relation between possible worlds that is reflexive, transitive and symmetric, and accessibility relation understood as universal accessibility, in which all worlds are accessible from one another.[15] The former, unlike the latter, allows inaccessible world, which can be seen if, for example, we posit "multiple systems of worlds, such that within each system all worlds are accessible from one another, but across systems no worlds are accessible from one another" (Picinnini 2017:85). Such a model represents accessibility relation understood as equivalence, and not as universal accessibility.

Bearing this in mind, it might be said that there is a sense in which many geometrical propositions can be considered (in line with Kant)[16] synthetic *a priori*: although sentences like "Sum of angles of a triangle is equal to the straight angle" are analytically true in Euclidean geometry, it is conceivable that they are false in some other systems of geometry (actually, the sentence "Sum of angles of a triangle is equal to the straight angle" is analytically false in hyperbolic and spheric geometries). If we recall the two notions of accessibility in S5 system for modal logic, we can say that although these sentences are necessarily true in at least one system of (geometrically possible) worlds, they are not true in all systems of (geometrically possible) worlds whatsoever.[17]

Now, it is an empirical question as to whether universal forces (mentioned above) exist in the actual world or not. At least, the dispute over their existence does not seem to be a merely verbal. However, information about them would be a nonmodal one, had we been able to found it out. On the other hand, information about the spatial properties of the world are likely *a priori* and modal. So, under assumption that universal forces exist and that we live in a non-Euclidean (Riemannian) space, statements like "Sum of angles of a triangle is greater than the straight angle" would be (in the twodimensionalist terminology) secon-

---

[15] Gultiero Piccinini (2017) used this distinction in his critique of Chalmers's zombie argument in philosophy of mind. Criticism of such a criticism is provided in Prelević 2017b.

[16] Kant (1998), of course, held that fundamental propositions of geometry are synthetic *a priori* for quite different reasons.

[17] Sentences like "Triangles have three edges and three vertices" are, of course. counted as *a priori* by modal rationalists since they are true in all scenarios (in all systems of worlds).

darily necessary and primarily possible.[18] This situation is much alike the standard Kripkean examples described at the very beginning, with the exception that, unlike in the case of the terms like "water", there is more than one primary intension of the terms like "triangle" (depending on which axiomatic system of geometry we use). If so, then pluralism in geometry can be easily accommodated to modal rationalist account of our modal knowledge.

3 Conclusion

Let us summarize. Modal rationalism is compatible with logical pluralism and metaphysical foundation of logic. That is because the space of conceivable worlds is narrower than the space of logically possible worlds, contrary to what is usually presupposed in the debates over the validity of the thesis (endorsed by modal rationalists) that conceivability entails metaphysical possibility. Like in the case of metaphysical modality, conceivability should not be conflated with logical possibility as such. Modal rationalism is also compatible with pluralism in geometry, given that competing systems of geometry are both primarily conceivable and primarily possible, but further empirical information is required for deciding which of them is counterfactually (secondarily) possible.

This is by no means the whole story about the plausibility of modal rationalism. One interesting question is to what extent this view is compatible with essentialism and a neo-Aristotelian account of metaphysical modality that many philosophers in the field are ready to endorse over last few decades. According to such an account, essence itself is not reducible to necessity but nonetheless it grounds metaphysical necessity,[19] which corroborates the claim that the epistemology of modality should be based on a corresponding epistemology of essence (see, for example, Hale 2013:§11). Much ink has been spilled by the adherents of com-

---

[18]These sentences are primarily possible if we take universal accessibility into account.

[19]Relatedly, Kit Fine (1994) provided well known examples purported to justify this thesis. For example, if existing Socrates belongs to existing singleton Socrates, then it is necessary that Socrates belongs to the singleton Socrates. Yet this by no means helps us, according to Fine, to understand Socrates's nature, and therefore it is not essential to Socrates to belong to the singleton Socrates.

peting views in the epistemology of modality (in particular, by those who endorse various versions of modal rationalism and modal empiricism) who have tried to handle various essentialist principles, such as the essentiality of origin, the essentiality of kind, and the like (see, for example, Roca-Royes 2011; Chalmers 2010; Prelević 2015b for more details). I think that modal rationalists can accommodate those principles by taking further metaphysical assumptions into account, but addressing this issue is beyond the scope of this paper.[20] However, considerations presented in this paper are fairly compatible with such endeavors.

Acknowledgements. I would like to thank Alexandre Costa-Leite for inviting me to contribute to this volume.This research was supported by Ministry of Education, Science and Technological Development of the Republic of Serbia (project: *Logico-epistemological Bases of Science and Metaphysics*, No. 179067).

References

[1] Aristotle. (1984). *Metaphysics* (trans. W. D. Ross). Princeton: Princeton University Press.

[2] Bealer, George. (2002). "Modal Epistemology and Rationalist Renaissance". In: T. Gendler and J. Hawthorne (eds.), *Conceivability and Possibility*. Oxford: Oxford University Press.

[3] Bensusan, Hilan, Costa-Leite, Alexandre and De Souza, Edélcio. (2015). "Logics and Their Galaxies". In: A. Koslow and A. Buchsbaum (eds.), *The Road to Universal Logic: Festschrift for the 50th Birthday of Jean-Yves Beziau* (vol. II). Birkhäuser.

[4] BonJour, Laurence. (1998). *In Defense of Pure Reason: A Rationalist Account of A Priori Justification*. Cambridge: Cambridge University Press.

[5] Chalmers, David. (1996). *The Conscious Mind: In Search of a Fundamental Theory*. Oxford: Oxford University Press.

---

[20] In Prelević (manuscript) it is argued that modal rationalism is compatible with the main essentialist views in metaphysics.

[6] Chalmers, David. (1999). "Materialism and Metaphysics of Modality". *Philosophy and Phenomenological Research* 59:473–496.

[7] Chalmers, David. (2002). "Does Conceivability Entail Possibility?". In: T. Gendler and J. Hawthorne (eds.), *Conceivability and Possibility*. Oxford:Oxford University Press.

[8] Chalmers, David. (2010). "The Two-Dimensional Argument against Materialism". In: D. Chalmers (ed.), *The Character of Consciousness*. Oxford:Oxford University Press.

[9] Chalmers, David. (2012). *Constructing the World*. Oxford: Oxford University Press.

[10] Chignell, Andrew. (2014). "Rational Hope, Possibility, and Divine Action". In: G. Michalson (ed.), *Kant's Religion within the Boundaries of Mere Reason: A Critical Guide*. Cambridge: Cambridge University Press.

[11] Costa-Leite, Alexandre. (2010). "Logical Properties of Imagination". *Abstracta* 6:103–116.

[12] da Costa, Newton, Béziau, Jean-Yves and Bueno, Otávio. (1995). "Paraconsistent Logic in a Historical Perspective". *Logique et Analyse*, 150–152:111–125.

[13] Dennett, Daniel. (1991). *Consciousness Explained*. New York: Back Bay Books/Little. Brown and Company.

[14] Estrada-Gonzáles, Luis. (2012). "Models of Possibilism and Trivialism". *Logic and Logical Philosophy* 21:175–205.

[15] Feferman, Solomon. (1962). "Transfinite Recursive Progressions of Axiomatic Theories". *Journal of Symbolic Logic* 27:259–316.

[16] Fine, Kit. (1994). "Essence and Modality". *Philosophical Perspectives* 8:1–16.

[17] Garcia-Carpintero, Manuel and Maciá, Josep (eds.). (2006). *Two-Dimensional Semantics: Foundations and Applications*. Oxford: Oxford University Press.

[18] Hale, Bob. (2013). *Necessary Beings - An Essay on Ontology, Modality and the Relation Between Them.* Oxford: Oxford University Press.

[19] Heyting, Arend. (1971). *Intuitionism: An Introduction.* Amsterdam: North-Holland.

[20] Horvath, Joachim. (2015). "Thought Experiments and Experimental Philosophy". In: C. Daly (ed.), *The Palgrave Handbook of Philosophical Methods.* Basingstoke:Palgrave Macmillan.

[21] Howard, Don. (2005). "Einstein as a Philosopher of Science". *Physics Today* 58:34–40.

[22] Kant, Immanuel. (1998). *Critique of Pure Reason.* Cambridge: Cambridge University Press.

[23] Kripke, Saul. (1980). *Naming and Necessity.* Cambridge: Harvard University Press.

[24] Łukasiewicz, Jan. (1968), "On Determinism," *The Polish Review* 13:47–61.

[25] Makinson, D. C. (1965). "The Paradox of the Preface". *Analysis* 25:205–207.

[26] Mallozzi, Antonella. (2018). "Two Notions of Metaphysical Modality". *Synthese* DOI:10.1007/s11229-018-1702-2 1–22.

[27] McGee, Vann. (1985). "A Counterexample to Modus Ponens". *The Journal of Philosophy* 82:462–471.

[28] Mortensen, Chris. (1989). "Anything Is Possible". *Erkenntnis* 30:319–337.

[29] Piccinini, Gualtiero. (2017). "Access Denied to Zombies", *Topoi* 36:81–93.

[30] Prelević, Duško. (2011). "Čalmersova odbrana argumenta na osnovu zamislivosti". *Theoria* 54:25–55.

[31] Prelević, Duško. (2013). *Modalna epistemologija i eksplanatorni jaz: Značaj argumenta na osnovu zamislivosti zombija* (PhD Thesis). Beograd: Univerzitet u Beogradu.

[32] Prelević, Duško. (2014). "Misaoni eksperimenti i argumenti na osnovu zamislivosti". *Treći program* 161–162:71–81.

[33] Prelević, Duško. (2015a). "Zombies Slap Back:Why the Anti-Zombie Parody Does Not Work". *Disputatio* 40:25–43.

[34] Prelević, Duško. (2015a). "Modal Empiricism and Knowledge of *De Re* Possibilities: A Critique of Roca-Royes' Account". *Organon F* 22(4):488–498.

[35] Prelević, Duško. (2017a). *Metafizičke osnove modalnog mišljenja*. Beograd: Srpsko filozofsko društvo.

[36] Prelević, Duško. (2017b). "Access Granted to Zombies". *Theoria* 63:58–68.

[37] Prelević, Duško. (manuscript). "The Explanatory Power of Modal Rationalism".

[38] Priest, Graham. (2016). "Thinking the Impossible". *Philosophical Studies* 173:2649–2662.

[39] Rasmussen, Douglas. (1983). "Logical Possibility: An Aristotelian Essentialist Critique". *The Thomist: A Speculative Quarterly Review* 47:513–540.

[40] Roca-Royes, Sonia. (2011). "Conceivability and De Re Modal Knowledge". *Nous* 45:22–49.

[41] Russell, Gillian. (2018). "Logical Pluralism: Could There Be No Logic?". *Philosophical Issues* 28:308–324.

[42] Russell, Gillian. (2019). "Logical Pluralism". *Stanford Encyclopedia of Philosophy*, https://plato.stanford.edu/entries/logical-pluralism/ Accessed: March 17, 2020.

[43] Salmon, Nathan. (1981). *Reference and Essence*. Princeton: Princeton University Press.

[44] Tahko, Tuomas. (2008). "A New Definition of *A Priori* Knowledge: In Search of a Modal Basis". *Metaphysica* 9:57–68.

[45] Tahko, Tuomas. (2009). "The Law of Non-Contradiction as a Metaphysical Principle". *Australasian Journal of Logic* 7:32–47.

[46] Vaidya, Anand. (2006). "The Metaphysical Foundation of Logic". *Journal of Philosophical Logic* 35:179–182.

[47] Vaidya, Anand. (2008). "Modal Rationalism and Modal Monism". *Erkenntnis* 68:191-212.

[48] Vaidya, Anand. (2010). "Understanding and Essence". *Philosophia* 38:811–833.

[49] Vaidya, Anand. (2015). "The Epistemology of Modality". *Stanford Encyclopedia of Philosophy*, Accessed: March 17, 2020.

[50] Zalta, Edward. (1988). "Logical and Analytic Truths that are not Necessary". *The Journal of Philosophy* 85:57–74.

Quasi-concepts of logic

Fabien Schang

Department of Philosophy, Federal University of Goiás, Brazil

Abstract

A analysis of some concepts of logic is proposed, around the work of Edelcio de Souza. Two of his related issues will be emphasized, namely: opposition, and quasi-truth. After a review of opposition between logical systems [2], its extension to many-valuedness is considered following a special semantics including partial operators [13]. Following this semantic framework, the concepts of antilogic and counterlogic are translated into opposition-forming operators [15] and specified as special cases of contradictoriness and contrariety. Then quasi-truth [5] is introduced and equally translated as a product of two partial operators. Finally, the reflections proposed around opposition and quasi-truth lead to a third new logical concept: quasi-opposition, borrowing the central feature of partiality and opening the way to a potential field of new investigations into philosophical logic.

1 Oppositions

The proper contribution of Edelcio de Souza with respect to logical oppositions has been through its application to logical systems [2], beyond mere formulas or philosophical concepts in a given logical system [3]. The concept of opposition comes from Aristotle's work and consists in logical relations between bivalent formulas, in such a way that each of these formulas is to be interpreted as either true or false. For this reason, oppositions should be explained in a semantic way. However, Edelcio and the co-authors claim that every logical opposition that does not relate propositions should not be explained in semantic terms ([2], 243):

"Because the vertices of the square (...) are not propositions we reconstruct the classical oppositions accordingly. We define them in terms of relations between logics –instead of logical values".

Anyway, the authors introduce the two central notions in a classical way. Letting $\Gamma \vdash_L \varphi$ the relation of consequence from any set of premises Γ to an arbitrary formula φ in a language L:

- $\bar{L}$ is called an *antilogic* of L if and only if it is not the case that φ is a logical consequence of Γ in $\bar{L}$: $\Gamma \not\vdash_{\bar{L}} \varphi$.

- $\tilde{L}$ is called an *counterlogic* of L if and only if it is the case that $\neg\varphi$ is a logical consequence of Γ in $\tilde{L}$: $\Gamma \vdash_{\tilde{L}} \neg\varphi$.

Beyond the bivalent stance, the aim of the present paper is to redefine oppositions between logics in semantic terms and to explore the possibility of non-standard oppositions.

On the one hand, such oppositions may be formulated in the Tarskian sense of semantic consequence as a relation of truth-preservation in a model, i.e., interpretations of formulas such that these are true (symbol: t) or false (symbol: f) in a model. Thus, $\Gamma \models_L \varphi$ means that any model w of $\psi \in \Gamma$ in L is also a model of φ in L: $\Gamma \models_L \varphi$, i.e. $v(w, \psi) = t \Rightarrow v(w, \varphi) = t$. Then any model w of $\bar{L}$ can be called an *antimodel* of L, and any model w of $\tilde{L}$ can be called a *countermodel* of L, such that:

- there exists a model w of $\psi \in \Gamma$ in L that is not a model of φ in $\bar{L}$: $\Gamma \not\models_{\bar{L}} \varphi$, i.e., $v(\psi, w) = t \not\Rightarrow v(w, \varphi) = t$.

- every model w of $\psi \in \Gamma$ in L is also a model of $\neg\varphi$ in L: $\Gamma \models_{\tilde{L}} \neg\varphi$, i.e., $v(w, \psi) = t \Rightarrow v(w, \neg\varphi) = t$.

On the other hand, the bivalent interpretation of formulas in the models entails that there is no logical difference between untruth and falsity. In other words, every antimodel of φ is a model at which φ is not true, that is, φ is false: $v(w, \varphi) \neq t = v(w, \varphi) = f$; and every countermodel of φ is a model at which $\neg\varphi$ is true, that is, φ is false: $v(w, \neg\varphi) = t$ means the same as $v(w, \varphi) = f$. The difference between both antilogic and counterlogic may be easily explained in terms of how many models there are for these: an antilogic has φ false at *some (but not all)* model of it at which ψ is true, whereas a counterlogic has φ false at *every* model of it at which ψ is true.

And yet, what of such higher-order oppositions in a many-valued system where bivalence does not obtain anymore? Answering to this question will be the central task of the present section, especially because bivalence is assumed in [2]. Our aim is to extend the notion of logical oppositions into non-bivalent or many-valued systems, accordingly. For this purpose, let us consider a general domain of valuation V_n including n logical values. Bivalence includes the class of logical systems where the $m = 2$ logical values are truth and falsehood. We assume that many-valuedness relates to any logical system whose domain of interpretation V_m includes more than 2 values, such that $m > 2$.

More generally, one way to characterize many-valuedness is by taking logical values to be ordered n-tuples of elements whilst keeping in mind that the basic values of logic are t and f. A characterization of such finitely n-valued systems consists in a 2^n-valued domain of values including n ordered elements and $2^m = n$ resulting logical values. Borrowing from various works from to Jaskówski [9] to Kapsner [10] through Shramko & Wansing [17], the following wants to focus on a specific case of structured logical values analogous to Belnap's 4-valued system First Degree Entailment [1]. Thus, let V_4 a 4-valued domain of structured logical values $X = \langle x_1, x_2 \rangle$. It includes $n = 2$ elements t and f such that, for any φ, $v_4(\varphi) = \langle x_1, x_2 \rangle$ and $x_i(\varphi) \mapsto \{1, 0\}$.

Given that logical values are structured objects in V_4, their characteristic valuation function proceeds as an ordered 2-uple $\mathbf{A}(\varphi) = \langle \mathbf{a}_1(\varphi), \mathbf{a}_2(\varphi) \rangle$ wherein $\mathbf{a}_1(\varphi) = x_1$ informs about whether φ is true, and $\mathbf{a}_2(\varphi) = x_2$ about whether φ is false. Correspondingly, we will rephrase the four logical values of V_4 by translating first their basic elements t and f in terms of structured values and, then, the combination of the latter.[1]

$v(\varphi) = t$ means that $\mathbf{a}_1(\varphi) = 1$, i.e., φ is true.
$v(\varphi) \neq t$ (or $v(\varphi) = \bar{t}$) means that $\mathbf{a}_1(\varphi) = 0$, i.e., φ is not true.
$v(\varphi) = f$ means that $\mathbf{a}_2(\varphi) = 1$, i.e., φ is false.
$v(\varphi) \neq f$ (or $v(\varphi) = \bar{f}$) means that $\mathbf{a}_2(\varphi) = 0$, i.e., φ is not false.

The logical values of V_4 can be considered as ordered structured pairs such that $B = \langle t, f \rangle$, $T = \langle t, \bar{f} \rangle$, $F = \langle \bar{t}, f \rangle$, and $N = \langle \bar{t}, \bar{f} \rangle$.[2]

$v(\varphi) = B$ means that $\mathbf{a}_1(\varphi) = \mathbf{a}_2(\varphi) = 1$, i.e., $\mathbf{A}(\varphi) = 11$.
$v(\varphi) = T$ means that $\mathbf{a}_1(\varphi) = 1$ and $\mathbf{a}_2(\varphi) = 0$, i.e., $\mathbf{A}(\varphi) = 10$.
$v(\varphi) = F$ means that $\mathbf{a}_1(\varphi) = 0$ and $\mathbf{a}_2(\varphi) = 1$, i.e., $\mathbf{A}(\varphi) = 01$.

---

[1] For a discussion about the meaning of such structured values and a doxastic interpretation of these, see e.g. [14].

[2] For sake of simplicity, the ordered pairs $\langle x, y \rangle$ will be rephrased as xy throughout the rest of the paper.

$v(\varphi) = N$ means that $\mathbf{a}_1(\varphi) = \mathbf{a}_2(\varphi) = 0$, i.e., $\mathbf{A}(\varphi) = 00$.

The semantic relation of consequence between a set of formulas Γ and a formula φ can also be rephrased in terms of structured logical values, such that $\Gamma \vdash_L \varphi$ means that, for every formula $\psi \in \Gamma$, $\mathbf{a}_1(\psi, L) = 1 \Rightarrow \mathbf{a}_1(\varphi, L) = 1$. The same does for the central notions of antilogic and counterlogic.

Antilogic: $\Gamma \vdash_{\bar{L}} \varphi$ if and only if it is not the case that $\Gamma \vdash_L \varphi$.
$\mathbf{a}_1(\Gamma, \bar{L}) = \mathbf{a}_1(\varphi, \bar{L}) = 1$ if and only if $\mathbf{a}_1(\psi, L) = 1$ and $\mathbf{a}_1(\varphi, L) = 0$.

Counterlogic: $\Gamma \vdash_{\tilde{L}} \varphi$ if and only if $\Gamma_L \vdash \neg\varphi$.
$\mathbf{a}_1(\Gamma, \tilde{L}) = \mathbf{a}_1(\varphi, \tilde{L}) = 1$ if and only if $\mathbf{a}_1(\psi, L) = \mathbf{a}_2(\varphi, L) = 1$.

Semantic consequence in a logical system can also be rephrased as a mapping function $\mathcal{F}_V$ on values such that, for a primary logical system L where truth is preserved from premisses Γ to consequence φ

$$\mathcal{F}_V(L) = t \mapsto t.$$

The corresponding antilogics and counterlogics can be redefined as follows, accordingly:

$$\mathcal{F}_V(\bar{L}) = t \mapsto \bar{t};$$
$$\mathcal{F}_V(\tilde{L}) = t \mapsto f.$$

Returning to the aforementioned paper [2], the authors gave a definition of the usual concepts of opposition whilst expressing these as set-theoretical relations of intersection $\cap$ between logical systems. Once again, we translate each of these into our semantic terms as follows: for every φ, the intersection $\vdash_{L_1} \cap \vdash_{L_2}$ is (not) empty if, and only if, φ's being true L_1 (does not) entail φ's not being true in L_2; and the intersection $\nvdash_{L_1} \cap \nvdash_{L_2}$, is (not) empty if, and only, φ's not being true L_1 (does not) entail φ's being true in L_2.[3]

---

[3] The second clause characterizing oppositions could be reformulated as a relation of union $\cup$ between any logical systems L_1, L_2, by virtue of the set-theoretical relation between intersection and union. Thus, $\nvdash_{L_1} \cap \nvdash_{L_2} = \emptyset$ means the same as $\vdash_{L_1} \cup \vdash_{L_2} \neq \emptyset$.

L_1 and L_2 are *contradictories* if and only if $\vdash_{L_1} \cap \vdash_{L_2} = \emptyset$ and $\nvdash_{L_1} \cap \nvdash_{L_2} = \emptyset$.
$\mathbf{a}_1(\varphi, L_1) = 1 \Rightarrow \mathbf{a}_1(\varphi, L_2) = 0$ and $\mathbf{a}_1(\varphi, L_1) = 0 \Rightarrow \mathbf{a}_1(\varphi, L_2) = 1$.

L_1 and L_2 are *contraries* if and only if $\vdash_{L_1} \cap \vdash_{L_2} = \emptyset$ and $\nvdash_{L_1} \cap \nvdash_{L_2} \neq \emptyset$.
$\mathbf{a}_1(\varphi, L_1) = 1 \Rightarrow \mathbf{a}_1(\varphi, L_2) = 0$ and $\mathbf{a}_1(\varphi, L_1) = 0 \nRightarrow \mathbf{a}_1(\varphi, L_2) = 1$.

L_1 and L_2 are *subcontraries* if and only if $\vdash_{L_1} \cap \vdash_{L_2} \neq \emptyset$ and $\nvdash_{L_1} \cap \nvdash_{L_2} = \emptyset$.
$\mathbf{a}_1(\varphi, L_1) = 1 \nRightarrow \mathbf{a}_1(\varphi, L_2) = 1$ and $\mathbf{a}_1(\varphi, L_1) = 0 \Rightarrow \mathbf{a}_1(\varphi, L_2) = 1$.

The fourth and ultimate case of *subalternation* differs from the preceding ones by being defined without the set-theoretical relation of intersection, in informal terms of 'sublogic'.

L_1 is *subaltern* to L_2 if and only if L_2 is a sublogic of L_1.

The latter is assumed to be known by the readers, in that it means a relation of consequence from the first system to the second one. That is:

$\mathbf{a}_1(\varphi, L_1) = 1 \Rightarrow \mathbf{a}_1(\varphi, L_2) = 1$ and $\mathbf{a}_1(\varphi, L_2) = 0 \Rightarrow \mathbf{a}_1(\varphi, L_1) = 0$

An alternative definition of subalternation has been proposed in [15], where oppositions are turned from relations into iterative functions. Thus, ψ is said to be 'subalternate' to φ if, and only if, ψ is the *contradictory of the contrary* of φ; and conversely, φ is 'superalternate' to ψ if, and only if, φ is the *contrary of the contradictory* of φ.
It would be interesting to see how such a functional interpretation of opposition may be implemented into the context of logical system [2]. Assuming that antilogicality and counterlogicality are special cases of contradictoriness and contrariety, respectively, then there is a discrepancy between the logical equations established in [15] and what the author said in their own symbols [2]. Thus,

(1) The antilogic of the antilogic of a given logical system L_1 is L_1 itself in [2]

$$\bar{\bar{L}} = L$$

which is confirmed in [15] by stating the contradictory of the contradictory of a given term φ is φ itself

$$cd(cd(\varphi)) = \varphi.$$

At the same time, (2) the counterlogic of the counterlogic of a given logical system L_1 does equate with L_1 itself in [2]

$$\tilde{\tilde{L}} = L$$

whereas the contrary of the contrary of a given term φ may differ from φ in [15]

$$ct(ct(\varphi)) \neq \varphi.$$

And (3) the counterlogic of the antilogic of a given logical system L_1 does equate with the antilogic of its counterlogic in [2]

$$\tilde{\bar{L}} = \bar{\tilde{L}}$$

whereas we have already seen that the contradictory of a contrary differs from the contrary of a contradictory in [14]. Indeed, the former iteration amounts to a case of subalternation

$$cd(ct(\varphi)) = sb(\varphi)$$

whereas the latter yields the converse case of superalternation

$$ct(cd(\varphi)) = sp(\varphi).$$

How to account for such a discrepancy, and what does it entail about the logical accuracy of [2] and [15]? In order to disentangle the situation, we have not only to prove that antilogicality relates to contradictoriness and counterlogicality to contradictoriness. But also, the calculus of opposition-forming operators set up in [15] leads to an important difference with respect to [2]. Indeed, such operators are not 'functions' in the strict mathematical sense of a bijection: one input value may have more than one contrary, subcontrary and subaltern (or superaltern), so that the above singular expression 'the contrary of' is misleading. Actually, it is possible to compute the output value of such opposite-forming operators only by means by a special semantics, namely: a 'bitstring semantics' in which terms do not receive a customary 'truth-value' but, instead, a Boolean bitstring characterizing their truth-conditions in a finite set of logical spaces. It turns out that this Boolean but *not* truth-functional semantics departs from the approach of [2]. On the one hand, it matches with [2] in that every logical system has one and only one *antilogic* as a counterpart of contradictoriness ([2], 245):

"It is clear that for each L there is exactly one $\bar{L}$", which can be explained set-theoretically once again:

$$\vdash_L \cap \vdash_{\bar{L}} = \emptyset;$$
$$\vdash_L \cup \vdash_{\bar{L}} = \wp(F) \times F.$$

On the other hand, it is shown in [15] that what the authors call 'counterlogic' is just a particular, truth-functional case of contrariness:

"It is clear that for each L, and for each negation operation, there is exactly one $\tilde{L}$."

The authors rightly assume that one and the same operator of negation occurs in L_1 and L_2, so that there can be only one system L_2 where φ is false whenever φ is true in L_1. A way to account for this unique case of contrariness occurs in algebraic terms of abstract operators [8,12,15]. In the second reference [12], for example, Piaget's INRC Group depicts the operation of reciprocity as mapping from an order set of conjunctive normal forms of literals *abdc* upon its reverse *cdba*. This helps why there cannot be but one of 'contrariness' once constructed in this bijective way. In [15], the same operation is applied to make sense of 'contrary' beliefs operators as ordered set of truth-conditions whilst noticing that there is one more than such one way to characterize contrariety.

And yet, one may imagine however more than one way of satisfying the clauses of antilogicality and counterlogicality once bivalence is not assumed. This requires to go beyond the Boolean approach, assumed both in [2] and [15]. For there may be more than one way of being true and false in V_4, for example, so that there may be more than one antilogic and counterlogic to an initial logical system L_1. Now going beyond bivalence is to go beyond the realm of 'classical' oppositions, which seems to lead to a *terra incognita* in the literature of logic. For what had been said thus far about 'non-classical oppositions'? Be this as it may, 'classical' oppositions may be characterized by two clauses such as *completeness* and *consistency*. Classicality is claimed and sustained in

[2,247] as follows:[4]

"It is not straightforward to present oppositional structures for any logic. We will proceed by introducing some restrictions. First, we restrict ourselves to logics which accept elimination of double negation in an obvious sense. Additionally, let L be a logic with negation. We say that L is *well-behaved* if and only if for every pair (Γ, φ), it is not the case that $(\Gamma \vdash_L \varphi$ and $\Gamma \vdash_L \neg\varphi)$".

Double negation relates to completeness: $\vdash_L \varphi$ if and only if $\vdash_L \neg\neg\varphi$, whist well-behavior has to do with consistency. Both properties and their opposite may be formulated as follows:

Consistency
$\Gamma \vdash \varphi \Rightarrow \Gamma \nvdash \neg\varphi$
Inconsistency
$\Gamma \vdash \varphi \nRightarrow \Gamma \nvdash \neg\varphi$
Completeness
$\Gamma \nvdash \varphi \Rightarrow \Gamma \vdash \neg\varphi$
Incompleteness
$\Gamma \nvdash \varphi \nRightarrow \Gamma \vdash \neg\varphi$

These metalogical properties characterize what is considered as the proper features of logical oppositions ([2], 427):

"We call a square *complete* if it is a square with all four oppositions: contradiction, contrariety, sub-contrariety and subalternation. A square is *standard* if it fits any family of concepts satisfying traditional oppositions. A square is *perfect* if it is complete and standard. Moreover, any square which is not complete or/and standard is called *degenerate square*."

Why sticking to such features, however? Let us consider in the following what non-standard squares should amount to, assuming that they might relate many-valued systems which are not well-behaved and do not accept elimination of double negation. In V_4, for example, logical systems may be incomplete or inconsistent whenever $\mathbf{A}(\varphi) = 00$ or $\mathbf{A}(\varphi) = 11$, respectively. Let us see what does follow from this non-bivalent situation: does it result in new kinds of oppositions? In order to answer this

---
[4]Note that classicality need not be a synonym of bivalence, given that there may be classical theorems that do not correspond to a bivalent domain (and conversely). See e.g. [15] about this point.

question, let us consider by now another issue which has been addressed by de Edelcio de Souza.

2 Quasi-truth

Indeed, one of de Souza's main contributions to the reflection in philosophical logic relates to the concept of *quasi-truth* [4], inherited from da Costa's seminal work. Roughly speaking, quasi-truth is to be viewed as a set of *partial* structures such that the predicates are seen as triples of pairwise disjoint sets $\{R^+, R^-, R^u\}$: the set of tuples which satisfies, does not satisfy and may satisfy or not a predicate in a given model. Our attention will be focused on the third subset R^u, since it stands for the 'partial' features of structures and leads to the notion of quasi-truth. R^u may be taken to be the set of undeterminate logical values, $\{11,00\}$, such that logical value of φ is neither determinately true nor determinately false. Although quasi-truth is usually interpreted into a 3-valued domain $V_{3+} = \{11, 10, 01\}$ or $V_{3-} = \{10, 01, 00\}$ –depending upon whether the additional third value is designated or not, it makes sense to consider as two proper cases of quasi-truth the situations in which there is evidence both for and against a given formula or neither for nor against, respectively. The concrete upshot is the same as the one when there is evidence neither for nor against the formula, in the sense that it leads to the same *practical* stance of indecision. Likewise, the coming 4-valued framework accommodates with the 3-valued definition of quasi-truth by treating gappy and glutty values (00 and 11) as two *pragmatic* variants of the same partial structure: underdetermined and overdetermined logical values amount to the same result of remaining undecided about φ, insofar as the logical value of formulas relate to what agents should do in the light of such informational data.

We propose to reconstruct both logical values and relations of opposition between logical systems into a common framework $\mathbf{AR}_{4[O_i]}$ [13]. It includes a number of logical systems distinguished by two sets of unary operators of *affirmation* $[O_i]$ and *negation* $[N_i]$. The language of $\mathbf{AR}_{4[O_i]}$ can be described by means of the usual Backus-Naur form:

$$\varphi ::= \quad [O_i]p \mid [O_i](\varphi \bullet \psi) \mid [O_i]\varphi \bullet [O_i]\psi \mid \neg_1[O_i]\varphi \mid [O_i]\neg_2\varphi$$

The lowercase variable i of $[O_i]$ means that there is a plurality of affirmative and negative operators in $\mathbf{AR}_{4[O_i]}$. Roughly speaking, both categories of operators constitute a variety of ways to restrict the logical values of formulas in V_4. Affirmative operators are not redundant by excluding logical values whilst always affirming their input value, whereas negative operators always exclude the input value. Their general definitions are the following, for any pairs of values $\{x_i, x_j\}$ in V_n:

Affirmative operators
$[A_i]\varphi : x_i \mapsto \overline{x_j}$

Negative operators
$[N_i]\varphi : x_i \mapsto \overline{x_i}$

An essential feature of $[A_i]$ and $[N_i]$ is that these are *partial*: they turn some, but not necessarily all input values into output values of the entire domain V_4.[5]

Given any domain of valuation V_n, there is a set of $i = 2^n - 1$ affirmative operators. In the present case of V_4, there are $2^4 - 1 = 15$ affirmative and negative operators which obey double negation in a metalogical sense of the word: $\bar{\bar{x}} = x$.

$[A_1]\varphi : t \mapsto \bar{f}$
$[A_2]\varphi : f \mapsto \bar{t}$
$[A_3]\varphi : \bar{t} \mapsto f$
$[A_4]\varphi : \bar{f} \mapsto t$
$[A_5]\varphi : t \mapsto \bar{f} \otimes f \mapsto \bar{t}$
$[A_6]\varphi : t \mapsto \bar{f} \otimes \bar{t} \mapsto f$
$[A_7]\varphi : t \mapsto \bar{f} \otimes \bar{f} \mapsto t$
$[A_8]\varphi : f \mapsto \bar{f} \otimes \bar{t} \mapsto t$
$[A_9]\varphi : f \mapsto \bar{f} \otimes \bar{f} \mapsto f$
$[A_{10}]\varphi : \bar{t} \mapsto f \otimes \bar{f} \mapsto t$
$[A_{11}]\varphi : t \mapsto \bar{f} \otimes f \mapsto \bar{t} \otimes \bar{t} \mapsto f$
$[A_{12}]\varphi : t \mapsto \bar{f} \otimes f \mapsto \bar{t} \otimes \bar{f} \mapsto t$
$[A_{13}]\varphi : t \mapsto \bar{f} \otimes \bar{t} \mapsto f \otimes \bar{f} \mapsto t$
$[A_{14}]\varphi : f \mapsto \bar{t} \otimes \bar{t} \mapsto f \otimes \bar{f} \mapsto t$
$[A_{15}]\varphi : t \mapsto \bar{f} \otimes f \mapsto \bar{t} \otimes \bar{t} \mapsto f \otimes \bar{f} \mapsto t$

$[N_1]\varphi : t \mapsto \bar{t}$
$[N_2]\varphi : f \mapsto \bar{f}$
$[N_3]\varphi : \bar{t} \mapsto t$
$[N_4]\varphi : \bar{f} \mapsto f$
$[N_5]\varphi : t \mapsto \bar{t} \otimes f \mapsto \bar{f}$
$[N_6]\varphi : t \mapsto \bar{t} \otimes \bar{t} \mapsto t$
$[N_7]\varphi : t \mapsto \bar{t} \otimes \bar{f} \mapsto f$
$[N_8]\varphi : f \mapsto \bar{f} \otimes \bar{t} \mapsto t$
$[N_9]\varphi : f \mapsto \bar{f} \otimes \bar{f} \mapsto f$
$[N_{10}]\varphi : \bar{t} \mapsto t \otimes \bar{f} \mapsto f$
$[N_{11}]\varphi : t \mapsto \bar{t} \otimes f \mapsto \bar{f} \otimes \bar{t} \mapsto t$
$[N_{12}]\varphi : t \mapsto \bar{t} \otimes f \mapsto \bar{f} \otimes \bar{f} \mapsto f$
$[N_{13}]\varphi : t \mapsto \bar{t} \otimes \bar{t} \mapsto t \otimes \bar{f} \mapsto f$
$[N_{14}]\varphi : f \mapsto \bar{f} \otimes \bar{t} \mapsto t \otimes \bar{f} \mapsto f$
$[N_{15}]\varphi : t \mapsto \bar{t} \otimes f \mapsto \bar{f} \otimes \bar{t} \mapsto t \otimes \bar{f} \mapsto f$

This language includes two main negations, the Boolean one $\neg_1$ and the Morganian one $\neg_2$, in addition with a set of binary connectives $\bullet = \{\wedge, \vee, \rightarrow\}$. Products $\otimes$ are idempotent, commutative, transitive

---

[5] Another way to characterize these operators is to take these as a combination of redundant and non-redundant mappings: they turn some (but not all) of their input values into some other output values.

and associative operators that merely add different mappings of the same kind to each other. For example, $[A_7]$ proceeds in such a way that every formula is unfalse whenever true *and* true whenever unfalse, whereas $[A_8]$ means that every formula is false whenever untrue *and* untrue whenever false. The single values occurring in boldface in the below matrix correspond to the outputs altered by the affirmative operators, the other ones remaining unchanged.

φ	$[A_7]\varphi$	$[A_8]\varphi$
11	**10**	**01**
10	10	10
01	01	01
00	**10**	**01**

Both $[A_7]$ and $[A_8]$ are *bivalence*-forming, or *normalization* operators: they reintroduce bivalence by restricting the output values in different ways, such that the resulting logical values are either 10 or 01. That is, every true formula is thereby not false and conversely. The aforementioned case of Boolean negation correspond to a single negative operator, that is:

$$\neg_1\varphi = [N_{15}]\varphi : t \mapsto \bar{t} \otimes \bar{t} \mapsto t \otimes f \mapsto \bar{f} \otimes \bar{f} \mapsto f.$$

At the same time, the structuration of such unary operators is such that it helps to see to what extent Morganian negation is not a 'pure' negation. Rather, it is case of 'mixed' operator conflating both affirmative and operators into mappings of the form $x_i \mapsto \overline{\bar{x_j}} = x_i \mapsto x_j$. The corresponding process is a *fusion* of the partial operators of affirmation and negation, thus resulting in 'affirmed negations' $[AN]$ or, equivalently. 'negated affirmations' $[NA]$:

$$\neg_2\varphi = [NA_{15}]\varphi = [AN_{15}]\varphi : t \mapsto f \otimes \bar{t} \mapsto \bar{f} \otimes f \mapsto t \otimes \bar{f} \mapsto \bar{t}.^{6}$$

---

[6]Fusion of partial operators differs both from their product $\otimes$ and the following operation of composition or iteration, $\circ$. It could be also shown that two other kinds of *redundancy*-making operators are equivalent with each other in $\mathbf{AR}_{4[O_i]}$, namely: $[NN]\varphi = [AA]\varphi$. The proof of such equivalences can be established as follows:
Proof.
$[NA]\varphi : x_i \mapsto [A]\overline{x_i} = x_i \mapsto \overline{\bar{x_j}} = x_i \mapsto x_j$.
$[AN]\varphi : x_i \mapsto [N]\overline{x_j} = x_i \mapsto \overline{\bar{x_j}} = x_i \mapsto x_j$.
Therefore $[AN]\varphi = [NA]\varphi$.

It turns out that antilogics and counterlogics are may be constructed by means of the unary operators of *Boolean* negation $\neg_1$ and *Morganian* negation $\neg_2$, following the definitions given in [13] and leading to the following truth-tables:

φ	$\neg_1\varphi$	$\neg_2\varphi$
11	00	11
10	10	01
01	01	10
00	11	00

According to this, Boolean negation $\neg_1$ turns logics L into *antilogics* $\tilde{L}$ whenever they turn true (or false) formulas into untrue (or unfalse) ones; and Morganian negation $\neg_2$ turn logics L into *counterlogics* $\check{L}$ whenever they turn true (or false) formulas into false (or true) ones. Antilogics correspond to situations in which a set of formulas belonging to L do not belong to another language $\bar{L}$, and this may be obtained by more than negative operator –not only $[N_{15}] = \neg_1$, but also every negative operator including the clauses of $[N_1]$ and $[N_2]$: $t \mapsto \bar{t} \otimes \bar{t} \mapsto t$. In the same vein, counterlogics correspond to situations in which the negations of a set of formulas belonging to L do belong to another language $\tilde{L}$, and this may be obtained by more than mixed operator –not only $[AN_{15}] = \neg_2$, but also every negative operator including the clauses of $[AN_1]$ and $[AN_2]$: $t \mapsto f \otimes f \mapsto t$.

Furthermore, it can be shown by now how the equations established in [2] may be validated or not according to the kind of partial operator selected in $\mathbf{AR}_{4[O_i]}$. The expressions 'antilogic of antilogic' and 'counterlogic of antilogic' correspond to cases of iteration or *composition* $\circ$, which are to be clearly distinguished from those of product $\otimes$ and mixed operators. Whilst the difference between product and composition can be easily shown by induction upon truth-tables,[7] it also helps to see that the following equations hold only when the corresponding operators proceed by iteration of specific operators –Boolean negation as an

---

$[AA]\varphi : x_i \mapsto [A]\overline{x_j} = x_i \mapsto \overline{\overline{x_i}} = x_i \mapsto x_i$.
$[NN]\varphi : x_i \mapsto [N]\overline{x_i} = x_i \mapsto \overline{\overline{x_i}} = x_i \mapsto x_i$.
Therefore $[AA]\varphi = [NN]\varphi$.

[7] Let $[A_3]$ and $[A_4]$ be two such partial operations. Then the following truth-tables show both that their product differs from their composition and that, unlike product, composition is not a symmetrical operation.

antilogic-forming operator and Morganian negation as a counterlogic-forming operator, once again.

$\bar{\bar{L}} = L$, that is, $[N_{15}][N_{15}]\varphi = \varphi$;

$\tilde{\tilde{L}} = L$, that is, $[AN_{15}][AN_{15}]\varphi = \varphi$;

$\tilde{\bar{L}} = \bar{\tilde{L}}$, that is, $[AN_{15}][N_{15}]\varphi = [N_{15}][AN_{15}]\varphi$.

Again, it must be recalled that all of these equations fail whenever antilogicality and counterlogicality are rephrased into $\mathbf{AR}_{4[O_i]}$ by partial operators which satisfy lesser semantic constraints whilst behaving as proper contradictory- and contrary-forming operators. This means that antilogic does not go on par with contradictoriness and counterlogic does not go on a par with contrariness –they are so only in a bivalent frame, where the unique negative operator is both Boolean and Morganian.

Coming back to the central section of the present issue, quasi-truth, it has been shown in [13] that the affirmative operators $[A_7]\varphi$ and $[A_8]$ are plausible candidates for being four-valued counterparts of the modalities of necessity and possibility in S5. Letting τ be a translation function from S5 to $\mathbf{AR}_{4[O_i]}$ and including a redundant-forming operator $[AA_{15}] = [NN_{15}]$ such that

$[AA_{15}]\varphi = [NN_{15}]\varphi : t \mapsto t \otimes \bar{t} \mapsto \bar{t} \otimes f \mapsto f \otimes \bar{f} \mapsto \bar{f}$.

It follows from this that

$\tau(\varphi, S5) = [AA_{15}]\varphi = [NN_{15}]\varphi$;
$\tau(\Box\varphi, S5) = [A_8]\varphi$;
$\tau(\Diamond\varphi, S5) = [A_7]\varphi$.

We are going to use the two many-valued translations of necessity and possibility in the following, in order to propose a many-valued counter-

φ	$[A_3]\varphi$	$[A_4]\varphi$	$[A_3]\varphi \otimes [A_4]\varphi$	$[A_3]\varphi \circ [A_4]\varphi$	$[A_4]\varphi \circ [A_3]\varphi$
11	11	11	11	11	11
10	10	10	10	10	10
01	01	01	01	01	01
00	01	10	11	10	01

part of *quasi-truth* in $\mathbf{AR}_{4[O_i]}$. On the other hand, it has been claimed in [5] that there is a connection between the concepts of quasi-truth and *contingency*, ∇. According to the author ([6],176),

"non-mathematical justifications are not able to lead to necessary but, rather, only to contingent truths. If there does not exist any demonstration about the truth of a proposition, then there is no certainty. Therefore, the proposition is not entitled to be acknowledged as true necessarily."

In other words, quasi-true formulas are those for which there is no conclusive evidence and that remain possibly false without being so determinately ([6], 180):

"Logics of justification – on its two approaches – can be used in order to define and think about the concept of quasi-truth. This was proposed by Newton da Costa in (1986) because, as a matter of fact, whenever we stand outside mathematics and logic we cannot talk exactly in terms of necessary truth, but only in terms of contingent truth, that is, quasi-truth."

Our main idea is to render da Costa & Bueno & Souza's insightful idea of quasi-truth as *partial structures* in semantic terms of quasi-truth as a *partial operator*, whereas some affirmative operators $[A_i]$ proceed as normalization-forming operators by restoring normal structures through partial ones. Assuming that quasi-truth proceeds as a contingency operator, and given our preceding translations of S5-modal necessity and possibility into $\mathbf{AR}_{4[O_i]}$, let us characterize quasi-truth QT as a conjunction of possibility and unnecessity.

Quasi-truth (as contingency)
$\nabla \varphi \Leftrightarrow \Diamond \varphi \wedge \neg \Box \varphi$
$\tau(QT(\varphi)) = [A_7]\varphi \wedge \neg_1 [A_8]\varphi.$[8]

---

[8] Only Boolean negation $\neg_1$ has a wide scope in $\mathbf{AR}_{4[O_i]}$, but note that the above translation of negated possibility would result in the same truth-table had the corresponding operator of negation been the Morganian one $\neg_2$ –due to the bivalent behavior of QT. Moreover, the logical constants of $\mathbf{AR}_{4[O_i]}$ have not been defined thus far, given that these are useless for the present purpose. However, contingency requires some words on conjunction since the latter makes part of its definition. So let $max(x,y)$ and $min(x,y)$ be the functions selecting the greater and lesser value among x and y, respectively, given that $1 > 0$. Then:
$$\mathbf{A}(\varphi \wedge \psi) = \langle min(\mathbf{a}_1(\varphi), \mathbf{a}_1(\psi)), max(\mathbf{a}_2(\varphi), \mathbf{a}_2(\psi)) \rangle.$$
See [12,13,15] for more information about these 4-valued logical constants.

φ	$[A_7]\varphi$	$[A_8]\varphi$	$\neg_1[A_8]\varphi$	$QT\varphi$
11	10	01	10	10
10	10	10	01	01
01	01	01	10	01
00	10	01	10	10

The above matrix accounts for quasi-truth as being false with every formula whose logical value is determinately true or determinately false –i.e., $\mathbf{A}(QT\varphi)) = 01$ whenever $\mathbf{A}(\varphi) \in \{10, 01\}$.

Such an operator may also be seen as a proper translation by satisfying the main negative features of quasi-truth, namely:

(i) $\not\models QT\varphi \to \varphi$
(ii) $QT\varphi, QT\neg\varphi \not\models \psi$
(iii) $QT\varphi \not\models \neg QT\neg\varphi.$[9]

Our final consideration will consist in combining the previous two issues of the paper, opposition and quasi-truth, in order to pave the way to a third new topic: *quasi-oppositions*. This will answer to the question about whether there could be further non-standard relations of opposition in a non-bivalent frame like V_4.

3 Quasi-oppositions

Following [15], we assume that consequence and opposition can be treated either as relations $R(x, y)$ or as operators $f(x) = y$ (without any specification about the nature of the objects x and y). Consequence $Cn(\Gamma, \varphi)$ has been studied since Tarski though several features like monotonicity, closure or structurality; and it has also be viewed as a possible operator mapping from given sets to close sets. Opposition $Op(\varphi, \psi)$ is traditionally considered as a relation between truth-values, and it has also been turned into an operator $op(\varphi) = \psi$ in the above reference. Given

---

[9]The translations of the formulas (i)-(iv) into $\mathbf{AR}_{4[O_i]}$ and their corresponding counter-models are the following, given the rules established in [14] and our previous definition of QT:

$\tau(i)$ $\not\models QT\varphi \to [AA_{15}]\varphi$ (counter-model: $\mathbf{A}(\varphi) = 00$.)
$\tau(ii)$ $QT\varphi, QT\neg_2\varphi \not\models \psi$ (counter-model: $\mathbf{A}(\varphi) = 11$.)
$\tau(iii)$ $QT\varphi \not\models \neg_1 QT\neg_2\varphi$ (counter=model: $\mathbf{A}(\varphi) = 11$.)

that logical oppositions are set of truth- and falsity-conditions between 'opposed' terms, truth-values constitute an essential feature in order to make sense of them. In the present context of a 4-valued domain, our main concern will be something like this: what sort of opposition is there between one formula which is neither-true-nor-false and another one which is both-true-and-false, for example?

One simple way to make an end to this discussion until its very opening is by applying the rationale urged by Roman Suszko, thereby rejecting the logical relevance of many-valuedness and reducing it to only two possible values: *designated*, or *not designated*. Thus, formulas are said 'designated' whenever they include the value of truth; they are 'not designated', otherwise. There are at least two ways not to follow this path, otherwise. Firstly, philosophical arguments –including those about quasi-truth, gave some reason to develop a set of many-valued inferences beyond Suszko's strictly bivalent policy. Following this stance introduced by Malinowski [11] and extended by Frankowski [7], there may be more than one way to characterize semantic consequence (or 'entailment') beyond the Tarskian classical pattern of truth-preservation. Here is a remainder of the four ways of dealing with consequence in a many-valued framework:

(Cn_t) $\quad \Gamma \models_t \varphi$ iff $\forall v[(\forall \psi \in \Gamma : v(\psi) \in \mathcal{D}^+) \Rightarrow v(\varphi) \in \mathcal{D}^+]$
(Cn_f) $\quad \Gamma \models_f \varphi$ iff $\forall v[(\forall \psi \in \Gamma : v(\psi) \notin \mathcal{D}^-) \Rightarrow v(\varphi) \notin \mathcal{D}^-]$
(Cn_q) $\quad \Gamma \models_q \varphi$ iff $\forall v[(\forall \psi \in \Gamma : v(\psi) \notin \mathcal{D}^-) \Rightarrow v(\varphi) \in \mathcal{D}^+]$
(Cn_p) $\quad \Gamma \models_p \varphi$ iff $\forall v[(\forall \psi \notin \Gamma : v(\psi) \notin \mathcal{D}^-) \Rightarrow v(\varphi) \in \mathcal{D}^+]$

In addition to the Tarskian pattern (Cn_t), the other three extensions depict semantic consequence as either a relation of non-falsity presentation (Cn_f), or a derivation of truth from non-refuted premises (Cn_q), or a derivation or mere plausibility from truth (Cn_p).

Following the developments around 4-valued inference by Blasio & Marcos & Wansing [4], three central issues will be approached in this last section: (a) What does truth and falsity mean into such a 4-valued frame? (b) How to systematize the kind of semantic consequence endorsed by Malinowski's line? (c) How to express the logical difference between the relations of consequence and opposition into one and the same framework?

With respect to (a), our 4-valued framework is such that the two main sets of logical values $\mathcal{D}^+$ and $\mathcal{D}^-$ will receive a special interpretation. For although these are generally taken to be exclusive from each other, the domain of values V_4 motivates another treatment. For let $\mathcal{D}^+ = \{11, 10\}$ be the subset of *designated* values that are cases of truth, and $\mathcal{D}^- = \{11, 01\}$ the subset of *antidesignated* values that are cases of falsehood. Then the glutty value 11 is both designated and antidesignated whereas the gappy value 00 is none, which entails that

$$\mathcal{D}^+ \cap \mathcal{D}^- \neq \emptyset$$
$$\mathcal{D}^+ \cup \mathcal{D}^- \neq \wp(F)$$

This means that $\mathcal{D}^-$ is not the mere complementary of $\mathcal{D}^+$, due to the overlapping relation of truth and falsity in V_4.

With respect to (b), one can make abstraction from the intuitive meaning of truth-values and conceive an exhaustive set of relations between designated and anti-designated sets. The reason why there are four kinds of entailment can be explained in a combinatorial way, given that it relies upon two clauses: belonging to the set of true formulas, and not belonging to the set of false formulas. This results in a set of $2^2 = 4$ possibles clauses for entailment, and we are going now to see how to extend this set to further semantic clauses. Starting from an initial set of two sets of formulas, i.e. designated and anti-designated, one can conceive of further relations between formulas and whose clauses of satisfaction do not consist in tracking truth whilst avoiding falsehood. Such is precisely the case with opposition, insofar as the latter essentially consists in tracking falsehood for a given formula whenever its 'opposed' term is true.

By thus introducing the additional two clauses of belonging to the set of false formulas and not belonging to the set of true formulas, it results in a set of $2^4 = 16$ kinds of relations. Letting $\mathcal{O}$ be a general meta-operator mapping between sets or their complementaries, two main interpretations of $\mathcal{O}$ will be naturally of interest in the following: consequence Cn, and opposition Op. Here is an exhaustive list of possible relations between subsets of values $\mathcal{D}^i = \{\mathcal{D}^+, \mathcal{D}^-\} \in V_4$:

- from D^+ onto D^+
  - (i) $\quad v(\varphi) \in D^+ \Rightarrow v(\psi) \in D^+$
  - (ii) $\quad v(\varphi) \notin D^+ \Rightarrow v(\psi) \notin D^+$
  - (iii) $\quad v(\varphi) \in D^+ \Rightarrow v(\psi) \notin D^+$

(iv) $\quad v(\varphi) \notin D^+ \Rightarrow v(\psi) \in D^+$

- from D^+ onto D^-

  (v) $\quad v(\varphi) \in D^+ \Rightarrow v(\psi) \in D^-$
  (vi) $\quad v(\varphi) \notin D^+ \Rightarrow v(\psi) \notin D^-$
  (vii) $\quad v(\varphi) \in D^+ \Rightarrow v(\psi) \notin D^-$
  (viii) $\quad v(\varphi) \notin D^+ \Rightarrow v(\psi) \in D^-$

- from D^- onto D^+

  (ix) $\quad v(\varphi) \in D^- \Rightarrow v(\psi) \in D^+$
  (x) $\quad v(\varphi) \notin D^- \Rightarrow v(\psi) \notin D^+$
  (xi) $\quad v(\varphi) \in D^- \Rightarrow v(\psi) \notin D^+$
  (xii) $\quad v(\varphi) \notin D^- \Rightarrow v(\psi) \in D^+$

- from D^- onto D^-

  (xiii) $\quad v(\varphi) \in D^- \Rightarrow v(\psi) \in D^-$
  (xiv) $\quad v(\varphi) \notin D^- \Rightarrow v(\psi) \notin D^-$
  (xv) $\quad v(\varphi) \in D^- \Rightarrow v(\psi) \notin D^-$
  (xvi) $\quad v(\varphi) \notin D^- \Rightarrow v(\psi) \in D^-$

With respect to (c), let us recall that the framework assumed in [2] was bivalent. This gave rise to a standard view of the square of opposition, in which whatever is not true is false and conversely. That is, in terms of structured values:

$\mathbf{a}_1(\varphi) = 1 \Rightarrow \mathbf{a}_2(\varphi) = 0$ and $\mathbf{a}_1(\varphi) = 0 \Rightarrow \mathbf{a}_2(\varphi) = 1$.

Such a normal or complete square may be depicted as follows, thereby fulfilling the clauses of consistency and completeness.

$$\begin{array}{ccc}
\mathbf{a}_1(\varphi) = 0 & \xrightarrow{ct} & \mathbf{a}_2(\varphi) = 0 \\
\downarrow sb & \begin{array}{c} cd \\ \times \\ cd \end{array} & \downarrow sb \\
\mathbf{a}_2(\varphi) = 1 & \xrightarrow{sct} & \mathbf{a}_1(\varphi) = 1
\end{array}$$

The situation is sensibly different into a 'non-standard square', that is, a non-bivalent set of relations where the aforementioned clauses are not followed:

$$\mathbf{a}_1(\varphi) = 1 \not\Rightarrow \mathbf{a}_2(\varphi) = 0 \text{ or } \mathbf{a}_1(\varphi) = 0 \not\Rightarrow \mathbf{a}_2(\varphi) = 1$$

So what should such a non-standard square look like? Given that the extension of logical values and their subsequent logical relations must complicate the resulting picture, one may begin answering to the above question by making a list of the possible relations of consequence and opposition. It appears that each of the four aforementioned relations of many-valued consequence corresponds to one case of the exhaustive list of the 16 $\mathcal{O}$-relations (i)-(xvi). Thus,

Many-valued consequence

(Cn_t)	$\varphi \in D^+ \Rightarrow \psi \in D^+$	(i)
(Cn_f)	$\varphi \notin D^- \Rightarrow \psi \notin D^-$	(xvi)
(Cn_q)	$\varphi \in D^+ \Rightarrow \psi \notin D^-$	(vii)
(Cn_p)	$\varphi \notin D^- \Rightarrow \psi \in D^+$	(xii)

Bueno & Souza [5] depicted quasi-truth in terms of partial structures whose final conclusion is open, which means that the formula into consideration may be true without being definitely so through the justification process [4]. For this reason, the above three non-Tarskian characterizations of consequence Cn_f, Cn_q, Cn_p may be taken to be various sorts of *quasi-consequence*. Likewise, the introduction of untrue and unfalse sets with $\mathcal{D}^+$ and $\mathcal{D}^-$ also seems to be in position make sense of the coming *quasi-oppositions*.

Roughly speaking, each case of 'quasi'-X is a situation in which the assessed object (proposition, concept, logical system, or whatever) is not X whilst being possibly so. Let us take the case of contrariness. According to the standard definition, any two objects are contrary to each other if, and only if, they cannot be true together in such a way that the second is false whenever the first is true. In a case of of *quasi-contrariness*, however, the second term is merely not true (or untrue) whenever the first is true. Assuming that being almost or being still in position to be (true or false) affords an intuitive meaning of the 'quasi'-phrase, here is the list of quasi-oppositions Op_f, Op_q, Op_p that correspond to the remaining cases of non-consequence relations (or operators) $\mathcal{O}$.

Many-valued opposition

Contrariness

(Ct_t)	$\varphi \in D^+ \Rightarrow \psi \in D^-$	(v)
(Ct_f)	$\varphi \notin D^- \Rightarrow \psi \notin D^+$	(x)
(Ct_q)	$\varphi \in D^+ \Rightarrow \psi \notin D^+$	(iii)
(Ct_p)	$\varphi \notin D^- \Rightarrow \psi \in D^-$	(xvi)

Contradictoriness

(Cd_t)	$\varphi \in D^+ \Rightarrow \psi \in D^-$ and $\varphi \in D^- \Rightarrow \psi \in D^+$	$(v) \otimes (ix)$
(Cd_f)	$\varphi \notin D^- \Rightarrow \psi \notin D^+$ and $\varphi \notin D^+ \Rightarrow \psi \notin D^-$	$(x) \otimes (vi)$
(Cd_q)	$\varphi \in D^+ \Rightarrow \psi \notin D^+$ and $\varphi \notin D^+ \Rightarrow \psi \in D^+$	$(iii) \otimes (iv)$
(Cd_p)	$\varphi \notin D^- \Rightarrow \psi \in D^-$ and $\varphi \in D^- \Rightarrow \psi \notin D^-$	$(xvi) \otimes (xv)$

Subcontrariness

(Sct_t)	$\varphi \in D^- \Rightarrow \psi \in D^+$	(ix)
(Sct_f)	$\varphi \notin D^+ \Rightarrow \psi \notin D^-$	(vi)
(Sct_q)	$\varphi \notin D^+ \Rightarrow \psi \in D^+$	(iv)
(Sct_p)	$\varphi \in D^- \Rightarrow \psi \notin D^-$	(xv)

Subalternation

(Sb_t)	$\varphi \in D^+ \Rightarrow \psi \in D^+$	(i)
(Sb_f)	$\varphi \notin D^- \Rightarrow \psi \notin D^-$	(xvi)
(Sb_q)	$\varphi \in D^+ \Rightarrow \psi \notin D^-$	(vii)
(Sb_p)	$\varphi \notin D^- \Rightarrow \psi \in D^+$	(xii)

It clearly appears that subalternation and consequence are one and the same logical relation (or operator), at least when these resort to the same non-standard kind Cn_x and Sb_x. This amounts to say that every such $\mathcal{O}$-mapping is a single case of opposition, reminding that subalternation can be parsed as the iteration of two simple opposite-forming operators [14].

Two future investigations might be pursued with respect to this new concept of quasi-opposition, provided that the latter turn out to be a relevant issue. One first work would have to do with the philosophical

applications to it into informal contexts use, just as q-entailment and p-entailment had been interpreted by their authors in terms of plausibility and degrees of truth [7,11]. Another work would be about a calculus of quasi-operators, thus extending the work devoted to consequence-forming operators [15].
Thanks already to Edelcio for opening the way towards these potential tools of logic.

References

[1] H. Belnap. "A useful four-valued logic". In M. Dunn (eds.), *Modern Uses of Multiple-Valued Logic*, Reidel, Boston, 1977: 8-37.

[2] H. Bensusan & A. Costa-Leite & E. G. de Souza. "Logics and their galaxies". In: *The Road to Universal Logic*, volume II, Springer, 2015: 243-252.

[3] R. Blanché. *Les structures intellectuelles. Essai sur l'organisation systématique des concepts*. Vrin: Paris, 1966.

[4] C. Blasio & J. Marcos & H. Wansing. "An inferentially many-valued two-dimensional notion of entailment". *Bulletin of the Section of Logic*, Vol. 46, 2017: 233-262.

[5] O. Bueno & E. G. de Souza. "The concept of quasi-truth". *Logique et Analyse*, Vol. 153-154, 1996: 183-199.

[6] A. Costa-Leite. "Lógicas da justificação e quase-verdade", *Principia: An International Journal of Epistemology*, Vol. 18, 2014: 175-186.

[7] S. Frankowski. "Formalization of a plausible inference". *Bulletin of the Section of Logic*, Volume 33, 2004: 41-52.

[8] W. H. Gottschalk. "The Theory of Quaternality". *The Journal of Symbolic Logic*, Vol. 18, 1953: 193-196.

[9] S. Jaskówski. "Recherches sur le système de la logique intuitionniste". In: *Actes du Congrès International de Philosophie Scientifique*. Partie 6: Philosophie des Mathématiques, Paris, 1936: 59-61. Tradução em S. MacCall (ed.): *Polish Logic, 1920-1939*, Oxford University Press, 1967: 259-263.

[10] A. Kapsner. *Logics and falsifications*. Cham: Springer, 2014.

[11] G. Malinowski. "That p + q = c(onsequence)", *Bulletin of the Section of Logic*, Vol. 36, 2007: 7-19.

[12] J. Piaget. *Traité de logique (Essai de logique opératoire)*, Paris, Dunod (2a ed.), 1949.

[13] F. Schang. "A general semantics for logics of affirmation and negation", draft.

[14] F. Schang. & A. Costa-Leite. "Une sémantique générale des croyances justifiées", *CLE e-prints*, Vol. 16, 2016: 1-24.

[15] F. Schang. "End of the square", *South American Journal of Logic*, Vol. 4, 2018: 1-21.

[16] F. Schang. "Epistemic pluralism", *Logique et Analyse*, Vol. 239, 2017: 337-353.

[17] Y. Shramko & H. Wansing. "Some useful 16-valued logics: How a computer network should think", *Journal of Philosophical Logic*, Vol. 34, 2005: 121-153.

www.ingramcontent.com/pod-product-compliance
Lightning Source LLC
Chambersburg PA
CBHW071328190426

3193CB00041B/922